D0890085

Neuronal Recognition

CURRENT TOPICS IN NEUROBIOLOGY

Series Editors:
Samuel H. Barondes
University of California, San Diego
La Jolla, California
and
Floyd E. Bloom
The Salk Institute
La Jolla, California

Tissue Culture of the Nervous System • 1973
Edited by Gordon Sato

Neuronal Recognition • 1976
Edited by Samuel H. Barondes

A Continuation Order Plan is available for this series. A continuation order will bring delivery of each new volume immediately upon publication. Volumes are billed only upon actual shipment. For further information please contact the publisher.

Neuronal Recognition

Edited by
Samuel H. Barondes
University of California, San Diego

Plenum Press · New York and London

Library of Congress Cataloging in Publication Data

Main entry under title:

Neuronal recognition.

(Current topics in neurobiology)
Includes bibliographies and index.
1. Neurons. 2. Cellular recognition. I. Barondes, Samuel H., 1933- [DNLM: 1. Nervous system – Physiology. 2. Neurons – Physiology. WL102 N4945]
QP363.N478 591.1'88 75-45291
ISBN 0-306-30885-1

A Division of Plenum Publishing Corporation
227 West 17th Street, New York, N.Y. 10011

Printed in the United States of America

Contributors

SAMUEL H. BARONDES — Department of Psychiatry
School of Medicine
University of California, San Diego
La Jolla, California

RICHARD P. BUNGE — Department of Anatomy
Washington University School of Medicine
St. Louis, Missouri

CARL W. COTMAN — Department of Psychobiology
University of California
Irvine, California

DOUGLAS M. FAMBROUGH — Department of Embryology
Carnegie Institution of Washington
Baltimore, Maryland

L. GLASER — Department of Biological Chemistry
Washington University School of Medicine
St. Louis, Missouri

G. GOMBOS — Centre de Neurochimie du CNRS
Strasbourg, France

D. I. GOTTLIEB — Department of Biological Chemistry
Washington University School of Medicine
St. Louis, Missouri

MARCUS JACOBSON — Department of Physiology and Biophysics
University of Miami School of Medicine
Miami, Florida

GARY S. LYNCH — Department of Psychobiology
University of California
Irvine, California

RICHARD B. MARCHASE — Department of Biology
The Johns Hopkins University
Baltimore, Maryland

R. MERRELL — Department of Biological Chemistry
Washington University School of Medicine
St. Louis, Missouri

I. G. MORGAN — Department of Behavioural Biology
Research School of Biological Sciences
Australian National University
Canberra, Australia

A. A. MOSCONA — Departments of Biology and Pathology
University of Chicago
Chicago, Illinois

KARL H. PFENNINGER — Section of Cell Biology
Yale University School of Medicine
New Haven, Connecticut

ROSEMARY P. REES — Department of Anatomy
Washington University School of Medicine
St. Louis, Missouri

STEVEN D. ROSEN — Department of Psychiatry
School of Medicine
University of California, San Diego
La Jolla, California

STEPHEN ROTH
Department of Biology
The Johns Hopkins University
Baltimore, Maryland

BRYAN P. TOOLE
Developmental Biology Laboratory
Departments of Medicine and Biological Chemistry
Harvard Medical School at Massachusetts General Hospital
Boston, Massachusetts

Preface

An outstanding characteristic of the nervous system is that neurons make selective functional contacts. Each neuron behaves as if it recognizes the neurons with which it associates and rejects associations with others. The specific interneuronal relationships that result define the innate neuronal circuits that determine the functioning of this system.

The purpose of this volume is to present some approaches to the problem of neuronal recognition. The volume has been somewhat arbitrarily divided into three sections. In the first section, the overriding theme is the degree of specificity of neuronal recognition. How specific is specific? Is the specificity so precise that the neurites of one neuron will only make synaptic contact with a unique target neuron? If less precise, within what range? Are the rules for specification that are operative in the embryo still operative at the same level of precision when connections regenerate in the mature organism? Are they still operative in dissociated tissue grown in culture?

The second section of this volume contains reviews of morphological studies of synaptogenesis and biochemical studies of synaptic components. Can the morphology of developing cellular contacts provide clues about selectivity? Can the chemical components of synaptic junctions be isolated and characterized? Do they include resolvable components that mediate neuronal recognition?

The third section contains studies seeking to identify the existence of specific molecules that might mediate cellular recognition. A major question here is whether molecules of this type even exist. Does the program that controls differentiation of the nervous system operate simply by regulating the time of maturation and outgrowth of neurites from individual neurons? Do these neurites then make contact on a first-come first-serve basis? Or are there specific molecules on the surfaces of neurons which bind selectively and with high affinity only to specific complementary molecules on appropriate target cells? If the latter, how specific is specific? Is synapse formation dependent on vast numbers of unique complementary pairs? Is it dependent on cell surface differences in *amounts* of complementary substances? Is this problem best studied with nervous system tissue or with other more homogeneous vertebrate tissues which appear to display cellular recognition properties? Are simple eukaryotes that display cellular recognition, such as cellular slime molds, useful experimental tools for studying this problem?

The authors of this volume provide tentative answers to many of these questions. It appears that the questions are neither premature nor hopelessly complex. Systems and strategies are being developed to resolve them. The problem of neuronal recognition is difficult—but solvable.

La Jolla, California SAMUEL H. BARONDES

Contents

6. Biochemical Studies of Synaptic Macromolecules: Are There Specific Synaptic Components?

I. G. MORGAN AND G. GOMBOS

III. TOWARD A MOLECULAR BASIS OF NEURONAL RECOGNITION

7. Cell Recognition in Embryonic Morphogenesis and the Problem of Neuronal Specificities

A. A. MOSCONA

8. An *in Vitro* Assay for Retinotectal Specificity

STEPHEN ROTH AND RICHARD B. MARCHASE

I

Specificity in Synaptic Development and Regeneration

1

Neuronal Recognition in the Retinotectal System

MARCUS JACOBSON

Map me no maps, sir, my head is a map, a map of the whole world.
Henry Fielding, 1730, *Rape upon Rape, or the Justice Caught in His Own Trap*, Act I, Scene v

1. INTRODUCTION

Regularity of association between cells is ubiquitous; it is the *sine qua non* of any organized system, but the nervous system shows a greater range and variety of regular and orderly cellular relationships than any other system. For example, glial cells and neurons are always found in association, and there are very many examples of exclusive association between neurons of specific types. If the same neurons are always found together, either lying in close proximity, forming nonsynaptic intercellular junctions, or making synaptic contact, the embryologist asks how such intercellular contacts developed. Is the intercellular relationship, regular as it may be, merely the result of a web of circumstances that cannot be traced directly to any single cell, or is the orderly relationship the direct result of the properties and activities of certain,

MARCUS JACOBSON · Department of Physiology and Biophysics, University of Miami School of Medicine, Miami, Florida.

identifiable cells? The question has some pragmatic interest—it is likely to be far more difficult to discover the mechanisms of cellular association if they are distributed through the developing system in time as well as space than if the mechanisms are intrinsic properties and functions of the associating cells. Such intrinsic mechanisms might therefore be expressed by cells that are experimentally isolated from the entire system *in vitro* or in cells that by transplantation are put into novel spatial and/or temporal contexts in the developing nervous system.

In summary, these are the experimental tests for intrinsic mechanisms of cellular association, and where such tests have not been applied there is always doubt about the mechanism, however obvious the fact of invariant association between neurons may be. So, before consideration of the retinotectal system, where the proper tests have shown that cells form invariant associations because they have intrinsic compatibility or affinity, it is well to give some examples of the more doubtful cases.

2. ALTERNATIVES TO NEURONAL RECOGNITION

There is no direct evidence for neuronal recognition if, by that term, we mean the specific recognition by one nerve cell of another in order to form a synaptic connection. The concept of neuronal recognition comes indirectly from evidence of intercellular recognition in other systems and can be inferred from the evidence that neurons form orderly and invariant associations. These observations, however impressive, need not necessarily be the result of specific recognition between those nerve cells that form invariant associations. Such associations might just as well have resulted from an initially random arrangement followed by selection of the most stable connections. The selection may be made, for example, on the basis of some principle of utility. There is, however, much evidence against initially random networks out of which organized circuits later develop as the result of use. The evidence against formation of organized neuronal circuits from an initially equipotential network is overwhelming (Sperry, 1965; Jacobson, 1970, pp. 315*ff*). Nevertheless, the cases in which the function of neuronal systems has been shown to change as a result of use or disuse (reviewed by Jacobson, 1970, pp. 336–342) involve mechanisms that seem to be quite different from those that initially establish the neural organization that must have developed before the nervous system can function properly. First, most neuronal systems are completely, or almost completely, organized prenatally before the animal is free to use its nervous system. Even in those systems in which some modification occurs as a result of experience and learning, as in the mammalian visual cortex, the basic structural patterns and functional activities develop prenatally, and we still have to discover how such basic neuronal circuits develop. Why, for example, does the retina always become connected to the appropriate parts of the brain rather than to other parts, and why do the retinal ganglion cells project to the visual centers in a spatially organized pattern, forming selective associations with particular types of neurons there?

Before considering to what extent recognition between neurons may be involved in the initial establishment of functional circuits, we have to consider other mechanisms that might accomplish the same end. Organized neuronal circuits could, in theory, develop as a result of a timetable of assembly. In such a timetable, the elements that are to form the appropriate connections are brought together by suitable schedules of the origin of the cells, growth of the fibers, and readiness of the elements for mutual interactions. The complexity of the final product tends to lead one into thinking that the orderly connectivity in the nervous system could develop only as the result of highly specific mutual affinities and disaffinities between neurons. However, it is conceivable that at the time when any pair of elements formed an association those were the only elements that were able to do so, the others either being absent or at inappropriate stages in their developmental timetable. This hypothesis is sufficiently important to warrant serious consideration, yet it has had very few experimental tests.

It must be borne in mind that the strength of cellular recognition mechanisms may vary from one case to another, and in some experiments the recognition may be altered by the experimental conditions, or other conditions may override the specificity of cellular recognition. Thus, in experiments in which motor nerves and muscles are mismatched, the connections between nerves and muscles seem to develop as the result of temporal coincidence rather than of specific recognition. Any motor nerve can be made to connect with any skeletal muscle either by deflecting the nerves or by transplanting the muscles. The selectivity that occurs in normal development seems to result entirely from spatial and temporal constraints on random connections between motor nerves and muscles. Yet developmental contingencies or accidents may not be the only way in which nerve and muscle form their associations. Cellular recognition may play some role not only in preventing connections between sensory nerves and muscles but even in achieving the correct matches between motor nerves and the appropriate muscles. The work of Mark and his associates has shown that although a given muscle will form connections with any motor nerve it will make functional connections preferentially with the appropriate nerve when that nerve is available, even in the presence of other motor nerves (Mark, 1969, 1974). Such affinity between motor nerves and muscle has so far been found in urodele amphibians and fish. It is possible that in mammals it either does not occur or is not strong enough to manifest itself under the experimental conditions.

To test whether a normal temporal pattern of development is essential for the formation of normal retinotectal connections, we have surgically altered the schedule of formation of connections between the eye and the brain. The results of these experiments discount timing as an important factor (Jacobson and Hunt, 1973; Hunt and Jacobson, 1973). Eyes removed from developing embryos of the clawed frog, *Xenopus*, were either grown *in vitro* for up to 2 weeks or grafted to the body for up to 2 months before being replaced in the eye socket and permitted to form connections with the midbrain tectum. In all cases, an orderly and complete retintotectal projection developed, although many cells had been added to both the retina and the tectum while they were separated and in spite of the fact that there was a change in the timing of arrival of retinal axons into the tectum (Jacobson and Hunt, 1973; Hunt and Jacobson, 1973). This result is consistent with earlier evidence that, in adult fish and amphibians, cutting the optic nerve results in regeneration and reestablishment of orderly retinotectal connections, although the regenerating nerve fibers probably do not reenter the tectum in the same temporal order as they did during normal development (Jacobson, 1961; Gaze and Jacobson, 1963). In a system that was assembled entirely by mechanisms of cellular recognition, the temporal order of assembly would not affect the final product. No such demonstration has yet been made: reaggregated nerve cells may show some semblance of orderly assembly often euphemistically described as "histotypic," although obviously abnormal. Cellular disaggregation, however "gentle," not only alters the timetable of normal development but probably also alters the cellular recognition mechanisms. Other heroic manipulations of the developing nervous system, such as exposure to chemical inhibitors, irradiation, virus infection, or surgery, have been used to alter development, but in all cases interpretation of the results is limited by lack of information about the total effect of the experimental procedures. The experiment may conceal mechanisms that operate in normal embryos as well as reveal mechanisms that may not exist under normal conditions.

Experimental tests of the timing hypothesis have the disadvantage that they may interfere, surgically or chemically, with normal development and may produce effects that have nothing to do with timing. On the other hand, inferences made from observations of normal development are weak because they can only show whether events in development are coincident, not whether they are causally related. Yet there have been several attempts to infer that the regularity of neuronal organization is the result of the temporal order of assembly of the system. For example, the conclusion that there is a temporal factor in the development of specific afferents to granule cells of the hippocampal dentate gyrus (Gottlieb and Cowan, 1972) is a case of that kind of inference. It is based on evidence that the ratio of crossed to uncrossed inputs to the hippocampal granule cells correlates with the time of origin of the granule cells: the earlier their origin the greater the proportion of uncrossed inputs. Thus at the time the first-formed granule cells are ready to receive synapses on the proximal portions of their dendrites the only afferents present are those from the same side of the hippocampus, with the fibers from the opposite side approaching the dentate gyrus only some days later. From such evidence it has been inferred that recognition between the specific afferents and the dentate granule cells is unnecessary; all that seems to be required is that the axons arrive at a time when the granule cell dendrites are ready to receive them. The obvious truth that both the pre- and postsynaptic components of any synapse must be present at the same time in the same place before they can form a synapse does not necessarily mean that the temporal coincidence is the cause of the formation of the synapse. The observations are equally consistent with a neuronal recognition mechanism, and for that mechanism to be effective it is also necessary for the components to come together and be ready to interact at the right time in development. I have labored this point only because the relationship between cause and effect has not been clearly defined in the analysis of the development of the nervous system. I am taking the commonsense view that a cause should be a necessary and sufficient condition that invariably accompanies an event—in this case, the formation of a synapse. In most cases, nerve cells come together during development without forming synapses. That the proper components are present at a certain time and place is a necessary but not sufficient condition for the formation of a synapse. It seems as if there must be some degree of mutual affinity between the neurons, or *neuronal recognition*, before the "correct" synaptic associations can develop. Neuronal recognition, therefore, as a mechanism of achieving association between the "correct" neurons, requires that the neurons be prelabeled before they form any association. This introduces the problem of the origin of the labels and their spatial deployment in sets of neurons and their expression during neuronal recognition in the formation of synaptic connections. Neurons, usually at some considerable distance apart initially, grow toward each other to form a synaptic connection; this raises the problem of the mechanisms of targeting of pre- upon postsynaptic elements in the development of neuronal connections and the precision of such neuronal targeting.

3. DEVELOPMENT OF SPECIFIC PROPERTIES IN NEURONAL SETS

Neuronal recognition as a mechanism of establishing the "correct" association between two sets of neurons—for example, between the retinal ganglion cells and the tectal cells—requires prior labeling of those neuronal sets. One problem therefore is how the sets of retinal ganglion cells acquire specific cellular labels and how the tectal cells acquire a set of matching labels so that retinal neurons can recognize tectal neurons. At the very least this demands matching of retinal and tectal neuronal sets, but there is evidence that recognition may occur between subsets or even between particular neurons.

In many parts of the central nervous system, one set of neurons projects its axons to form connections with one or more neuronal sets so that the spatial order of the elements in one set is preserved in the spatial order of the connections to the other set or sets. Let us assume that the elements of each neuronal set must develop specific properties, spatially deployed as a map in the neuronal set and expressed during development as the formation of a topographically ordered projection from that set to other sets in which matching maps of properties have developed. Because the material nature of such properties or maps is unknown, it is convenient to refer to the *locus specificity* of neurons, defined as the property of an element of a neuronal set, determined by its position in the set, predisposing it to connect selectively with an element at the corresponding position in other sets.

Neuronal sets develop independently, in parallel, in different places in the nervous system (Fig. 1). How does each set develop all the necessary locus specificities deployed throughout its neuronal population and with the axes of each set aligned with those of other sets and with the embryonic axes? The evidence is that all sets derive locus specificity from a common source of positional information available throughout the embryo. It can be shown that neuronal sets can obtain positional information not only in their normal situation in the nervous

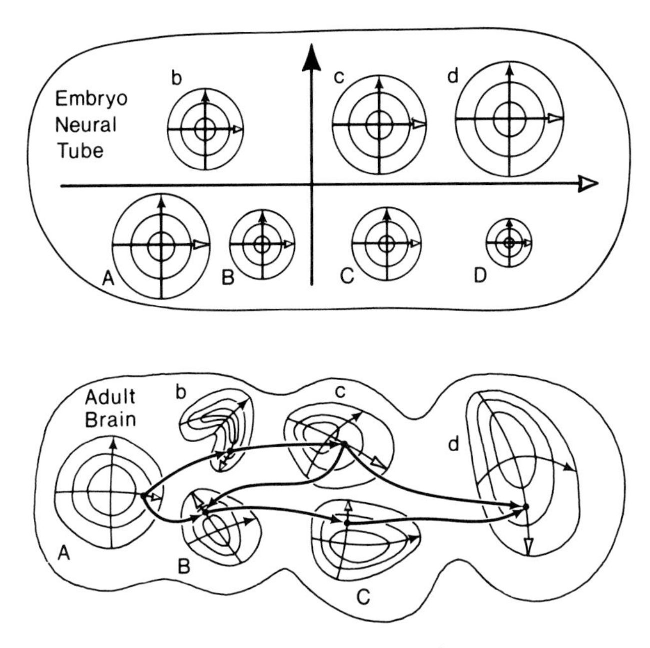

Fig. 1. Development of neuronal sets. In the upper diagram, seven neuronal sets are shown in various places in the embryonic neural tube. They are aligned with one another and with neural tube as a whole because they derive their AP and DV axes from an external source of axial cues or positional information. Within each set, the cells acquire position-dependent properties, termed *locus specificities*, with reference to the axes laid down in the neuronal set. The lower diagram shows that the sets change their size and shape, and may even disappear completely (set D), but connections between the sets are formed by linking neurons that have the same locus specificity in different sets.

system but even when they are grafted to other parts of the embryo outside the nervous system (Fig. 2).

The hypothesis proposed here is that positional information is transferred to neuronal sets and within the cells comprising the set by means of intercellular communication channels. The mode of transfer between cells has not been discovered. It may occur via intercellular junctions such as gap junctions, or by exocytosis and pinocytosis, or by exchange of cell surface materials between cells that are in contact. While these channels of information transfer are open, the set may obtain new positional information, and thus, if it is displaced during growth of the embryo or by experimental surgery, it can obtain positional information from the external reference source in order to replace its reference axes with axes that are realigned with those of other

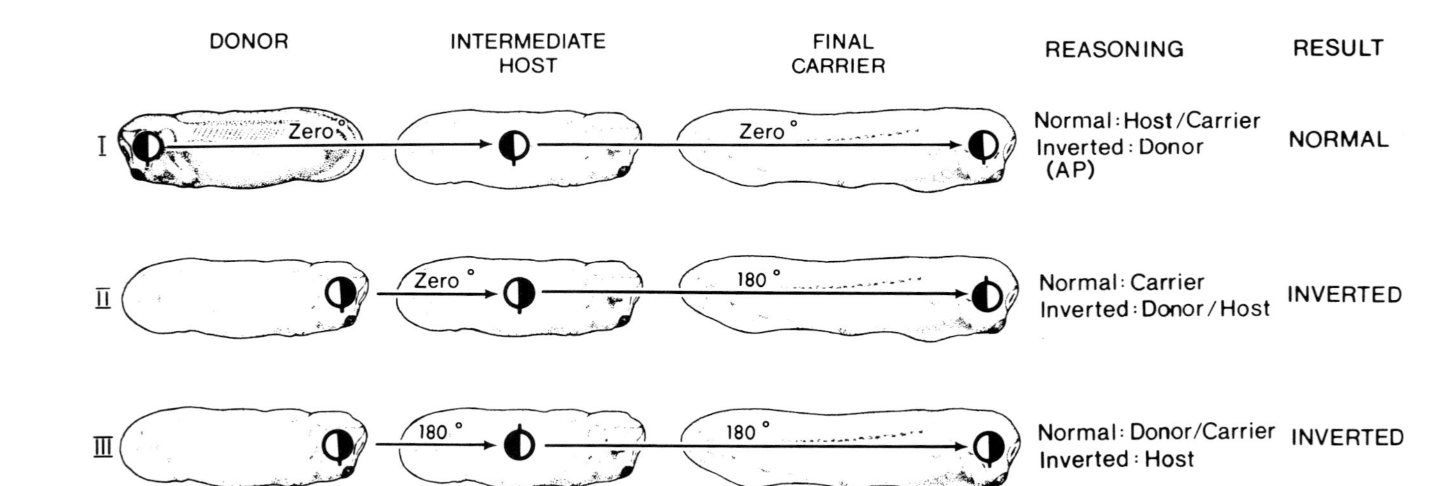

Fig. 2. An eye that spends the critical period (stages 29–31 in *Xenopus* embryos) while implanted on the body of a host embryo, and is later returned to the eye socket, forms a retinotectal map whose axes are derived from the eye's orientation on the body of the host. From Hunt and Jacobson (1972).

neuronal sets and with the embryo as a whole (Jacobson, 1968*a*; Hunt and Jacobson, 1972). This regulatory capacity is lost at some stage in normal development. At that stage the neuronal set becomes refractory to the influence of external positional information and the set is said to be *specified*. It has been shown that this refractoriness is not due to unavailability of positional information, for the positional information from the external reference source remains available to other sets of neurons that are not refractory to it (Hunt and Jacobson, 1973). It bears repeating that many neuronal sets, in different parts of the embryonic neural tube, share the same external source of positional information and thus develop internal reference axes that are aligned with the AP and DV axes of the embryo.

At the time of specification, which occurs in the embryo before the development of connections between neuronal sets, the set becomes uncoupled from the external source of positional information, but cells that may continue to be added to the set derive positional information from within the set. Nevertheless, although neuronal sets become different from one another in size and shape, they conserve the positional information that they all obtained from the shared external source and use the information in the development of locus specificities in all the cells that finally comprise the set. These locus specificities are expressed in the formation of connections between neuronal sets. An element carrying a particular locus specificity in one set—for example, the retina—will be able to connect with an element carrying the same locus specificity in other sets—for example, the tectum, pretectal nuclei, and lateral geniculate nucleus. A neuron in one set will then be able to connect in a point-to-point way with the matching neurons in other sets, with the proviso that the developmental anatomy permits such linkages to occur (Fig. 1).

Most of the evidence for the thesis that I have put forward is available elsewhere (Jacobson, 1968*a*,*b*, 1970; Hunt and Jacobson, 1972, 1973; Jacobson and Hunt, 1973). I will give a very brief synopsis of the most important pieces of evidence and concentrate mainly on the experiments that bear directly on the problem of information transfer in developing neuronal sets.

In order to study the developmental origins of the locus specificities in the set of retinal ganglion cells, our experimental strategy was to reposition one eye of the embryo of the clawed frog, *Xenopus*, at various stages of development before the eye had started forming nervous connections with the brain. The eye either was inverted within the eye socket, or was transplanted to another position on the body, or was explanted into tissue culture for some time. In all cases, the eye was finally reimplanted in the eye socket, either misoriented or correctly aligned with the embryonic axes, in order to assay the resulting retinotectal projection.

The basic experiment, in which the eye was simply excised and reimplanted in different orientations in its own socket, at different stages of embryonic development, showed that the set of retinal ganglion cells undergoes a change of state, termed *specification*, at stages 29–31 in *Xenopus* embryos. Inversion of the eye before stage 28 results in replacement and realignment of the retinal axes so that the resulting retinotectal projection is normally aligned with the embryo, even though the eye is anatomically inverted. Inversion of the eye after stage 31 results in an inverted retinotectal projection. The system therefore undergoes a change of state in about 7 h, at embryonic stages 29–31, which alters its response to misalignment. These basic experiments (Jacobson, 1968*a*) were followed by others showing that before specification the eye has stable axes about which retinal locus specificities are deployed. Thus eyes from donor embryos at stages 22–26 were grown *in vitro* until they developed the appearance of stage 32 eyes and were then reimplanted in stage 32 carrier embryos. The resulting retinotectal maps were misaligned when the eye was misaligned and normal when the eye was normally aligned, showing that the retinal locus specificities were derived from the original donor and not from the carrier embryo (Hunt and Jacobson, 1973). We also showed that the conditions necessary to replace the retinal axes of an inverted eye are not confined to the eye socket but also exist on the body, so that axial replacement occurs in unspecified eyes grafted upside down on the flank of the embryo (Fig. 2).

Specification is not due to changes in the extraocular conditions necessary to replace the retinal axes in an inverted eye but is due to a refractoriness within the eye itself (Hunt and Jacobson, 1973). Thus, in *Xenopus* embryos, the axes of a stage 26 eye are replaced after the eye is grafted upside down into the eye socket of a stage 32 embryo, but the axes of a stage 32 retina are not replaced when the eye is inverted in a stage 32 embryo, or when a stage 32 eye is grafted upside down into a stage 28 embryo. The specified eye cannot normally have its retinal axes replaced, although fragments of an eye that has already been specified may revert to an unspecified state (Hunt and Jacobson, 1973, 1974). However, it is not possible to hasten the time of specification in the retina by transplanting an eye at stage 26 to a stage 29–31 embryo—the eye undergoes the change of stage on its own-developmental schedule (Jacobson and Hunt, 1973). This result should be considered in conjunction with the experiments showing that eyes can be prematurely specified by treatment of the eyes at stages 24–25 with the ionophore X537A (Jacobson, 1976*a*).

There are many possible mechanisms that may be responsible for the change from replaceable to fixed axes in neuronal sets. Failure of axial replacement in misaligned or inverted neuronal sets may be the result of absence of the signals arising outside the neuronal set, absence of channels through which the signals are transmitted to and within the neuronal set, failure of reception or transduction of the signals or any other deficiency, or loss of response by the neuronal set. The evidence, although inconclusive, suggests that the change to a refractory state is due to an uncoupling of the neuronal set as a whole from the external source of positional signals.

The evidence is against specification resulting from changes in the extraocular conditions. We can rule out mechanisms in which signals arise, inhibitory controls disappear, or conditions change transiently at the time of specification because we have shown that the extraocular conditions necessary for axial replacement in the retina of inverted eyes persist long after the normal time of specification.

There are three important facts to take into account when considering the intraocular conditions at the time of specification. First, at the time of specification the first retinal cells withdraw from the mitotic cycle, and these cells later differentiate as ganglion cells in the center of the retina (Jacobson, 1968*b*; Kahn, 1973). Second, intercellular gap junctions are frequently seen between the retinal cells before embryonic stage 28 but disappear between the cells in the center of the retina at stages 29–31 (Dixon and Cronly-Dillon, 1972; Jacobson, unpublished observations). These gap junctions may reappear if the eye is cut in half at stage 32 (Jacobson, unpublished). In such half-eyes, the retinal axes can be realigned after the stage at which the eye has normally become refractory to axial realignment (Jacobson and Hunt, 1973). Third, at stages 29–31 the retina contains less than 1% of the total number of retinal cells present in the adult. Neurons that are generated after the time of specification must derive locus specificity from within the eye, for if the eye is inverted at stage 31, when it contains less than 1000 ganglion cells, the entire retinotectal projection is inverted in the adult, whose retina contains hundreds of thousands of ganglion cells. This shows that the ganglion cells that originated after the eye was inverted did not take heed of the eye's relationship to the body axes but derived locus specificity from reference axes within the eye.

To determine how information is transmitted to the newborn retinal cells, it is necessary to determine the pattern of retinal histogenesis and the lineage of the ganglion cells. Until stages 29–30, the retina consists entirely of stem cells, but at that stage, when the total number of cells is less than 1000, the first postmitotic cells appear. These cells are destined to differentiate as ganglion cells, for if the eye is cumulatively labeled with tritiated thymidine, starting at stages 29–30, the only unlabeled cells are some ganglion cells at the center of the retina (Jacobson, 1968*b*). After stage 30, until adulthood, the ganglion cell population increases linearly by about 300 cells per day, and the total cell content of the retina increases linearly by about 3000 cells per day (Jacobson, 1976*b*). Assuming a generation time of 12 h (Jacobson, 1968*b*), there are about 1500 stem cells in the retina, of which 150 give rise to ganglion cells. These stem cells lie at the periphery of the retina and divide asymmetrically to produce another stem cell, situated peripherally, and a postmitotic cell that lies central to the stem cell, between it and the previous generation of ganglion cells. The retinal ganglion cells form in rings, with the younger generations lying more peripherally (Straznicky and Gaze, 1971). By means of electron microscopy, it has been shown that there are no gap junctions between the older ganglion cells whereas gap junctions are seen only between cells at the retinal periphery (Dixon and Cronly-Dillon, 1972; Jacobson, unpublished). However, it is not clear from the electron micrographs whether these gap junctions are exclusively between stem cells, between stem cells and their most recent progeny, or between those and the earlier generation of young ganglion cells. Knowing which retinal cells are coupled would help to establish the direction of flow of information during retinal histogenesis.

Because histogenesis continues after the stage of development at which the neuronal set becomes specified, the problem is how locus specificities are acquired by the newly formed neurons that develop after the change to a refractory state of the set as a whole. There must be a source of positional information within the set that can be transferred to the newly formed neurons originating later in development. In principle, there are two ways in which such information transfer may be accomplished in the histogenesis of neuronal sets: (1) by a cell lineage mechanism in which a singularity of the stem cell is transmitted at each division to its daughter cells or (2) by a cell interaction mechanism in which the newborn cell derives positional information from neighboring cells. Evidence that a cell lineage mechanism does not operate in the histogenesis of the vertebrate retina is that after removal of half of the embryonic retina, including about half the stem cells, a normal, apparently complete, retinotectal projection develops from the residual retinal cells (Hunt and Jacobson, 1973, 1974). However, it is not yet certain whether positional information is transferred from the stem cells to their progeny during the latter's terminal cell cycle or whether the information is transferred from the central retinal cells to the peripheral, newly born generation of cells.

4. TARGETING OF THE RETINAL AXONS ONTO TECTAL NEURONS

How axons from one neuronal set, such as the set of retinal ganglion cells, grow along the correct pathways and connect with the appropriate cells in other neuronal sets is too complex a problem to deal with as a whole, but must be reduced to more tractable parts. I will not discuss here how axons select the correct pathway to grow to their targets or how the axons stop growing after arriving at those targets. I will limit the discussion to the problem of how the retinal axons might associate selectively with the correct elements in the optic tectum. In theory, there are two main ways in which the retinal axons might arrange themselves in a retinotopically organized map within the boundaries of the midbrain tectum and other visual centers. In the first of these ways, the retinal axons might sort themselves out without reference to any other cells along the pathway or within the tectum. Merely by axon–axon interactions and preferential association between axons arising close together in the retina, the axons could sort out to form a retinotopically organized map within the boundaries of the tectum. In this mechanism the tectal cells need have no specific labels. However, if the retinal axons orient themselves in some invariant relationship to the tectum, there must be some tectal markers with which the set of retinal axons as a whole is aligned. The question then arises whether a single tectal marker is sufficient to align the retinal axon population as a whole or whether the retinal axons assume the same spatial order at their endings in the tectum as they had at their origin in the retina, because both the retina and the tectum have matching maps of position-dependent properties. The rule for the development of such a retinotectal map would thus be that each retinal element can form a stable connection only with a corresponding tectal element (Sperry, 1963). However, as we have previously discussed at considerable length (Hunt and Jacobson, 1974), this does not mean that the tectal element that corresponds with any given retinal element should always be at the same position in the tectum under all experimental conditions or at all stages of development. An element may retain its identity even though it moves during development or is surgically transplanted to a different position within the map, or an

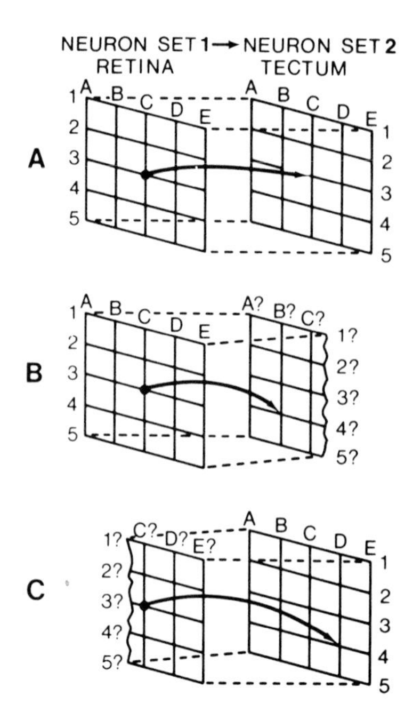

Fig. 3. Relations between complete and partial neuronal sets. (A) Mapping of neuronal set 1 (retina) onto set 2 (tectum), with elements linked at position C3 in both sets. (B, C) What might be found some time after removal of part of the retina or tectum. The relations appear to have changed, but one cannot know what changes have occurred because the identities of elements at various positions cannot be determined from the map.

element so moved may change its identity to one consistent with its final position in the map. These alternatives can only be resolved by the appropriate experimental tests, to be described below. During normal development of the retinotectal projection, the map cannot show which elements are present at any stage of development. It also cannot indicate whether an element at any position at one time during development has the same identity as the element found at the same position at another time. Although the map can show that a projection is continuous and orderly, it cannot show whether it is complete or incomplete. This problem becomes particularly troublesome in assessing the effects of experimental lesions on the retinotectal projection (Fig. 3). The main issue at present is whether the retina maps on the tectum element by element as Sperry's theory demands, or, alternatively, whether the entire retina matches the entire tectum as a system (Gaze and Keating, 1972), or even whether mapping may be accomplished entirely by interactions between the retinal axons themselves, which would require only a single point of tectal reference for alignment with the tectum (Levine and Jacobson, 1974).

In the following discussion of these modes of retinotectal mapping, it should be noted than an element of a set of neurons need not be defined anatomically: an element may be a single neuron or a group of neurons lying close together, and the anatomical identity of the element in any set cannot be determined *a priori* but must be discovered empirically. The three alternative modes of mapping retina on tectum are in fact quite different, so that it should be possible to design experiments to distinguish between them. However, there are serious limitations both in the concept of retinotectal matching and in the actual practice of retinotectal mapping that have to be taken into consideration in designing experiments to test between the various alternative modes of mapping retina on tectum. We have discussed these methodological problems elsewhere (Hunt and Jacobson, 1974). In brief, the retinotectal map merely shows that there is an orderly deployment of retinal axons within the boundaries of the tectum but does not show whether any particular retinal axon invariably associates with the same tectal element under all experimental conditions. Because the identity of the tectal or retinal elements cannot be derived from the map, the rules for mapping or matching retinal and tectal elements also cannot be derived from such maps. The map gives the positions of retinal and tectal elements in a spatial coordinate system with reference to the anatomical boundaries of the retina and the tectum and shows a continuous and orderly deployment of retinal axonal endings within the tectal boundaries. By stimulating retinal neurons and recording postsynaptically from tectal cells, we have shown that the retinotectal map is projected onto the tectal elements (Skarf and Jacobson, 1974). We infer that a particular retinal element always associates with a tectal element at the same position in the map in normal adult animals. This inference is consistent with all the theories mentioned above, so that other tests have to be used to distinguish between them. Five experimental strategies have been used to try to discover the rules for mapping the retina on the tectum. One may summarize the results of these experiments by saying that when the evidence permits any conclusion it is in favor of matching the retina and tectum element by element, but where the evidence has been adduced in favor of matching system by system it can be shown that the evidence is intrinsically incapable of being interpreted unambiguously. The five experimental approaches to the problem are as follows:

1. Mapping the retinotectal projection during normal development to determine the changes that occur at different times in development (Gaze *et al.*, 1972, 1974; Chung *et al.*, 1974).
2. Studying changes in the retinotectal map produced by removing parts of the retina and/or tectum (Gaze and Sharma, 1970; Yoon, 1971, 1972*a*,*b*, 1973; Sharma, 1972; Jacobson and Levine, 1975*a*).

3. Studying changes in the retinotectal map produced by altering the number of axons projecting from the retina to the tectum without surgically damaging either the retina or the tectum (Jacobson and Gaze, 1965).
4. Comparing the tectal projection of a surgically altered retina with the projection of a normal retina into the same tectum (Jacobson and Hunt, 1973; Hunt and Jacobson, 1974).
5. Studying the retinotectal map after rearrangement of the elements of the tectum (Yoon, 1973; Levine and Jacobson, 1974; Jacobson and Levine, 1975*a*,*b*).

Two kinds of results have been obtained from these various experimental strategies, and we may consider both kinds in terms of sets of neurons connected in a point-to-point manner as shown in Fig. 3. In the first kind of result, the elements of each set retain their expected associations after various surgical operations (Attardi and Sperry, 1963; Jacobson and Gaze, 1965; Levine and Jacobson, 1974; Jacobson and Levine, 1975*b*), showing, for example, that an element in the retinal set will recognize the same element in the tectal set after the latter has been moved to another position in the tectum (Jacobson and Levine, 1975*b*). In the second kind of result, after removal of part of one or both sets an element in the retinal set is found to connect in the tectal set with a different element postoperatively from the element with which it associated preoperatively (Gaze and Sharma, 1970; Sharma, 1972; Yoon, 1971, 1973; Jacobson and Levine, 1975*a*). In theory, there are many mechanisms that might subserve such cases of plasticity. For example, such changes may conceivably be due to lax constraints on the association between presynaptic and postsynaptic elements, so that failure to find the matching element results in another one's being selected. The same final configuration would result if, after removal of an element bearing a particular specificity, the latter was adopted by one of the remaining elements. All the logical alternative mechanisms must be considered when trying to deduce the mechanism of a change of association between retina and tectum following surgical intervention.

Consideration of the development of the retinotectal connections requires that each element have at least two attributes, namely a position-dependent property and a function (growth of axons and dendrites, selection of a synaptic site, formation of a functional synapse) which must be performed to connect elements in the retina with the corresponding elements in the tectum. Then, if an element at position C3 in the retina normally connects with an element C3 in the tectum, we assume that the elements have position-dependent property pC3 and function fC3. However, if experimental surgery results in an apparent change in the matching between retina and tectum, we cannot assume that the elements have retained their preoperative properties and operations. There are six alternative ways in which the system may respond to experimental intervention:

1. New elements may be formed to replace those that have been lost or damaged. This is particularly likely to occur in developing systems, and the proper controls have to be done to show whether cell replacement has occurred.
2. The elements themselves may have moved to new positions in one or both sets. For example, in Fig. 3B, an element in the tectum at position C3 might have moved to position B4. This cellular migration might have occurred with or without changes in the functions and/or properties of the elements.
3. The properties of the elements might have changed in one or both sets. For example,

$$\begin{array}{ll} & pC3 \rightarrow pB4 \\ \text{or} & pB4 \rightarrow pC3 \\ \text{or} & pC3 \rightleftharpoons pB4 \end{array}$$

4. The functions of the elements might have changed in one or both sets. For example,

$$\begin{array}{ll} & fC3 \rightarrow fB4 \\ \text{or} & fB4 \rightarrow fC3 \\ \text{or} & fB4 \rightleftharpoons fC3 \end{array}$$

5. The functions and operations might have changed.
6. The surgical operation might have altered the rules for mapping retina into tectum. For example, if the rule is that each retinal element matches only one tectal element, the rule may change to matching several elements in the retina with several in the tectum, or to any other rule that results in an orderly map.

Consideration of these alternatives is no mere academic diversion. Chemical or surgical interference with the nervous system, especially during development, is very likely to produce one or more of the changes just enumerated. These considerations invite a critical reappraisal of all the studies of retinotectal matching. As a start in this direction, Hunt and Jacobson (1974) found that while several ways of matching retina and tectum had been theorized in the literature the actual mechanism had not yet been discovered, and that in view of technical limitations the mechanism could not have been discovered by the experimental strategies used. For example, consider the experiments in which the retinal projection becomes compressed into the remaining half of the tectum after removal of the other half, or in which the projection from a remaining half-retina expands to fill the entire tectal space (Gaze and Sharma, 1970; Yoon, 1971, 1972*a*,*b*, 1973; Sharma, 1972). We do not know which of the six previously enumerated mechanisms might have resulted in the reorganization of retinotectal maps. Yoon (1972*a*,*b*) has favored "respecification" of tectal cells, but that term might be applied equally to all the possible changes in properties and functions of retinal and tectal elements that are produced by experimental surgery. In any case, one cannot infer the rules of mapping retina onto tectum from such experiments.

Studying the changes in retinotectal maps during development might appear to be a more promising way of understanding how an individual retinal element recognizes its tectal counterpart. However, in practice, the results have been disappointing. To illustrate the limitations of this approach, consider a retinal axon which originates from the center of the retina and ends in the center of the tectum in normal adult animals. During development this axon may be found to terminate at different tectal positions at different stages of development, finally coming to rest at its adult position, as Gaze *et al.* (1972, 1974) have observed. The question which cannot be answered from such evidence is whether an axon arising from a given retinal position associates, at different times, with the same tectal elements that move from the front to the back of the tectum during development or whether the axon in question associates with a succession of different tectal elements, each of which remains at a fixed position in the tectum. In the latter case, do the different neurons with which the axon associates during development have different specificities or do they all have the same specificity? Finally, does the specificity of the axonal ending change or remain constant during development? The question of whether an axon is permanently connected to a neuron which is moving in the tectum or the retinal axon forms synaptic connections successively with a series of neurons cannot be answered by the mere demonstration of the presence of synaptic connections between axons and tectal neurons (Chung *et al.*, 1974; Scott, 1974). A patent failure to ask the right questions distinguishes discussions of such experiments. The simple conclusion that retinal axons are not constrained to form permanent associations with tectal cells with which they are specifically matched is an indication of the restricted view adopted in interpreting such results.

Because all the experiments mentioned above fail to give any information about the properties and operations that underlie the specific association between retinal and tectal elements, they are inherently incapable of determining whether one retinal axon recognizes and connects with a single specific tectal target or whether, failing to find that target, it can connect elsewhere in the tectum. Also, none of the experiments mentioned above deals with the question of whether individual retinal and tectal elements are matched, whether the retina and tectum match as wholes, or whether the retinotectal projection develops without recognition of tectal cells by retinal fibers.

The most direct evidence of neuronal recognition in the retinotectal system is the observation that the retinal axons regenerate to the tectal elements with which they were previously associated, even after those elements have been translocated in the tectum (Yoon, 1973; Levine and Jacobson, 1974) or when some tectal elements have been duplicated by interchange of noncorresponding pieces between the left and the right sides of the tectum (Jacobson and Levine, 1975*a*,*b*). In such experiments

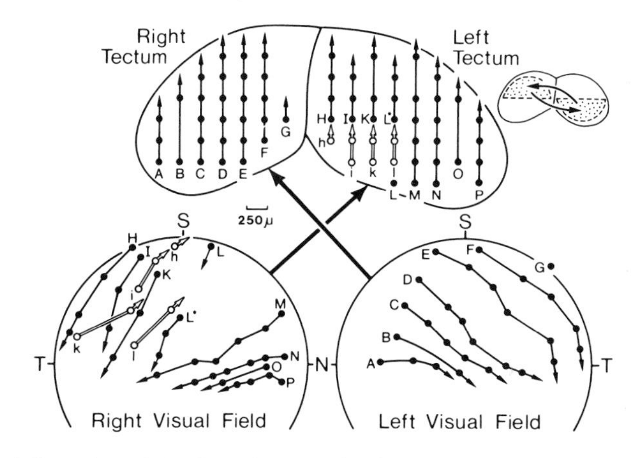

Fig. 4. Stability of positional markers in the frog's optic tectum is shown by the targeting of optic fibers onto a translocated tectal graft. Visuotectal projection 64 days after left–right exchange of pieces of tectum (stippled areas), with 180° rotation of each tectal graft. Each letter and dot on the tectum represent the position at which a microelectrode recorded action potentials from several units in response to a 1° spot of light at the correspondingly designated position in the visual field. The full extent of the visual field from which a stimulus evoked any response was about 3 times the diameter of the dot. The projection from the left visual field to the right tectum is complete and orderly, and does not differ from the normal, as discussed by Jacobson and Levine (1975*a*). The projection from the right visual field to the left tectum has a scotoma in the superior nasal quadrant of the field, whereas part of the superior temporal portion of the field projects in duplicate: normally to the back of the left tectum (rows H, I, K, L, L*, shown by black dots) and inverted to the front of the left tectum, to the position of the graft (rows h, i, k, l, shown by white dots).

(Fig. 4), restoration of the association between retinal elements and misplaced tectal elements shows that part of the set of retinal elements matches part of the tectal set without reference to the remainder of the elements in either the retinal or the tectal set. To determine whether recognition occurs between individual retinal and tectal elements is not at present feasible, as it would necessitate transplantation of a single tectal element to see whether it continued to serve as a target for the corresponding retinal axons. The best that can now be done is to look at the discrimination made by individual retinal fibers in selecting the appropriate tectal site—for example, at the boundaries between the tectal transplant and surrounding tectum. In the most favorable cases, it has been found that the retinal axons can discriminate between the matching and nonmatching tectal elements that are separated by about 50 μm across the boundary between the tectal transplant and surrounding tectum (Jacobson and Levine, 1975*b*).

5. REFERENCES

Attardi, D. G., and Sperry, R. W., 1963, Preferential selection of central pathways by regenerating optic fibers, *Exp. Neurol.* **7**:46.

Chung, S. H., Keating, M. J., and Bliss, T. V. P., 1974, Functional synaptic relations during the development of the retino-tectal projection in amphibians, *Proc. R. Soc. London Ser. B* **187**:449.

Crossland, W. J., Cowan, W. M., Rogers, L. A., and Kelly, J. P., 1974, The specification of the retino-tectal projection in the chick, *J. Comp. Neurol.* **155**:127.

Dixon, J. S., and Cronly-Dillon, J. R., 1972, The fine structure of the developing retina in *Xenopus laevis, J. Embryol. Exp. Morphol.* **28**:659.

Gaze, R. M., and Jacobson, M., 1963, A study of the retino-tectal projection during regeneration of the optic nerve in the frog, *Proc. Soc. London Ser. B* **157**:420.

Gaze, R. M., and Keating, M. J., 1972, The visual system and "neuronal specificity," *Nature (London)* **237**:375.

Gaze, R. M., and Sharma, S. C., 1970, Axial differences in the reinnervation of the goldfish optic tectum by regenerating optic nerve fibers, *Exp. Brain Res.* **10**:171.

Gaze, R. M., Chung, S. H., and Keating, M. J., 1972, Development of the retinotectal projection in *Xenopus, Nature (London) New Biol.* **236**:133.

Gaze , R. M., Keating, M. J., and Chung, S. H., 1974, The evolution of the retinotectal map during development in *Xenopus, Proc. R. Soc. London Ser. B* **185**:301.

Gottlieb, D. I., and Cowan, W. M., 1972, Evidence for a temporal factor in the occupation of available synaptic sites during the development of the dentate gyrus, *Brain Res.* **41**:452.

Hunt, R. K., and Jacobson, M., 1972, Development and stability of positional information in *Xenopus* retinal ganglion cells, *Proc. Natl. Acad. Sci. USA* **69**:780.

Hunt, R. K., and Jacobson, M., 1973, Specification of positional information in retinal ganglion cells of *Xenopus*: Assays for analysis of the unspecified state, *Proc. Natl. Acad. Sci. USA* **70**:507.

Hunt, R. K., and Jacobson, M., 1974, Neuronal specificity revisited, *Curr. Top. Dev. Biol.* **8**:203.
Jacobson, M., 1961, The recovery of electrical activity in the optic tectum of the frog during early regeneration of the optic nerve, *J. Physiol. (London)* **157**:27P.
Jacobson, M., 1962, The representation of the retina on the optic tectum of the frog: Correlation between retinotectal magnification factor and retinal ganglion cell count, *Q. J. Exp. Physiol.* **47**:170.
Jacobson, M., 1968*a*, Development of neuronal specificity in retinal ganglion cells of *Xenopus, Dev. Biol.* **17**:202.
Jacobson, M., 1968*b*, Cessation of DNA synthesis in retinal ganglion cells correlated with the time of specification of their central connections, *Dev. Biol.* **17**:219.
Jacobson, M., 1970, *Developmental Neurobiology*, Holt, Rinehart and Winston, New York.
Jacobson, M., and Gaze, R. M., 1965, Selection of appropriate tectal connections by regenerating optic nerve fibers in adult goldfish, *Exp. Neurol.* **13**:418.
Jacobson, M., and Hunt, R. K., 1973, The origins of nerve cell specificity, *Sci. Am.* **228**:26.
Jacobson, M., and Levine, R. L., 1975*a*, Plasticity in the adult frog brain: Filling the visual scotoma after excision or translocation of parts of the optic tectum, *Brain Res.* **88**:339.
Jacobson, M., and Levine, R. L., 1975*b*, Stability of implanted duplicate tectal positional markers serving as targets for optic axons in adult frogs, *Brain Res.* **92**:468.
Kahn, A. J., 1973, Ganglion cell formation in the chick neural retina, *Brain Res.* **63**:285.
Levine, R., and Jacobson, M., 1974, Deployment of optic nerve fibers is determined by positional markers in the frog's brain, *Exp. Neurol.* **43**:527.
Mark, R. F., 1969, Matching muscles and motoneurones: A review of some experiments on motor nerve regeneration, *Brain Res.* **14**:245.
Mark, R. F., 1974, *Memory and Nerve Cell Connections*, Clarenden Press, Oxford.
Scott, T. M., 1974, The development of the retinotectal projection in *Xenopus laevis*: An autoradiographic and degeneration study, *J. Embryol. Exp. Morphol.* **31**:409.
Sharma, S. C., 1972, Reformation of retinotectal projections after various tectal ablations in adult goldfish, *Exp. Neurol.* **34**:171.
Skarf, B., and Jacobson, M., 1974, Development of binocularly driven single units in frogs raised with asymmetrical visual stimulation, *Exp. Neurol.* **42**:669.
Sperry, R. W., 1963, Chemoaffinity in the orderly growth of nerve fiber patterns and connections, *Proc. Natl. Acad. Sci. USA* **50**:703.
Sperry, R. W., 1965, Embryogenesis of behavioral nerve nets, in: *Organogenesis* (R. C. DeHaan and H. Ursprung, eds.), pp. 161–186, Holt, Rinehart and Winston, New York.
Straznicky, K, and Gaze, R. M., 1971, Growth of the retina in *Xenopus laevis*: an autoradiographic study, *J. Embryol. Exp. Morph.* **26**:69.
Yoon, M., 1971, Reorganization of retinotectal projection following surgical operations on the optic tectum in goldfish, *Exp. Neurol.* **33**:395.
Yoon, M., 1972*a*, Reversibility of the reorganization of retinotectal projection in goldfish, *Exp. Neurol.* **35**:565.
Yoon, M., 1972*b*, Transposition of the visual projection from the nasal hemiretina onto the foreign rostral zone of the optic tectum in goldfish, *Exp. Neurol.* **37**:451.
Yoon, M., 1973, Retention of the original topographic polarity by the 180° rotated tectal reimplant in young adult goldfish, *J. Physiol. (London)* **233**:575.

2

Specificity of Nerve–Muscle Interactions

DOUGLAS M. FAMBROUGH

1. INTRODUCTION

Developmental neurobiology today can be described as the field of scientific research in which investigators are attempting to explain the organization of extremely complex adult nervous systems in terms of a minimum number of "simple" mechanisms of cellular interaction and the timing of these interactions during development. Even the briefest description of a nervous system will acknowledge the proper connectivity of functional elements. Within a species the organization of the nervous system is extremely similar and not very different from that of any closely related species. This uniformity of organization is ensured by cellular mechanisms that determine the establishment and maintenance of proper connections. In this chapter, I will examine the evidence related to mechanisms of interaction between motoneurons and skeletal muscle fibers in the vertebrates and will discuss what inferences and deductions might be made from the evidence concerning the specifity of such interactions.

DOUGLAS M. FAMBROUGH · Department of Embryology, Carnegie Institution of Washington, Baltimore, Maryland.

2. NORMAL ADULT NEUROMUSCULAR CONNECTIONS

2.1 *Structure of Neuromuscular Junctions*

A good place to begin any discussion on neuromuscular interactions is at a neuromuscular junction. The basic structural features of neuromuscular junctions are similar for all vertebrates, although the finer details vary considerably. There are many good reviews of various aspects of this topic (Andersson-Cedergren, 1959; Bone, 1964; Cöers, 1969; Couteaux, 1960, 1963; Hess, 1970; Sandbank and Bubis, 1974; Zacks, 1964).

Much of our understanding of the structure and function of neuromuscular connections has come from observations on the frog sartorius and cutaneus pectoris muscles. Light microscopic observations (summarized in Peper *et al.*, 1974), reveal that the axon of each approaching motoneuron is myelinated all the way to the area of nerve and muscle contact. Then the terminal, unmyelinated portion of the axon usually branches, and these terminal branches run along the muscle surface, usually parallel to the long axis of the muscle fiber. These terminal branches lie in shallow grooves on the muscle surface and are overlain and partially enclosed by accessory cells termed *Schwann cells.*

Stained for the enzyme acetylcholinesterase, the frog neuromuscular junction appears as a thin line with many deeper-stained crossbars spaced about 1 μm apart. These areas of denser staining correspond to positions of infolding of the postsynaptic membrane. The precise localization of acetylcholinesterase at neuromuscular junctions has been the subject of hundreds of studies. This matter is still not completely resolved, but it does appear that the enzyme is present in much higher concentration at the neuromuscular junction than in adjacent regions of the muscle surface and that the most probable locations of the esterase molecules are in weak association with the postsynaptic membrane or in the basement membrane (Betz and Sakmann, 1971, 1973; Hall and Kelly, 1971; Salpeter, 1967) which exists just outside the muscle plasma membrane, including the invaginations called *junctional folds.* In either case, the banded pattern of esterase staining is explained in terms of an esterase activity distributed over the postsynaptic surface so that the staining product of the enzymatic actions accumulates in the folds.

Ultrastructural observations on thin sections through frog neuromuscular junctions have added a great deal to this description (Birks *et al.*, 1960; Couteaux and Pécot-Dechavassine, 1970; Dreyer *et al.*, 1973; Heuser and Reese, 1973; Rosenbluth, 1974). Of particular relevance to the present discussion are the following observations: (1) The *postsynaptic surface* consists of two sorts of areas. One is the juxtaneural surface, which is separated from the nerve terminal by about 600 Å. It is characterized by the intimate association of plasma membrane and some cytoplasmic material which binds heavy metal ions and thus appears especially dark in electron micrographs. Such structures are called *postsynaptic thickenings*. The other area of postsynaptic surface consists of folds which occur at rather regular intervals of 0.3–2.0 μm along the major axis of the neuromuscular junction and extend about 1 μm down into the muscle fiber. The deep portion of these folds may be branched or curved. The fairly regular spacing between junctional folds is unrelated to spacings in the sarcomeres of subjacent myofilaments. (2) The *nerve terminal* contains ultrastructural features which are aligned opposite the postsynaptic folds. In longitudinal sections through the neuromuscular junction, these structures generally appear as slight protrusions of the presynaptic membrane opposite each postsynaptic junctional infolding. At such points, there is a slight accumulation of electron-dense material, and the synaptic vesicles cluster near such points. These configurations have been implicated as points of transmitter release and are sometimes referred to as *active zones*. Three-dimensional reconstructions of release sites indicate that these areas are bars of presynaptic density opposite the junctional folds and are thus perpendicular to the long axis of the terminal. Each bar is about as wide as the terminal (1 μm), and at each bar there are about 40 synaptic vesicles aligned in two rows, one on either side of the presynaptic density.

Studies of frog neuromuscular junctions, using the freeze-cleave technique, have confirmed and extended the observations made on thin sections. These studies (Peper *et al.*, 1974; Heuser *et al.*, 1974) have identified intramembrane particles in both pre- and postsynaptic membranes at positions corresponding to the release sites and to the areas of postsynaptic densities. The presynaptic intramembrane particles exist as two parallel rows of doublet particles. It has been suggested that such particles may include the molecular components of the presynaptic membrane that are involved in the entry of calcium ions during depolarization of the nerve terminal, a prerequisite to the release of acetylcholine.

Intramembrane particles are densely clustered in the areas of postsynaptic density in the postsynaptic membrane. There is an especially dense accumulation at the lips of the junctional folds. It has been shown in the neuromuscular junctions of the mouse diaphragm that these areas contain large numbers of acetylcholine receptors (Albuquerque *et al.*, 1974; Daniels and Vogel, 1975; Fertuck and Salpeter, 1974). Thus these intramembrane particles may include the molecular machinery involved in transduction of chemical to electrical signals.

When we extend our consideration of synaptic structure to neuromuscular junctions in other species and to tonic (graded-contraction) muscles as well as phasic (twitch), we find that not all the major features of the neuromuscular junction in frog twitch muscle are really general ones. In many vertebrates the organization of the neuromuscular junction is not as a series of linearly arranged units (as in frog) but as a more compact structure. It is much more difficult to obtain an appreciation of the arrangement of these structures in the latter case, for sections through the junction may reveal different details depending on their alignment. Nevertheless, three-dimensional reconstructions of neuromuscular junctions of mouse intercostal muscle have been made (Andersson-Cedergren, 1959). Information about the arrangement of junctional folds can even be obtained from high-resolution light microscopic observation of end plates in teased muscle fibers after staining for acetylcholinesterase (e.g., see Csillik, 1965). Junctional folds are basically platelike invaginations of the postsynaptic membrane, oriented perpendicular to the longer axis in any area of synaptic gutter.

In the mammalian (mouse, rat, human) neuromuscular junction, where some freeze-cleave studies have been done (Rash and Ellisman, 1974; Rash *et al.*, 1975), there are structures homologous to all of the elements of the frog neuromuscular junction, but the organization is one in which "active zones" are far less abundant than are postsynaptic folds. It appears that active zones occur opposite junctional folds, but the data are scant on this point.

2.2 *Stability of Neuromuscular Junctions*

The connectivity of motoneurons and skeletal muscle fibers appears to be extremely stable in normal animals. Based on observations of the innervation of limb muscles of the cat, Barker and Ip (1966) originally proposed that there might be a continual turnover of neuromuscular junctions during adult life. Further analysis of the innervation of the soleus and peroneus digiti quinti muscles of cats ranging in age from 1 to 19 years, reported by Tuffery (1971), has demonstrated that individual neuromuscular junctions only become more complex during adult life, due to the sprouting of a process usually from the last node of Ranvier and the formation of synaptic connections between the termini of such sprouts and the muscle fiber to which the original neuronal process remains connected. The major change in the neuromuscular system with age is the loss of muscle fibers altogether, presumably following death of the innervating motoneuron. During such a sequence of events it would be expected that some of the muscle fibers denervated by the death of a motoneuron would become innervated by collateral sprouts from other axons, as happens following experimental partial denervation of skeletal muscle (see Edds, 1955; Guth and Brown, 1965*b*).

2.3. *Inferences about Neuromuscular Interactions*

For the present purposes it is not so important to discuss the possible functions of the various structures characteristic of the frog neuromuscular junction (see Hubbard, 1973) as it is to observe that a number of such distinctive structures exist and that they are assembled in such a way that nerve terminal and postsynaptic elements are *in register* along the synapse. Even before recent reconstructions and freeze-cleave images sharpened our perception of synaptic structures, the alignment of the elements had led Couteaux (1963) to speculate that the presynaptic element exerts inductive influence which leads to the genesis of aligned postsynaptic structures. Based on the relative completeness of corresponding pre- and postsynaptic structures, Peper *et al.* (1974) have suggested that the inductive influence might be a *post*synaptic influence on nerve terminal differentiation and organization. Of course, it has always been a lamented aspect of microscopy that matters of temporal or causal relationship cannot be reliably inferred without some other information. But whichever way the system operates, a transsynaptic organizing influence seems to be needed to explain the formation and alignment of pre- and postsynaptic structures.

3. DEVELOPMENT OF NEUROMUSCULAR CONNECTIONS

3.1. *Development of Skeletal Muscle*

It has long been known that muscle can develop in the absence of nervous influence. This was definitively shown for the tail and limb musculature of the larvae of several species of frogs by Harrison (1904). His experiments consisted of removing the caudal portion of the neural tube in very young larvae, before the outgrowth of nerve fibers or the development of the lateral musculature of the tail. These larvae developed in an essentially normal manner except for some minor distortion due to the removal of tissue. Cross-striated muscle developed in the aneural parts. The autonomous differentiation of skeletal muscle in tissue culture was achieved shortly thereafter. In fact, in his original description of the technique of tissue culture, Harrison (1910) described the development of skeletal muscles in explants containing portions of myotome or lateral mesoderm, although these cultures also contained developing neurons. Largely as the result of culture techniques for the growth and differentiation of muscle free from virtually all other cell types, it has been possible to learn a great deal about the degree of muscle differentiation, which can and probably does occur before innervation.

Skeletal muscle fibers form *in vitro* and *in vivo* by the fusion of mononucleated myogenic cells into multinucleated cells called *myotubes* (reviewed by Fischman, 1970; Yaffe, 1969). The fusion process itself has been shown to be a tissue cell type specific phenomenon, and in culture it is possible to obtain species hybrid myotubes by coculture of myogenic cells from different creatures—for example, between rat and chick, cow and rabbit (Yaffe and Feldman, 1965). Of more direct importance to the questions of the specificity of neuromuscular interactions is the high degree of functional differentiation which occurs in developing muscle. Almost all of the characteristics of the adult, innervated muscle are generated in cell cultures of muscle alone. As we will discuss later, the interaction between muscle and nerve results more in quantitative changes and organizational changes than in qualitative ones. The functional components capable of subserving neuromuscular transmission and muscular contraction are elaborated during muscle differentiation.

As discussed above, the postsynaptic surface at the neuromuscular junction is endowed with at least two very important functions: sensitivity to the neurotransmitter, acetylcholine, and the ability to rapidly hydrolyze acetylcholine. Molecules subserving these functions appear very early in the development of skeletal muscle. Fambrough and Rash (1971) have demonstrated that in tissue cultures of rat myogenic cells all multinucleate products of the fusion of myogenic cells are able to respond to extracellular acetylcholine, with characteristic depolarizing responses. Even some mononucleate cells have this capacity, and when cell fusion is inhibited by lowering the calcium ion concentration of the medium the development of acetylcholine sensitivity occurs without cell fusion. This line of investigation has been extended by Patterson and Prives (1974; Prives and Patterson, 1975), who have shown that in the absence of calcium several muscle cell surface characteristics are expressed on the same time schedule and to the same extent as if the formation of multinucleate myotubes had been allowed to take place.

Other properties of skeletal muscle develop with a much slower time course in fusion-arrested cells, but nevertheless the organization of contractile proteins and of the tubular systems of the muscle occurs. Obata (1974) has also shown that the correct chemosensitivity develops in muscle cultures and furthermore that inappropriate chemosensitivity—e.g., to GABA and noepinephrine and glutamate—does not develop. While this at first seems to be a test of a very unlikely possibility, it turns out that some chemosensitivities of cultured cells are surprisingly unlike those of the tissue or origin. The biological meaning of these inappropriate chemosensitivities is not known.

An especially interesting aspect of the chemosensitivity of embryonic skeletal muscle to acetylcholine is the distribution of sensitivity. Whereas in innervated adult skeletal muscle the chemosensitivity is extremely high at the neuromuscular junction and very much lower (often immeasurable) at other points on the muscle fiber surface, in embryonic skeletal muscle there is a generalized sensitivity of the cell surface to acetylcholine (Diamond and Miledi, 1962). Vogel *et al.* (1972) and Sytkowski *et al.* (1973) have investigated this aspect of chemosensitivity using autoradiography of whole myotubes after incubation of cultures with medium containing [^{125}I] α-bungarotoxin. They report that at later times in culture the distribution of acetylcholine receptors, as judged from grain distributions in autoradiography, is very nonuniform, with small areas of very high receptor accumulation occurring on most of the fibers. These sites of receptor accumulation were also found by Fischbach and Cohen (1973) as areas of high chemosensitivity along the length of muscle fibers in culture. The obvious questions raised by these observations are whether or not the accumulation of receptors at loci of high receptor density is homologous to the accumulation of ACh receptors at neuromuscular junctions, whether, indeed, these points of high chemosensitivity are truly areas of high packing density of receptors into the surface membrane, and whether such receptor accumulations occur *in vivo* as part of the muscle's autonomous preparation for innervation. The author is inclined to feel that the answer to the last question will be no, but he is prepared to be proven wrong. There are basically two reasons for this inclination. First, innervation probably occurs *in vivo* on very immature skeletal muscle and has been noted to occur on small myotubes in rat nerve–muscle cultures (Robbins and Yonezawa, 1971), whereas receptor accumulations in culture appear rather late and are characteristic of chick cell cultures but are not often noted on rodent muscle in culture. Second, during reinnervation of denervated adult skeletal muscle, neuromuscular junctions can form rapidly at positions on the muscle surface, where, in the absence of available nerves, foci of accumulated receptors do not seem to occur (Hartzell and Fambrough, 1972). On the other hand, motoneurons characteristically regenerate to innervate former areas of postsynaptic surface (see below); thus a preference for a preformed postsynaptic surface may exist. However, even if one were able to demonstrate that, in culture, motoneurons preferentially form synaptic connections with areas of preformed receptor accumulation on muscle fibers, this would not prove that such areas were available for innervation during embryogenesis.

Much less is known about the behavior of acetylcholinesterase prior to innervation of muscle. The problem is complicated in several ways. First, most of the experiments on the distribution of cholinesterases have not distinguished amond the various enzymes capable of hydrolysis of acetylcholine analogues (and there seem to be several such enzymes). Second, most of the work has involved histochemical localization of esterase activities. Thus there is always the question of whether the reaction products are deposited at the sites of hydrolysis. One has only to review the literature on the distribution of acetylcholinesterase at the neuromuscular junction to get a firm conviction that this problem is a severe one. Several investigations have revealed the presence of strong cholinesterase activity in a large portion of the myoblast population prior to cell fusion (Fluck and Strohman, 1973; Tennyson *et al.*, 1973; Wilson *et al.*, 1973). Wilson *et al.* (1973) have found that a surprisingly large fraction of the acetylcholinesterase synthesized in myogenic cell cultures is eventually found in the medium outside the cells. They have some data to suggest that the same phenomenon occurs in embryonic chicks *in ovo*. While it is clear that some molecular species of acetylcholinesterase are produced by uninnervated skeletal muscle, such molecules are not accumulated to any great extent on the cell surface. Fischback (1972) and several other investigators have shown that the physiological responses of myotubes to acetylcholine are not enhanced by treatment with anticholinesterases, unlike the case at the neuromuscular junction (see Hubbard, 1973). Cultures do not show a strong cell surface stain for cholinesterases, and no one has reported accumulations of high esterase activity comparable to the focal accumulations of acetylcholine receptors in cultured myotubes.

Propagated action potentials also develop in skeletal muscle cultured in the absence of innervation, and spontaneous contractions are routinely noted in well-developed cultured muscle (see Nelson, 1975). In general, the maximum rate of rise and the half-decay time of action potentials are much slower than in mature, innervated fibers *in vivo*. Also, several qualitative differences between these and adult muscle action potentials have been discovered. In the case of rat muscle (Harris and Marshall, 1973) (but not chick; Harris *et al.*, 1973), the embryonic action potentials are only partially sensitive to tetrodotoxin, a potent inhibitor of the opening of active sodium channels in innervated adult muscle (see Redfern and Thesleff, 1971). Possibly unrelated to the tetrodotoxin-resistant action potentials, there is a pronounced calcium component to the action potential (Kidokoro, 1973) and in the chick a chloride spike has been found (Fukuda, 1974).

There are two reasons for summarizing some of the autonomously developing properties of skeletal muscle in the context of neuromuscular specificity. First, these properties characterize the postsynaptic component of the interaction, suggesting ample possibilities for "recognition" by motoneurons. Second, when some of the details of the functional characteristics of the embryonic states are brought up, the question arises as to whether these embryonic muscle properties are subserved by the same molecular species which are present in the adult innervated muscle or by a set of perhaps homologous but distinct molecular species (e.g., see Harris *et al.*, 1973). Is the broadly distributed embryonic acetylcholine receptor a *substrate* for formation of a postsynaptic surface, or does the nerve "induce" the formation of an entirely new species of receptor molecule? Similarly, does the embryonic acetylcholinesterase become utilized in the neuromuscular interaction or is its function not concerned with neuromuscular interaction? Thus the question of recognition during synaptogenesis in the motor system and the question of genetic regulation in embryonic and adult cells may represent two avenues toward an analysis of the same problem.

3.2. *Development of Motoneurons*

Much less is known about the differentiation of motoneurons than about the differentiation of skeletal muscle. Whereas in the case of muscle there are many distinctive morphological and physiological traits, specific markers for the development of motoneurons are not available. In fact, there is not a single adequate criterion for the *identification* of a motoneuron other than its position and size in the spinal cord and the fact that its axon exits via one of the ventral roots and connects to a number of muscle fibers. Once a motoneuron is isolated in tissue culture, there is no adequate way to identify it. It is a minority cell in the spinal cord, and it is not the only type which can synthesize acetylcholine. Its pharmacological properties are also probably not distinctive. Even if it can form neuromuscular connections with muscle fibers in culture, there is still the possibility that a particular neuron was not supposed to be a motoneuron according to the developmental plan of the embryo. As discussed later, there are several demonstrations that other types of neurons can at least occupy positions of synapsis with skeletal muscle fibers and in some cases mediate synaptic transmission. Unfortunately, a simple, reliable histochemical reactions for choline acetyltransferase activity does not exist. With such a reaction we could at least determine when acetylcholine synthesis might begin *vis-à-vis* the time of first neuromuscular interaction. It is known that motoneurons have a high acetylcholinesterase activity in their somata before neuromuscular interaction begins and that this activity shows no marked change at the time of interaction (Atsumi, 1971*b*).

There are some other facts pertinent to the specificity of neuromuscular interaction: (1) several times as many potential motoneurons form in the spinal cord as are finally operative in adult animals (Hughes, 1968), the remainder atrophy and disappear; (2) motoneurons develop along the developing spinal cord in a positional pattern characteristic of the species and may vary from uniform distribution along the cord to largely concentrated at cord positions from which major innervation to the limbs will emerge; (3) the initial distribution and outgrowth of motoneurons are intrinsic properties of the spinal cord, independent of the presence or absence of appropriate peripheral fields such as limb buds.

The pattern of distribution of motoneurons in the developing chick spinal cord has been examined by Levi-Montalcini (1950) and Hamburger (1952). At early stages in development the motor columns are continuous, but after 4 days there is progressive elaboration of more motoneurons at spinal cord levels opposite the developing limbs and a loss of cells at other spinal levels. These changes in the distribution of potential motoneurons were shown by Wenger (1951) to be due to the autonomous development of the spinal cord, for when levels of spinal cord which normally innervate limbs were transplanted to positions where the peripheral field was greatly diminished the elaboration of motoneurons in those spinal cord areas continues. Thus development of large pools of motoneurons at limb levels is not dependent on the periphery. The uniform distribution of cells in the early spinal cord is also found in mammals, and presumably the same mechanisms are operating in the subsequent development of motoneuron pools there (see discussion by Hughes, 1968). Experimental manipulations in mammalian embryos are extremely difficult, so experiments comparable to those of Wenger have not been reported. Hamburger (1946) has also shown that the development of motoneurons in the chick spinal cord is not influenced by nerve fibers from higher levels in the cord: tantalum foil was used to isolate segments of spinal cord from both lower and upper inputs, and a subsequently normal developmental pattern was found. Finally, Piatt (1940) found that the outgrowth of nerves from the spinal cord was nonspecific with respect to the periphery, outgrowth occurring whether or not the appropriate target tissue was there.

The pattern of nerve branching in the limbs is also not dependent on the existence of the specific target cell types in the limb. Taylor (1943) found that the pattern of nerve branching in the developing limbs of *Rana pipiens,* with separation of the mixed nerve trunk into its sensory and predominantly motor branches, occurred before skeletal muscle differentiated and *very* long before limb morphogenesis was complete. Similar results have been reported by Hughes (1965) for the developing nerve pathways in *Xenopus.*

Hughes and Prestige and coworkers (Prestige and Hughes, 1967; Hughes, 1965, 1968; Prestige, 1967; Reier and Hughes, 1972; Prestige and Wilson, 1972) have examined the pattern of motoneuron and dorsal root ganglion cell degeneration during the time which encompasses the formation of neuromuscular connections in amphibians. They have found a correlation between the degeneration of nerve fibers and the time of progressive specialization in the innervation of muscles. At early times after the first limb movements, the connectivity of nerves with muscles seemed to be less precise than in the adult, and stimulation of different nerves resulted in more widely spread contractions than should occur with correct nerve–muscle connections. However, this could be caused by incompletely separated and differentiated muscles (see Dunlap, 1966). During the ensuing period of progressive specialization of innervation, the turnover of nerve fibers was maximal. These findings together with other data (see below) suggest that innervation of skeletal muscle involves an early period of "competition" between different neurons for the establishment of permanent connections with muscle fibers. Just what parameters are involved in determining the outcome of such competition is at the heart of the question of nerve-muscle specificity during development. The answer is not known, but some other information related to the question has come from morphological observations on the establishment of neuromuscular connections during embryogenesis and from studies on the reinnervation of denervated muscles.

Taken together, the data on the development of nerve and muscle indicate that the early stages, including genetic expression of most of the characteristics of mature cells, occur without synaptic contact. When nerve and muscle do come into contact, recognition might reasonably be expected to utilize one or more of the distinctive properties of the respective differentiated cells. Of course, there is no guarantee that this is the case. Synaptic connections form between motoneurons and skeletal muscle fibers, while the sensory neurons which follow the same pathways from spinal cord toward periphery mostly divert to the surface to become cutaneous nerves. A small number of sensory fibers interact with specialized muscle fibers to from the muscle spindles which participate in proprioception and others become associated with tendons. Just what cues are involved in the final choice of destination for the growing nerves is wholly unknown. A glimpse into the nature of the scientific problems is afforded by experiments on the behavior of sensory nerves which have been routed into skeletal muscle (discussed in connection with reinnervation experiments below).

3.3. *Neuromuscular Interactions During Synapse Formation*

In the rest of this chapter, three levels of specificity will be examined: (1) which muscle a nerve goes to, (2) which fibers in that muscle are innervated by which nerves, and (3) where on a muscle fiber a neuromuscular junction forms. While the main concern is how this all comes about during development, many experiments on embryos are exceedingly difficult. Therefore, investigators have frequently used regeneration and reinnervation as a model for development.

In this section, we will explore the events surrounding the formation of neuromuscular junctions *in vivo* and *in vitro*. First we will discuss the morphological and physiological correlates of the formation and maturation of neuromuscular connections. Then we will discuss single and multiple innervation of muscle fibers and the problem of position of neuromuscular junction along muscle fiber length. In later sections, we will discuss histochemically and physiologically distinct muscle fiber types and associated motoneuron types and the results of experimental cross-innervation of skeletal muscle with foreign motor, autonomic, and sensory nerves.

a. Morphological and Histochemical Studies on the Formation of Neuromuscular Junctions. Studies have been made of the ultrastructural correlates of the formation of neuromuscular connections in chick sartorius muscle (Hirano, 1967; Bennett and Pettigrew, 1974*a*) and rat muscles (Teravainen, 1968; Kelly and Zacks, 1969; Bennett and Pettigrew, 1974*a*). The observations are similar. Very primitive neuromuscular junctions can be found occasionally in muscle at the time at which the first spontaneous movements begin. The first signs of neuromuscular junctions include a "thickening" in the submembranous cytoplasm under the postsynaptic membrane and the presence of a small nerve terminal, containing a few clear synaptic vesicles about 500–800 Å diameter and some dense-core vesicles. The pre- and postsynaptic elements are separated by the myotube's basement membrane, and the apposition of nerve and muscle appears tenuous. The large synaptic cleft might even be an artifact of sample preparation in a very fragile, loosely packed tissue. At early stages no junctional folds are found in the postsynaptic surface. These develop slowly, so that junctions fully comparable in general features to adult neuromuscular junctions are found only 10 or more days postnatally in the rat. Similar sequences of events have been reported for rat spinal cord and muscle grown *in vitro* (Yonezawa *et al.*, 1973) and for neuromuscular junctions forming on regenerating skeletal muscle in *Triturus* limb (Lentz, 1969). In addition to the ultrastructural studies there have been a larger number of studies of synaptogenesis using light microscopic methods, particularly silver staining for neuronal processes and histochemical staining for acetylcholinesterase. It is the consensus opinion that histochemically demonstrable acetylcholinesterase appears slightly later than neuromuscular contact and approximately coincident with or later than the onset of spontaneous movement. This has been found in developing and regenerating lizard tail (Liu and Maneely, 1968), in tail musculature and transverse ventralis muscle of the larvae of *Xenopus laevis* (Lewis and Hughes, 1960) and *Rana catesbeiana* (Letinsky, 1974), in various rat embryo muscles (Teravainen, 1968; Zelena and Szentagothai, 1957; Bennett and Pettigrew, 1974*a*), in chick embryo hind limb muscles (Hirano, 1967; Mumenthaler and Engel, 1961) and intercostal muscle (Atsumi, 1971*a*), in regenerating salamander limb (Lentz, 1969), in rat and mouse explant cultures developing *in vitro* (Yonezawa *et al.*, 1973), and in cultured combinations of rodent embryonic spinal cord and muscle of various species of adult mammals (Peterson and Crain, 1972).

Biochemical analyses of chick nerve–muscle cultures have suggested that the culture of spinal cord cells in the presence of muscle results in elevated levels of choline acetyltransferase, the enzyme catalyzing acetylcholine synthesis in neurons (Giller *et al.*, 1973), and that an elevation of skeletal muscle acetylcholinesterase results from coincubation of skeletal muscle with spinal cord explants (Oh *et al.*, 1972). An earlier study, testing the possibility of substrate induction, had failed to demonstrate any effect of acetylcholine itself on acetylcholinesterase levels in muscle cultures (Goodwin and Sizer, 1965). In the light of the finding that much of the acetylcholinesterase of skeletal muscle in culture is lost to the medium (Wilson *et al.*, 1973), the significance of either positive or negative findings is not yet clear. Nevertheless, it seems likely that coculture of muscle and spinal cord cells will be a valuable tool in further analysis of the factors involved in the interactions of nerve and muscle during synapse formation.

b. Physiological Studies on the Formation of Neuromuscular Junctions. In line with ultrastructural observations on embryonic tissue, physiological studies have demonstrated the very low content of releasable quanta in primitive neuromuscular connections (Fischbach, 1972; Robbins and Yonezawa, 1971), reflecting probably the small volume and sparse vesicle content of new nerve terminals. The frequency of spontaneous miniature end-plate potentials is low (Diamond and Miledi, 1962; Letinsky, 1974) and the amplitude distribution may be skewed with both smaller and larger potentials than expected, suggesting that while acetylcholine is released in packets as at adult neuromuscular junctions the number of acetylcholine molecules per packet might vary considerably. The suggestion has also been made repeatedly that the distribution and time course of spontaneous miniature end-plate potentials could indicate multiple innervation of the muscle fiber. The same data could also be interpreted as reflecting the release of packets of acetylcholine at variable distances from the chemosensitive muscle surface.

There are physiological data for both neonatal rat muscles (Redfern, 1970; Bennett and Pettigrew, 1974*a*) and kitten muscles (Bagust *et al.*, 1973;) that single muscle fibers are often innervated by two or more motorneurons and that multiply innervated fibers are more densely innervated in chick and salamander (Bennett and Pettigrew, 1974*a*; Dennis, 1975). By intracellular recording and electrical stimulation of the phrenic nerve with stimuli of increasing strength, Redfern (1970) showed that the muscle response was at first a simple end plate potential of constant latency but with stronger stimuli became a complex potential of two to four components, each component with its characteristic latency. This was explained by the hypothesis that the muscle fiber was innervated by several different neurons. A weak stimulation of the phrenic nerve activated only one of the axons supplying innervation of the muscle, resulting in a simple response. Stronger stimuli activated, two, three, and sometimes even four different axons. Variation in conduction velocity resulted in the arrival of nerve impulses at the various endings on the muscle at different times after the stimulus. An alternate means by which to judge multiple innervation of a muscle is to look for nonadditivity in the muscle tension produced by stimulation of the two parts of a divided nerve to a muscle vs. stimulation of the entire nerve (Brown and Matthews, 1960). Bagust *et al.* (1973) have demonstrated this effect in young kittens. Bagust *et al.* (1973) found that when the nerve to the 3-day-old kitten soleus and flexor hallucis longus was split into equal or unequal parts and these were stimulated separately, the sum of the muscular tension produced by these separate stimuli was up to 20% greater than when the entire nerve was stimulated at once. The effect diminished with age of the animals and was essentially gone by 6 weeks of age. This sort of analysis complements intracellular recording data. Neither set of data is a foolproof demonstration of multiple innervation by more than one neuron. Guth and Brown (1965*b*) (see also Brown and Matthews, 1960) have pointed out technical problems with the tension measurements showing anomalous results in rat soleus muscle, and the intracellular recording data can be explained in other ways, although the alternatives do not seem very likely. Studies on teased and silver- or gold-stained adult mammalian skeletal muscles have repeatedly demonstrated the rarity of multiple innervation (e.g., Guth and Brown, 1965*b*; Tuffery, 1971). There is a low incidence (about 1–2%) of muscle fibers with two widely separated neuromuscular junctions. Otherwise, the rule is one neuromuscular junction between one nerve process and a muscle fiber in adult mammalian muscle. An exception to this rule is the cricothyroideus muscle of the cat, which contains individual muscle fibers which Hunt and Kuffler (1954) have shown are innervated by axons from both the superior laryngeal nerve and the pharyngeal plexus.

c. Morphological Evidence for Multiple Innervation in Embryonic and Neonatal Skeletal Muscles. Although thin sections through embryonic and neonatal junction sometimes reveal several nerve terminal profiles (e.g., see Hirano, 1967; Bennett and Pettigrew, 1974*a*), without serial reconstruction it is not possible to say whether these profiles are of one or of several different nerve processes.

Multiple innervation of extrafusal muscle fibers frequently occurs in the lower vertebrates. Hess (1970) has summarized much of the data on these muscles. Multiple innervation is characteristic of "slow" or "tonic" skeletal muscles, those which respond to neuronal activation with local, graded contraction of slow time course. The best-studied examples are the anterior latissimus dorsi of the chicken and the slow fibers of the iliofibularis of the frog. However, there are some "fast" or "phasic" muscles of lower vertebrates which also have multiple innervation, the best known being the frog sartorius, where most or all of the fibers are dually innervated.

In a series of experiments by Zelena and coworkers (Hnik *et al.*, 1967; Jirmanova *et al.*, 1971; Zelena and Jirmanova, 1973; Zelena *et al.*, 1967), later confirmed and extended by Bennett and coworkers (Bennet *et al.*, 1973*c*; Bennett and Pettigrew, 1974*b*), the question of specific interactions in the development and innervation of the slow anterior latissimus dorsi (ALD) and fast posterior latissimus dorsi (PLD) muscles of the chick has been explored. *In vivo* the ALD develops much faster than the PLD and becomes multiply innervated while the PLD is focally innervated. When myogenic cells from the respective muscle tissues are cultured *in vitro,* however, no such difference in developmental schedule is found (Gutmann *et al.*, 1969). When the nerves to the two muscles are severed and the proximal stumps are sutured to the incorrect distal nerve part, the two muscles become innervated by the incorrect nerves. However, in older birds the muscles retain their normal single vs. multiple pattern of innervation, although the structure of individual neuromuscular junctions is altered. When the operations are performed in newly hatched chicks, on the other hand, conversion of the pattern of innervation occurs.

Gordon and Vrbova (1974) have also examined the innervation of these muscles, but with a very different experimental approach. They found that when eggs were injected with curare, a potent inhibitor of neuromuscular transmission, the pattern of innervation changed in a very interesting way. The rather regularly spaced nerve terminals on the multiply innervated ALD became much more closely spaced. Their explanation for this change is that the curare decreases the magnitude of the end-plate potentials, resulting is a situation where muscle surface some distance from the site of end-plate potential generation is hardly depolarized at all. Somehow, they postulate, the resistance to innervation displayed in the normal situation is dependent on frequent depolorization of the surface membrane to a certain critical level. When this level is not reached—because of a more localized depolarization than normally occurs—regions of muscle surface previously resistant to innervation become accepting. Whether or not this explanation proves correct, the experimental observations themselves are most exciting, for they represent a new experimental means of exploring the question of single vs. multiple innervation.

d. The Locus of Neuromuscular Interaction. In mammalian muscles the neuromuscular junction usually is very close to the center of the muscle fiber's long dimension (Tiegs, 1953). Thus in very thin muscles and in muscles where the fibers insert into tendons at approximately the same point, as in the diaphragm, all the neuromuscular junctions will lie in a narrow band of muscle tissue about midway between insertions into the tendons. This handy geometry has been exploited in various biochemical, pharmacological, anatomical, and physiological studies, making the rat diaphragm one of the most investigated of mammalian muscles. Two other much-studied muscles are the soleus and the extensor digitorum longus muscles of the lower hindlimb. Both of these have long straplike tendons which are continuous with sheaths of connective tissue on either side of the muscle. The muscle fibers insert into these straplike tendons at various levels so that the long axis of individual muscle fibers is somewhat oblique to the overall muscle long axis and the muscle fibers are not in register. Hence the neuromuscular junctions, which again tend to occur at the midpoint of each fiber, are not in register. (This shortcoming, from an experimental point of view, is balanced by several attractive features, including ease of dissection and relative homogeneity of motor unit type—a matter to be discussed later.)

As mentioned above, some mammalian muscles are multiply innervated (Cöers, 1969). These include laryngeal and facial muscles and some muscles of somitic origin such as the extrinsic ocular muscles, which possess scattered innervation. In the lower vertebrates, both scattered innervation of slow muscle and myoseptal innervation (i.e., innervation at the very ends of muscle fibers) as well as equatorial innervation are found. Myoseptal innervation is found in fish body musculature and the musculature of some reptiles and the primitive somitic muscles of frogs and toads (Cöers, 1969). The distribution of acetylcholine receptors on such muscles has been determined by Anderson and Cohen (1974), using fluorescent-labeled α-bungarotoxin as a specific probe. As expected, there is a very high accumulation of cholinergic receptors coincident with the position of high acetylcholinesterase activity at the sites of innervation.

e. Development of Neuromuscular Connections in the Absence of Function. One of the most amazing facts about the formation of neuromuscular junctions is that there is apparently no *requirement* that neuromuscular transmission occur during the formation of junctions. This was first shown by Harrison in 1904. He raised the larvae of three frog species from very early stages in an environment containing an anesthetic ("chloretone"). The tadpoles developed in this environment up through stages when they would normally be swimming about. When Harrison transferred the totally paralyzed animals to water, the anesthetic was quickly diluted out and the animals immediately displayed all of the normal behavior repertoire for their stage of development. Thus neuromuscular connections and central connections formed correctly in the absence of any functional validation. Crain and Peterson (1971) have demonstrated the formation of neuromuscular junctions in tissue culture in the presence of curare or hemicholinium-3 (which blocks choline uptake by nerve terminals), and Cohen (1973) has also shown the formation of functional neuromuscular connections in explant cultures of *Xenopus* spinal cord and myotomes in the presence of tubocurarine.

Neuromuscular junctions have been found to form in chick embryos poisoned with botulinum toxin (which irreversibly blocks neuromuscular transmission), and neuromuscular junctions form in mice with hereditary motor end-plate disease, characterized by a failure of neuromuscular transmission, and in muscular dysgenic mice, characterized by a hereditary failure of skeletal muscle contraction. An interesting "model" for synaptogenesis has been developed (Harris *et al.*, 1971; Steinbach *et al.*, 1973) in which a myogenic cell line in tissue culture interacts with a "neuronal" cell line, neuroblastoma. No true synapses are formed, but at points of contact between processes of the neuroblastoma and the myotubes high sensitivity to acetylcholine develops. Steinbach *et al.* (1973) have shown that such points of high sensitivity form even when the neuroblastoma is incapable of synthesizing acetylcholine or when the acetylcholine receptors on the myotubes are temporarily blocked by cobra α-toxin.

3.4. *Inferences about Nerve–Muscle Specificity Based on Developmental Studies*

The data on the physiological and morphological correlates of formation of neuromuscular junctions suggest several things about nerve–muscle interactions. It appears that both pre- and postsynaptic organization are effected as a result of the direct short-range interaction nerve ending and muscle fiber. The basic building materials for the interaction may well be present in the nerve ending and the muscle fiber, although these matters are far from being convincingly settled. The evidence suggests that very rudimentary neuromuscular connections are effective and that the movements of embryos (which are necessary for the normal morphogenesis of joints, for instance) are directed by nervous activity. Neuromuscular junctions of the adult form are generated fairly slowly.

Two pieces of evidence suggest that competition takes place between motoneurons for innervation sites within a single muscle. First, there is physiological evidence for multiple functional nerve endings from different neurons upon single figers, despite the morphological evidence for a single point of nerve–muscle contact at this time. The multiple nerve ending profiles seen in thin sections through immature neuromuscular junctions could be at least partially due to multiple innervation at the forming junction. Indeed, Bennett and Pettigrew (1974*a*) have found histological evidence for this. Second, there are more potential motoneurons produced in the developing spinal cord than actually become motoneurons in the adult. Hughs and coworkers (see Hughes, 1968) have estimated that the "overproduction" of motoneurons could be as much as tenfold in the urodele embryo. Most of the potential motoneurons die during development. This could be due to failure of these neurons to make adequate functional connection with the periphery. There is precedent for this in the central nervous system. For example, in the isthmooptic nucleus of the chick about a twofold excess of neurons is originally formed. Those which make successful connection with cells in the retina survive, the others degenerate. Partial ablation of the retina leads to corresponding loss in cells of the isthmooptic nucleus (Cowan and Wenger, 1968; Crossland *et al.*, 1974).

It seems unlikely that the early "competition" between neurons for connections to skeletal muscle fibers is part of a process of "hunting" for specific loci. The net result of these developmental events is the establishment of motor units, functional units of the motor system in which a single neuron is permanently is synapsis with a set of skeletal muscle fibers in a muscle. The muscle fibers of a motor unit are scattered through much of the cross-sectional territory of the muscle, quite opposite from the situation, for example, between retinal ganglion cells and the optic lobes or tectum in fish, birds, and amphibians, where a rigid point-to-point connectivity probably exists. Regeneration experiments, to be discussed below, support this supposition.

The positions of neuromuscular junctions (myoseptal, scattered, and equatorial) on skeletal muscle fibers seem to be due to subtle differences in the geometry of the encounter between growing nerve processes and to differences in subsequent growth parameters of the muscles. (Bennet and Pettigrew, 1974*a*,*b*). In the case of equatorial junctions, for instance, the early formation of neuromuscular junctions on small myotubes or short myofibers is followed by growth of the muscle fibers predominantly at their ends or equally along their whole length (see Mackay and Harrop, 1969; Williams and Goldspink, 1971). Experiments of Gordon and Vrbova (1974) suggest that neuromuscular transmission itself might be involved in finalizing the morphological type of innervation pattern. The experiments of Zelena and coworkers have shown that in the case of the chick ALD and PLD muscles the innervation pattern (multiple or single) is first determined by the identity of the nerve which participates in the formation of the neuromuscular junctions. At later times, however, when these muscles which had previously received scattered or equatorial innervation are denervated and then reinnervated with nerves of the other sort, the original type of innervation pattern is preserved in the cross-innervated condition. Thus, once a pattern of innervation is established, it is fairly permanent. We will have much more to say about this matter latter on:

it does not necessarily constitute irreversible determination of the quality of the muscle *vis-à-vis* innervation.

In sum, while many of the aspects of synaptogenesis could conceivably involve specific interactions between specific motoneurons and muscle fibers, most of the details of the innervation process can be accounted for, at least theoretically, without the necessity of presuming such a high degree of specificity. There is, of course, great specificity of interaction of motoneurons with skeletal muscle fibers while the sensory fibers reach their appropriate destinations. Some further specificity in the establishment of neuromuscular connections must exist such that fibers in certain muscles interact with their motoneurons to become slow muscle fibers (with multiple innervation in the case of lower vertebrates) while other muscles develop in such a way as to contain mixed populations or relatively pure populations of one muscle fiber type or another with the characteristic innervation pattern for the fiber type. If the muscle characteristics are determined by the interaction with specific types of motoneurons, then specific growth of motoneurons of slightly different physiological and "inductive" properties to the specific muscles during development must occur. Finally, of course, there is abundant evidence that functionally correct neuromuscular connections form during embryogenesis with little indication that such connectivity involves functional validation. The transient existence of multiple innervation in mammalian muscles during early life could be a correlate of exploratory activity, but it is doubtful that this activity involves massive exploration of incorrect muscles by the developing processes of motoneurons.

4. INNERVATION OF SKELETAL MUSCLES BY THEIR OWN AND FOREIGN NERVES

When peripheral nerves are cut, the damaged neurons show a remarkable capacity to regrow and reestablish connections with their target tissues (see Guth, 1956). It is also possible to surgically reroute the regenerating nerves to inappropriate targets and to observe the interactions which occur. This area of experimentation has been a very active one for many years, partly because of the medical importance of the subject. Both the clinical and basic research literature were comprehensively reviewed by Sperry (1945) from the point of view of whether or not the correct peripheral connections are reestablished and whether or not there is reorganization in the central nervous system to compensate for any incorrect reinnervation. Guth (1956) has reviewed the literature from 1929 to 1955 on the process of peripheral nerve regeneration, and more recent reviews by Guth (1968) and by Close (1972) cover, respectively, the trophic interactions between nerves and muscles and the effects of nerve cross-union on the physiological and biochemical properties of muscle fibers. I will try to touch upon the highlights of these areas as they relate to the specificity of nerve–muscle interactions and to present some of the most exciting and/or enlightening bits of information to appear in the recent literature.

4.1. Locus of Reinnervation

a. Reinnervation at Old Postsynaptic Sites. When a peripheral nerve is cut, there is a period of degeneration during which the proximal portion of the axons, still connected to neuronal cell bodies in the spinal cord and dorsal root ganglion, retract somewhat. Shortly thereafter, new processes sprout out from the ends of the axons and grow back toward the periphery, following the path of the degenerating distal portions of the old nerve fibers. This path of regeneration leads the motoneurons back to the denervated skeletal muscle fibers—in fact, back to the area of former neuromuscular junctions. Functional neuromuscular connections are soon established. Gutmann and Young (1944) showed that this will occur even if the muscle (rabbit small peroneal muscles) has been denervated for a long time (up to 17 months). With their techniques, it was not possible to determine *exactly* where the new junctions occurred relative to the old ones. More recent investigations, including several employing the electron microscope, have demonstrated that the reinnervation of skeletal muscle fibers preferentially occurs exactly at the position of previous innervation (Aitkin, 1965; Bennett *et al.*, 1973*a*; Guth and Brown, 1965*a*; Gonzenbach and Waser, 1973; Iwayama, 1969; Jirmanova and Thesleff, 1972; Koenig and Pecot-Dechavassine, 1971; Lüllmann-Rauch, 1971; Saito and Zacks, 1969). At times it appears that connective tissue deposits at the old postsynaptic surface prevent complete reestablishment of neuromuscular contact, in which case the nerve terminal may innervate only a part of the old subsynaptic membrane, and some new areas of postsynaptic surface may be generated. This tendency toward reinnervation to surviving postsynaptic structures has been well documented for several muscles in chickens, rats, mice, and rabbits, and data on the frog sartorius reinnervated by its normal nerve (Miledi, 1960) or by autonomic preganglionic fibers (Landmesser, 1972) suggest that the same phenomenon occurs in the frog as well.

Reinnervation of old postsynaptic surface by the regenerating motor axons implies that the postsynaptic strutures survive for rather long periods of time in the absence of innervation. There is a fairly large body of ultrastructural data on this point, the bulk of it confirming this assumption. In the frog sartorius muscle, Birks *et al.* (1960) found that the postsynaptic folds and thickenings remained unaltered up to 130 days after denervation while the nerve terminal degenerated within 1 week after denervation. Reger (1959) and Nickel and Waser (1968) and Bauer *et al.* (1962) found that the postsynaptic structure in various mouse muscles could be found up to 3 or 4 months after denervation, and Miledi and Slater (1968, 1970) found the same for rat diaphragm. Bourgeois *et al.* (1972) by electron microscopic autoradiography found that clusters of acetylcholine receptors were still present up to 140 days after denervation of *Electrophorus electroplax,* although the presence or absence of ultrastructural correlates of such clusters was not reported. Degenerative changes in postsynaptic structure occur much more rapidly in the newt (Lentz, 1969), where powers of regeneration of tissues are very impressive. In addition to these ultrastructural studies, there are a large number of reports on the survival of postsynaptic acetylcholinesterase activity following denervation. Again most of the work is on rat, mouse, and frog muscles (see Filogamo and Gabella, 1967). In the rat tibialis anterior muscle, postsynaptic acetylcholinesterase activity has been demonstrated up to 8 months after denervation (Eränkö and Teräväinen, 1967). Only one study has investigated the "functional efficiency" of acetylcholinesterase at denervated postsynaptic sites: in the frog, *Rana esculenta,* by 3 months after denervation the amount of postsynaptic cholinesterase was established to be about 50% of that at innervated postsynaptic sites by histochemical techniques, but this residual cholinesterase was not effective in altering the time course or response of the postsynaptic surface to iontophoretically applied acetylcholine (Pecot-Dechavassine, 1968).

Filogamo and Gabella (1967) have performed some experiments on the metabolic basis for the maintenance of acetylcholinesterase at denervated postsynaptic sites. Using the irreversible inhibitor of acetylcholinesterase activity, diisopropylfluorophosphate (DFP), they blocked all of the esteratic sites at control and denervated neuromuscular junctions in the chick limb and then followed by histochemical methods the appearance of junctional cholinesterase at the formerly blocked sites. New esterase activity began to appear in 2–3 days at the innervated junctions, and by 5–7 days the esterase activity was normal. Esterase activity did not return at the denervated endplate up to 90 days after DFP treatment. However, when reinnervation of the denervated sites was allowed, the esterase activity began to accumulate again as it did at the innervated end plates. A smilar kind of experiment was also done by Sonesson and Thesleff (1968) on rat tibialis anterior muscle, but a slightly different result was obtained. When reinnervation did not occur, there was still a partial recovery of acetylcholinesterase activity in the denervated postsynaptic structure. The experiments of Sonesson and Thesleff suggest that the junctional cholinesterase is at least partially the synthetic product of the skeletal muscle fiber. The experiments of Filogamo and Gabella suggest that cholinesterase *activity* of the postsynaptic membrane is not involved in the mechanism of neuronal recognition during reinnervation. The experimental approach of Filogam and Gabella and Sonesson and Thesleff would seem to be a good one to take in further exploration of the sufficient conditions for neuronal recognition of old postsynaptic structures.

b. Innervation at Other Loci. While there is no evidence that functional acetylcholine receptors are required during the formation of neuromuscular junctions, there is a general sentiment among researchers in this area that the presence of acetylcholine receptors on the skeletal muscle fiber surface is indeed always associated with the ability of the skeletal muscle fiber to receive innervation. In his beautifully clear presentation of experiments on the innervation of skeletal muscle, Elsberg (1917) demonstrated that innervated rabbit thigh muscles were refractory to innervation by the implanted nerve stump of a foreign nerve. However, denervated thigh muscle readily accepted the foreign innervation, whether the foreign nerve was implanted in the muscle long before denervation or many weeks afterward. It was discovered many years later that denervation results in the appearance of acetylcholine receptors all over the skeletal muscle fiber surface (Axelsson and Thesleff, 1959). In much more recent experiments it has been shown that aneural portions of muscle, formed by cutting muscles in two and leaving all of the innervation associated with one portion (Miledi, 1962; Koenig, 1963), can also accept innervation. This portion likewise becomes sensitive to acetylcholine upon disconnection from the innervated part of the muscle (Katz and Miledi, 1964). Other procedures which lead to supersensitivity to acetylcholine—botulinum toxin treatment (Thesleff, 1960), treatment with local anesthetics (Jansen *et al.*, 1973), and local trauma (Miledi, 1963)—also result in a removal of the normal refractoriness of the muscle to innervation. In these various cases new neuromuscular junctions can form on the muscle surface.

Since the acetylcholine receptor is a major component of the postsynaptic membrane, as we have already mentioned, this component can be considered an important *structural* as well as *functional* element of the synapse. Thus formation of neuromuscular junctions in the absence of acetylcholine receptors might seem at first to be a ridiculous thought, but whether or not it is ridiculous really depends on just how neuromuscular junctions are formed, and we know precious little about that. One possibility is that the acetylcholine receptors present on the muscle surface prior to innervation are used in the building of the postsynaptic membrane. Another possibility is that neuromuscular contact induces the formation of end-plate specific acetylcholine receptors without reference to the presence of or the synthetic machinery for production of extrajunctional acetylcholine receptors. If the latter case is correct, then the presence of acetylcholine receptors prior to innervation might be just an extremely distracting coincidence or perhaps a biologically important chemical identification tag for noninnervated skeletal muscle but no more than that. These are intriguing possibilities and part of what makes the study of nerve–muscle interactions all the more tantalizing as experimental approaches seem nearly within our grasp.

4.2. *Specificity of Innervation of Muscle Fibers During Reinnervation*

a. Fiber-Type Specificity in Reinnervation. As mentioned above, most muscles are composed of several morphologically, histochemically, and even physiologically distinguishable fiber types. In the lower vertebrates two very different types are present: one phasic or fast and one tonic or slow. Often the slow fibers are distinguished by a high myoglobin content, as is familiar to nearly everyone in the "dark" and "light" meat of chickens and fish. In mammals nearly all skeletal muscle is of the single-innervation phasic or twitch type, and the distinctions "fast" and "slow" refer to twitch speed. Correlations among myoglobin content, twitch speed, and other parameters such as pH sensitivity of myofibrillar ATPase and mitochondrial content vary from one species to the next, so that the situation is extremely confusing to nearly everyone. Some of the complexity is sorted out in tabulations by Close (1972) and Guth and Samaha (1969). As most elegantly demonstrated by Burke *et al.* (1973), all of the muscle fibers belonging to a single motor unit are of the same fiber type and these fibers tend to be scattered through much of the volume of the muscle rather than grouped into a neat bundle (Burke and Tsairis, 1973). Reinnervation experiments reveal that the arrangement of the muscle fibers of a motor unit in the muscle mass and the major distinctions between muscle fiber types are characteristics imposed on the muscle by the identity of the innervating motoneurons and the manner in which the innervation develops. Two types of experiments establish these points. First, as pointed out originally by Romanul and Van der Meulen (1967), when a muscle is reinnervated even by its own nerve, the different histochemical types of fibers occur in groupings of like fibers rather than distributed through the muscle cross-section in the usual random "checkerboard" fashion. This change in the spatial distribution of fiber types was interpreted as being due to the reinnervation of groups of neighboring muscle fibers by the ingrowing processes from single neurons. Second, when the nerves to a pair of muscles which differ greatly from each other in their composition of fiber types are cut and cross-sutured to the wrong distal nerve stumps, the muscles become reinnervated by the wrong nerves, and as a result the histochemical and physiological properties of the muscle fibers are converted to ones similar to those characteristic of the fibers to which the nerves normally connect. To what extent the changes are the results of alterations in the patterns of muscular activity and to what extent the changes are due to humorally induced alterations in genetic expression of the muscle fibers is not clear. Studies by Kuno *et al.* (1974) have demonstrated that while large changes occur in the physiological properties of cat skeletal muscle fibers after cross-union of slow and fast motoneurons there are not any corresponding large changes in the physiological properties of the motoneurons themselves.

Miledi and Stefani (1969), examining the reinnervation of the rat soleus muscle after sciatic nerve section, found that most of the reinnervated fibers displayed altered electrophysiological properties which indicated that the normally "slow" soleus had been innervated predominantly by "fast" fibers. The normally "fast" extensor digitorum longus muscle was also reinnervated predominantly by "fast" fibers. Miledi and Stefani concluded that nonselective reinnervation of muscle fibers had occurred and that the preponderance of "fast" nerve fibers in the sciatic nerve resulted in a preponderance of "fast" innervation. Bernstein and Guth (1961) have also examined the reinnervation of rat soleus (see below) and found no evidence for specific re-formation of synaptic connections. Similar conclusions were reached by Yellin (1967) based on histochemical analysis of reinnervated muscles, and similar electrophysiological results were obtained by Elul *et al.* (1968) examining reinnervation of frog fast and slow muscle fibers.

There are a few reports of selective reinnervation of fast and slow muscle fibers by corresponding nerve fibers (Feng *et al.*, 1965; Hoh, 1971). Close and Hoh (1968) have shown that slow fibers in the muscles of the toad, *Bufo marinus,* will accept fast nerve innervation when a pure fast nerve is forced into the muscle and further that nerve crosses do not result in transformation of nerve or muscle properties. Hoh (1971) found that when slow and fast nerves are cut and their proximal and distal stumps are tied together and resutured to form a nexus the slow muscle fibers become innervated by slow nerves again and fast fibers by fast nerves. Hoh postulates two sorts of specificity, one involving a sorting out of nerve fibers in growth toward the correct muscle and a second involving specific recognition between muscle fiber type and nerve fiber type. Hoh further suggests that the reinnervation of fast muscle specifically by fast nerve represents a case of correct reinnervation where there is no multiple innervation. However, as mentioned above, the correct reinnervation of fast muscle by fast nerve may be related more to the preponderance of fast nerve fibers and their greater regenerative ability than to specificity of interactions.

While these studies on the relation between motoneuron and muscle fiber type have not yet fully solved the questions of the mechanism by which nerve influences muscle properties, they do tell us some important things about specificity of neuromuscular connections. First, it is very likely that the shades of difference in properties of the muscle fibers in different motor units in an all-phasic muscle are not intrinsic, independently developed characteristics of a subset of muscle fibers which then result in innervation by a corresponding particular subset of motoneurons with complementary physiological attributes. Rather, all indications are that the initial innervation of the muscle is random with respect to the exact connectivity between nerve processes and muscle fibers in a given muscle. Some mechanism, presumably competitive interaction between neuronal processes during innervation, results in the attainment of a reasonably constant motor unit size for a given muscle. Second, given the liklihood that the differentiation of motor units with different physiological and biochemical properties is the result of innervation by motoneurons with somewhat different properties themselves, there must exist some mechanism in the embryo for *reproducibly* routing motoneurons with particular properties to specific muscles so that, for example, the rat soleus receives no "fast" innervation and the extensor digitorum longus almost exclusively "fast" innervation while the diaphragm receives about 60% "slow," 20% "intermediate," and 20% "fast." The alternative to specific routing of varieties of motoneurons to particular muscles is random outgrowth of motoneurons followed by some later specification of motoneuron types *after* muscular innervation, involving adjustments in the central nervous system. The idea that nerves regrow specifically to the muscles they originally innervated and the idea that central nervous system readjustments might occur after innervation to ensure the establishment of a coordinated motor system have a long and colorful history. We will deal in part with them next.

b. Whole Muscle Specificity During Reinnervation. While it appears that usually within a particular muscle the connectivity between motoneurons and muscle fibers involves no specific recognition beyond that of general cell type (nerve–muscle) and while it is definitely possible to "force" a muscle to become innervated by the wrong motor nerve, is there any evidence that during large-scale reinnervations of musculature correct nerve–muscle connections are reestablished? In some cases the answer to that question seems to be clearly no. In other cases the answer is apparently yes! Let us consider the negative cases first. As thoroughly reviewed by Sperry (1945), there is a wealth of evidence against the formation of correct neuromuscular connections during nerve regeneration in mammals and against any compensatory changes taking place in neuronal circuitry of the central nervous system to yield correct coordination and reflex activity. Apparent cases of correct reinnervation can be accounted for by "trick" movements and other subtle adjustments in behavior which can give the appearance of normal movement even to a rather careful observer. Weiss and Taylor (1944) examined the possibility that regenerating nerves might show some tropism but found that rat nerves growing in forked arteries demonstrate no preference for one branch or the other, even when one branch ends in a blind termination or contains degenerating nerve tissue such as the distal nerve stump. And Weiss and Hoag (1946) found no selective regrowth of nerves to their muscles in rats. However, there have been a number of cases, going back to Elsberg (1917), in which competition between two regenerating nerves for innervation of a denervated muscle was won by the correct nerve. In such cases it has not been possible to determine whether the mechanism for this outcome involved some specific recognition phenomenon or just a slight advantage given to the unwittingly correct nerve: for example, the "correct" nerve might happen to have an innate capacity to recover from the trauma of nerve section and begin to regenerate faster than the other nerve. Bernstein and Guth (1961) demonstrated that the nerves which normally connect with the soleus and plantaris muscles of the rat hindlimb fail to reestablish specific connections during reinnervation. They took advantage of the fact that two spinal roots, L4 and L5, contribute quite differently to the innervation of soleus and plantaris. Measuring isometric tensions when stimulating either L4 or L5, they found that the tension ratio (L4 stimulated/L5 stimulated) was much higher for plantaris than soleus. After nerve crush or transection and resuture, the muscles were reinnervated but the differential tension ratios were gone. Other cases of unperturbed reinnervation of slow muscle by fast nerve have been mentioned above.

c. Cases of Apparent Reestablishment of Specific Neuromuscular Con-

nections. There are a number of cases of complete functional recovery after denervation in lower vertebrates (newts, salamanders, and teleost fish) (see Mark, 1969), which suggest either that nerves grow back to their appropriate muscles or that reinnervation is random but changes in the central nervous system result in normal motor output. These include recovery of normal eye movements and normal fin movements after section of the oculomotor nerve or the brachial plexus in teleost fish. When the nerves to the protractor or retractor muscles of the pectoral fin were carefully crossed with little chance of growth to the correct muscles, reversed fin movements were observed and these were maintained without recovery of normal movement for up to 18 months, proving that no reorganization of neuronal circuits in the central nervous system could occur to account for the recovery of normal function seen when regeneration of the nerves occurred without experimental misguidance.

Grimm (1971) and Cass *et al.* (1973) have examined the return of normal limb coordination in the axolotl, *Ambystoma mexicanum*, after reinnervation with and without experimental nerve cross-suturing. Grimm made nerve crosses between the nerves to the flexor and extensor muscles of the forearm. After regeneration, stimulation of nerves close to muscles suggested that incorrect connections had been made (hence the crossing was successful), but stimulation of the nerves at positions proximal to the original nerve cross resulted in contractions of the correct original muscles. No anatomical base for these observations was noticed, but it was concluded that somehow the regenerating nerve fibers must have "recrossed" and reinnervated their original muscles. To accomplish this, the nerves would have to overcome substantial geometric obstacles. Cass *et al.* (1973) examined the basis for recovery of normal hindlimb movement after nerve crossing and after partial denervation and reinnervation of the limb. As expected, the nerve cross did not lead to abnormal limb movement. Each nerve after regeneration activated the original muscle. To get at the basis of this, Cass *et al.* examined a slightly simpler situation which led them to propose that normal movement was the result of growth of nerves randomly to muscles followed by supression of synapses from the inappropriate nerves. They cut one of the nerves to the limb (the sixteenth) and observed the subsequent innervation of the denervated area by collateral sprouting from the intact seventeenth nerve. By 3 days after denervation, the seventeenth nerve has taken over about 0.5 mm of muscle territory ordinarily innervated by the sixteenth nerve, and progressively over the next several weeks the seventeenth nerve innervated more and more of the denervated area. Electrophysiological analysis of the situation showed that the formerly denervated muscle fibers now responded to seventeenth-nerve stimulation with long-latency end-plate potentials which were smaller than normal. After about 30 days, correct innervation began again as the sixteenth nerve regenerated and the seventeenth nerve apparently ceased functioning. It appeared that the sixteenth nerve was "out-competing" the seventeenth for innervation of its original target. But what had really happened to the collateral connections between seventeenth nerve and the muscles now activated by the sixteenth nerve? Did they physically retract or were they just "turned off"? To test one possibility, the sixteenth nerve was sectioned a second time. This time the seventeenth nerve innervated the "denervated" muscle within 3 days (as opposed to 3 weeks the first time) and the end-plate potentials were large and did not have a long latency. This suggested the presence of mature but previously silent synapses from the seventeenth nerve. Apparently, collaterals from the seventeenth nerve *remained* in close association with the muscle innervated by the sixteenth nerve, although functional neuromuscular connections had not been detected until the sixteenth nerve was cut. Despite the absence of functional connections, evidence for anatomical connections to the incorrect muscle was found. When the sixteenth nerve was cut and 4 days later the adjacent muscle innervated by the seventeenth nerve was examined by electron microscopy, it was found that about 25% of the synapses (apparently silent synapses from the sixteenth nerve) showed signs of nerve terminal degeneration. In the newly denervated muscle, about 75% of the nerve terminals were degenerating, the other 25% being nonfunctional synapses from the seventeenth nerve. Thus it was proposed that each muscle was normally partially "innervated" by *nonfunctional* synapses from the incorrect nerve. Electrophysiological analyses of these connections have not yet been reported.

Mark and coworkers (Marotte and Mark, 1970*a*,*b*; Mark and Marotte, 1972; Mark *et al.*, 1970, 1972) have reported much more extensively on another experimental situation in which nonfunctional synapses are postulated to occur between incorrectly connected nerve and muscle. The experiments involve innervation of the extraocular muscles of the eye of the carp and goldfish, *Carassius carassius* and *Carassius auratus*. There are two major muscles which control the rotation of the eye in the plane of its flat front surface: the superior and inferior oblique muscles. There are innervated, respectively, by motoneurons from the fourth and third cranial nerves. The muscles contract in response to vestibular and visual input to the motoneurons and maintain the eye in approximately constant horizontal orientation regardless of whether the fish is oriented head up or head down in the water. By some surgical maneuvers, a *functional* crossing of nerve III to the superior oblique muscles was achieved (although the incorrect eye rotations were always much smaller than the normal). This was measured by noting the reflex counter-rotations of the eye in response to positioning the fish head up and head down. During reinnervation of a cross-innervated superior oblique muscle by its correct fourth nerve, the incorrect reflex rotation of the eye due to the incorrect foreign innervation stops within 1–2 days of the first observed return of correct reflex counter-rotations. Electron microscopic examination of the synapses in the cross-innervated and correctly innervated superior oblique muscle at various times after the original surgery (Marotte and Mark, 1970*b*; Mark *et al.*, 1972) gave no indication of degenerating synapses during the period of competitive reinnervation. Thus the original nerve apparently silenced the foreign innervation without affecting the morphology of the synapses. It is unfortunate that neuromuscular junctions in this case are atypically simple in structure, so there are fewer morphological criteria to use in the analysis of structure.

In related study (Mark and Marotte, 1972) it was found using electrophysiological and anatomical analyses that after section of nerves III and IV, regenerating processes from both nerves regrew following the old pathway of the third nerve but that the pathway of the fourth nerve was followed only by fibers with firing properties correct for fourth nerve fibers.

The reinnervation of goldfish extraocular muscles has been examined by Scott (1975). She was able to confirm in some cases the reversal in reflex counter-rotation upon cross-innervation, and the suppression of this reflex upon reinnervation by the correct nerve, but when the superior oblique muscle was tested *in vitro* for contraction in response to nerve stimulation, stimulation of either the third or fourth nerve gave rise to muscle contraction. Testing by comparing the isometric tension developed by stimulation of the nerves separately or together revealed a fairly large degree of dual innervation, and this was confirmed in 15 fibers in five muscles by intracellular recording from single muscle fibers which responded to stimulation of either nerve. Her studies suggest that behavioral repression of incorrect muscle movement during competitive reinnervation might have some explanation other than repression of synaptic transmission at the level of the neuromuscular junctions.

This matter of nonfunctional synapses has added a new twist to considerations of nerve–muscle specificity and, as Mark has stressed (1970, 1974), potentially has extremely great significance in studies of central nervous system activity. Mark has proposed a role of physiological (but not morphological) synaptic repression and activation in the processes of memory and learning. If such "silent synapses" do indeed exist, then anatomical data would have to be interpreted with even more caution than presently exercised when inferences about neuronal circuitry are being drawn. This area of research begs for additional comfirmation and analysis, for what at the outset might have appeared to be an unusual regenerative capacity peculiar to a few lower vertebrates may rather offer major insights into the structure and function of our own nervous systems as well.

As Mark (1969) has pointed out, all of the cases of recovery of correct function after reinnervation or even cross-innervation involve skeletal muscle fibers which are multiply innervated. The ability of the muscle to accept multiple inputs may allow an ensuing competition (of whatever nature) between correct and incorrect nerves leading to restoration of normal function. In the case of singly innervated muscles, it may be that the first nerve to establish synaptic contact with the old postsynaptic structure impedes reinnervation by the original nerve. Experiments indicate that, even in mammalian muscles, a foreign nerve cannot necessarily achieve permanent and exclusive connection with the muscle (Tonge, 1974*a*; Frank *et al.*, 1974).

That multiple innervation is not a sufficient condition for restoration of normal function has been demonstrated in avian muscle by Bennett *et al.* (1973*c*), who have shown that the multiply innervated anterior latissimus dorsi of the chicken can maintain dual innervation by fast and slow motoneurons.

d. Cross-innervation of Muscle by Sensory Nerves. There are several reports of the "innervation" of skeletal muscle by sensory neurons. In such experiments it has usually been the peripheral or dendritic portion of the sensory neurons which has been routed into denervated muscle. This sort of "innervation" does not result in the formation of synapses or in stabilization of the muscle against denervation atrophy. Weiss and Edds (1945), for example, examined the regeneration of sensory fibers into the quadriceps muscle of the rat. They found that the sensory fibers arborized richly through the muscle and were still present at 8 months, but these fibers afforded no trophic support to denervated fibers in the muscle and stimulation of dorsal roots did not result in any muscle contraction. Why these fibers persist in the muscle and whether or not these fibers could function as sensory fibers are unknown. Zalewski (1970) reexamined this matter, crossing *central* (or axonal) fibers of the vagal nodose ganglion into the sternomastoid muscles of adult rats. He found no functional reinnervation or trophic support of the denervated muscle. There is one report of successful innervation of muscle by the central fibers of sensory neurons (Vera and Luco, 1967). In a single cat a cross-union of central fibers of the nodose ganglion into the longus captis muscle resulted over a year later in functional connections.

e. Cross-innervation of Muscle by Autonomic Cholinergic and Adrenergic Nerves. Zalewski (1970) found that cross-union of preganglionic sympathetic nerves to sternomastoid muscle failed to result in the formation of neuromuscular connections or trophic support of the denervated muscle. On the other hand, there are several reports of successful reinnervation of skeletal muscle by preganglionic nerve fibers of the autonomic nervous system. The most thoroughly documented of these is the innervation of frog sartorius muscle by the gastric vagal nerve after transplantation of the sartorius to the thoracic region (Landmesser, 1971, 1972). This innervation stabilized the muscle, preventing atrophy and preventing the changes in electrical properties which normally follow denervation. Some of the physiological properties of neuromuscular transmission were quite different from normal motor innervation: end-plate potentials had a low quantal content, and long-lasting facilitation of neuromuscular transmission occurred. The nerve maintained a high threshold for stimulation and a low conduction velocity, indicating that the nerve properties were not altered by innervating muscle. According to Landmesser (1971), most of the earlier reports of innervation of skeletal muscle by preganglionic autonomic nerve fibers are equivocal demonstrations, with the exception of that by Langley and Anderson (1904), since the particular nerves used in the crosses probably contained some motor fibers. Electron microscopic analysis of the cross-innervated frog sartorius indicated that the autonomic nerve fibers had formed synaptic contacts with the old postsynaptic sites. Bennett *et al.* (1973*b*) have studied the innervation of rabbit diaphragm by preganglionic fibers of the thoracic vagus nerve. They too feel that most of the earlier demonstrations in mammals, except possibly that of Guth and Frank (1959) on the innervation of rat diaphragm by the vagus nerve, could be explained by aberrant motoneurons effecting the reinnervation. In the study by Bennett and coworkers, the diaphragm was shown to be innervated exclusively by nonmyelinated fibers and the nerve terminals occurred at the old postsynaptic sites. Spontaneous and evoked potentials were recorded, and occasional multiple potentials suggesting polyneuronal innervation were found. The reinnervated diaphragm retained cholinesterase activity at the neuromuscular junctions and regained normal muscle fiber diameters.

The cross-innervation of cat skeletal muscle fibers with adrenergic (postganglionic autonomic) nerves does not result in the establishment of functional neuromuscular connections (Mendez *et al.*, 1970; Luco and Luco, 1971), but the investigators found that the presence of the adrenergic nerves in the muscle had an antifibrillary effect, suggesting some trophic support of the denervated muscle.

f. Innervation of Structures Other Than Skeletal Muscle Fibers by Motoneurons. There are a few reports of successful innervation of autonomic ganglion neurons and smooth muscle by motoneurons. McLachlan (1974) has demonstrated partial innervation of superior cervical ganglion cells by the motor nerve to the sternohyoid muscle. The study focused on the physiological properties of the ganglion cells, which were not affected either by denervation or by reinnervation with foreign nerve.

4.3. *Inferences about Nerve–Muscle Specificity Based on Reinnervation Experiments*

The experiments of nerve regeneration with and without nerve cross-union have given us fairly substantial evidence with regard to several aspects of specificity. These experiments treat specificity at three different levels. In reverse order from that of presentation above, they are (1) the level of the individual muscle (do the correct nerve bundles grow back to their denervated muscles?) (2) the finer level of specificity in the interaction of nerve and muscle fiber types within a single muscle, and (3) the specificity for the reestablishment of neuromuscular junctions at the old postsynaptic sites. In addition, there is the question of absolute vs. preferential specificity of interactions.

a. Absolute vs. Preferential Specificity of Interactions. In all cases where it appears that sufficient effort has been made, investigators have found that cholinergic nerves and muscles can be *forced* to form functional connections without regard to the identity of nerve and muscle, identity of neuron and muscle fiber type, or previous site of synaptic connection. Thus there is no evidence for absolute specificity of interaction between motoneurons and skeletal muscle fibers.

b. Preferential Innervation of Old Postsynaptic Sites. All evidence suggests a very strong preference on the part of the regenerating nerve for a return to the precise positions of former neuromuscular connection. It is intriguing that even preganglionic cholinergic nerves will innervate skeletal muscle fibers at this position. In one case the preferential reinnervation of old postsynaptic sites has been shown not to involve the residual activity of acetylcholinesterase at those positions (Filogamo and Gabella, 1967). Developmental studies suggest that other electrophysiological functions are not involved in synaptogenesis.

While it is not yet determined whether this is so during reinnervation as well, Dennis and Miledi (1974) have found that the re-formation of nerve–muscle connectivity occurs before synaptic transmission is again possible, due to failure of nerve action potentials to invade the nerve terminal during the early stages of re-formation of neuromuscular junctions. Various other studies have suggested that the restoration of properties characteristic of innervated muscle as opposed to denervated muscle begins before synaptic transmission is reestablished and that miniature end-plate potentials may occur as a sign of reinnervation before synaptic transmission is possible (Bennett *et al.*, 1973*c*; Miledi, 1960; Stefani and Schmidt,1972; but see Tonge, 1974*b*). It seems unlikely, although not inconceivable, that the specific reestablishment of neuromuscular connections at previous synaptic sites is due to some structural congruence between molecules of the postsynaptic surface and receptors on the growing nerve fiber tips. It seems more likely that some humoral agent is released at the neuromuscular junction which stabilizes nerve and muscle connection by halting the extension of nerve processes and inducing the elaboration of nerve terminal structure. Evans (1974) has obtained some evidence for distinctive permeability properties of the postsynaptic structure of denervated muscle. The experiments and considerations summarized by Edds (1955) make it unlikely that neuromuscular junctions are sites for the release of long-range neurotrophic substances.

c. Specificity in the Reestablishment of Synaptic Connections Between Particular Motoneuron and Skeletal Muscle Fiber Types. In the literature on mammalian skeletal muscle reinnervation there is little or no suggestion that any specificity exists, preferential or absolute, in the reestablishment of synaptic connections. If any preference exists, it seems to be for the survival and regrowth of fast nerve fibers. In other cases it is difficult to judge to what degree the experimental results are due to uncontrollable advantages one nerve might have over another in regeneration in experimental situations. Such advantages might or might not be of biological significance. The early results of Elsberg (1917), as he carefully points out, could be an example of the latter.

Of the other data on reestablishment of fiber type specific neuromuscular connections, those of Hoh (1971), using the toad, are most convincing. This case warrants additional investigation. We would like to know, for example, if morphologically normal but synaptically silent or inefficient synapses involving fast motoneurons exist on reinnervated slow muscle fibers. In all cases of reinnervation, but particularly where the possibility of competition between correct and incorrect nerve fibers might occur, the most detailed time study of the reinnervation process is necessary to allow judgment as to the mechanism of the process and the degree of preference afforded correct over incorrect connections. Where investigators have tried to vary time of arrival of foreign vs. original nerve, the most important variable in which nerve makes the most synapses is which nerve arrives first (Frank *et al.*, 1974; Tonge, 1974*a*).

d. Specificity in the Reestablishment of Synaptic Connections Between Specific Muscles and Their Nerves. In the mammals there is little or no evidence for correct regrowth of nerves to their muscles. Thus reinnervation never leads to proper restoration of muscular coordination (Sperry, 1945; Mark, 1969). As for the lower vertebrates, some cases of restitution of normal coordination after nerve section seem explainable only in terms of correct regrowth of nerves. However, in all these cases there remains the possibility that the nerves also establish morphologically normal but physiologically silent synapses with incorrect muscles. These cases beg for closer physiological and morphological investigation, for their general relevance to neurobiology could be enormous.

ACKNOWLEDGMENTS

I wish to thank Mrs. Ann N. Murphy for typing the manuscript. I would also like to thank Dr. S. A. Scott for her critical review of the manuscript.

5. REFERENCES

Aitkin, J. T., 1965, Problems of reinnervation of muscle, in: Degeneration Pattern in the Nervous System (M. Singer and J. P. Schade, eds.), *Progr. Brain Res.* **14**:232.

Albuquerque, E. X., Barnard, E. A., Porter, C. W. and Warnick, J. E., 1974, The density of acetylcholine receptors and their sensitivity in the post-synaptic membrane of muscle endplates, *Proc. Natl. Acad. Sci. USA* **71**:2818.

Anderson, M. J., and Cohen, M. S., 1974, Fluorescent staining of acetylcholine receptors in vertebrate skeletal muscle, *J. Physiol. (London)* **237**:385.

Andersson-Cedergren, E., 1959, Ultrastructure of motor endplate and sarcoplasmic components of mouse skeletal muscle fiber as revealed by three-dimensional reconstruction from serial section, *J. Ultrastruct. Res.* (Suppl. 1).

Atsumi, S., 1971*a*, The histogenesis of motor neurons with special reference to the correlation of their endplate formation. I. The development of endplates in the intercostal muscle in the chick embryo, *Acta Anat.* **80**:161.

Atsumi, S., 1971*b*, The histogensis of motor neurons with special reference to the correlation of their endplate formation. III. The development of motor neurons innervating the intercostal muscle in the chick embryo, *Acta Anat.* **80**:504.

Axelsson, J., and Thesleff, S., 1959, A study of supersensitivity in denervated mammalian skeletal muscle, *J. Physiol. (London)* **147**:158.

Bagust, J., Lewis, D. M., and Westerman, R. A., 1973, Polyneuronal innervation of kitten skeletal muscle, *J. Physiol. (London)* **229**:241.

Barker, D., and Ip, M. C., 1966, Sprouting and degeneration of mammalian motor axons in normal and deafferented skeletal muscle, *Proc. Roy Soc. London Ser. B* **163**:538.

Bauer, W. C., Blumberg, J. M., and Zacks, S. I., 1962, Short and long-term unltrastructural changes in denervated mouse endplates, in: *Proceedings of the IV International Congress of Neuropathology* (G. Thieme, ed.), pp. 16–18, Stuttgart.

Bennett, M. R., and Pettigrew, A. G., 1974*a*, The formation of synapses in striated muscle during development, *J. Physiol. (London)* **241**:515.

Bennett, M. R., and Pettigrew, A. G., 1974*b*, The formation of synapses in reinnervated and cross-reinnervated striated muscle during development, *J. Physiol. (London)* **241**:547.

Bennett, M. R., McLachlan, E. M., and Taylor, R. S., 1973*a*, The formation of synapses in reinnervated mammalian striated muscle, *J. Physiol. (London)* **233**:481.

Bennett, M. R., McLachlan, E. M., and Taylor, R. S., 1973*b*, The formation of synapses in mammalian striated muscle reinnervated with autonomic pregangliomic nerves, *J. Physiol. (London)* **233**:501.

Bennett, M. R., Pettigrew, A. G., and Taylor, R. S., 1973*c*, The formation of synapses in reinnervated and cross-reinnervated adult avian muscle, *J. Physiol. (London)* **230**:331.

Bernstein, J. J., and Guth, L., 1961, Nonselectivity in establishment of neuromuscular connections following nerve regeneration in the rat, *Exp. Neurol.* **4**:262.

Betz, W., and Sakmann, B., 1971, "Disjunction" of frog neuromuscular synapses by treatment with proteolytic enzymes, *Nature (London) New Biol.* **232**:94.

Betz, W., and Sakmann, B., 1973, Effects of proteolytic enzymes on function and structure of frog neuromuscular junctions, *J. Physiol. (London).* **230**:673.

Birks, R., Huxley, H. E., and Katz, B., 1960, The fine structure of the neuromuscular junction of the frog, *J. Physiol. (London)* **150**:134.

Bone, Q., 1964, Patterns of muscular innervation in the lower chordates, *Int. Rev. Neurobiol.* **6**:99.

Bourgeois, J.-P., Ryter, A., Menez, A., Fromageot, P., Bouquet, P., and Changeux, J.-P., 1972, Localization of the cholinergic receptor protein in *Electrophorus* electroplax by high resolution autoradiography, *FEBS Lett.* **25**:127.

Brown, M. C., and Matthews, P. B. C., 1960, An investigation into the possible existence of polyneuronal innervation of individual skeletal muscle fibers in certain hindlimb muscles of the cat, *J. Physiol. (London)* **151**:436.

Burke, R. E., and Tsairis, P., 1973, Anatomy and innervation ratios in motor units of cat gastrocnemius, *J. Physiol. (London)* **234**:749.

Burke, R. E., Levine, D. N., Tsairis, P., and Zajac, F. E., 1973, Physiological types and histochemical profiles in motor units of the cat gastrocnemius, *J. Physiol. (London)* **234**:723.

Cass, D. T., Sutton, T. J., and Mark, R. F., 1973, Competition between nerves for functional connections with axolotl muscles, *Nature (London)* **243**:201.

Close, R. I., 1972, Dynamic properties of mammalian skeletal muscles, 1972, *Physiol. Rev.* **52**:129.

Close, R. I., and Hoh, J. F. Y., 1968, Effects of nerve cross-union on fast-twitch and slow-graded muscle fibers in the toad, *J. Physiol. (London)* **198**:103.

Cöers, C., 1969, Structure and organization of the myoneural junction, *Int. Rev. Cytol.* **22**:239.

Cohen, M. W., 1973, The development of neuromuscular connexions in the presence of D-tubocurarine, *Brain Res.* **41**:457.

Couteaux, R., 1960, Motor endplate structure, in: *Structure and Function of Muscle,* Vol. 1 (G. H. Bourne, ed.), pp. 337–380, Academic Press, New York.

Couteaux, R., 1963, The differentiation of synaptic areas, *Proc. R. Soc. London Ser. B* **158**:457.

Couteaux, R., and Pécot-Dechavassine, M., 1970, Vesicules synaptiques et poches au niveau des "zones actives" de la junction neuromusculaire, *C. R. Acad. Sci.* **271**:2346.

Cowan, W. M., and Wenger, E., 1968, The development of the nucleus of origin of centrifugal fibers to the retina in the chick, *J. Comp. Neurol.* **133**:207.

Crain, S. M., and Peterson, E. R., 1971, Development of paired explants of fetal spinal cord and adult skeletal muscle during chronic exposure to curare and hemicholinium, *In Vitro* **6**:373.

Crossland, W. J., Cowan, W. M., Rodgers, L. A., and Kelley, J. P., 1974, The specification of the retina–tectal projection in the chick, *J. Comp. Neurol.* **155**:127.

Csillik, B., 1965, *Functional Structure of the Post-synaptic Membrane in the Myoneural Junction,* Akademiai Kiado, Budapest.

Daniels, M. P., and Vogel, Z., 1975, Immunoperoxidase staining of α-bungarotoxin binding sites in muscle endplates shows distribution of acetylcholine receptors, *Nature (London)* **254**:339.

Dennis, M., 1975, Physiological properties of junctions between nerve and muscle during salamander limb regeneration, *J. Physiol. (London)* **244**:683.

Dennis, M. J., and Miledi, R., 1974, Non-transmitting neuromuscular junctions during an early stage of end-plate reinnervation, *J. Physiol. (London)* **239**:553.

Diamond, J., and Miledi, R., 1962, A study of foetal and new-born rat muscle fibers, *J. Physiol. (London)* **162**:393.

Dreyer, F., Peper, K., Akert, K., Sandri, C., and Moor, H., 1973, Ultrastructure of the "active zone" in the frog neuromuscular junction, *Brain Res.* **62**:373.

Dunlap, D. G., 1966, The development of the musculature of the hindlimb in the frog, *Rana pipiens, J. Morphol.* **119**:241.

Edds, M. V., 1955, Collateral regeneration in partially reinnervated muscles of the rat, *J. Exp. Zool.* **129**:225.

Elsberg, C. A., 1917, Experiments on motor nerve regeneration and the direct neurotization of paralyzed muscles by their own and foreign nerves, *Science* **45**:318.

Elul, R., Miledi, R., and Stefani, E., 1968, Neurotrophic control of contracture in slow muscle fibers, *Nature (London)* **217**:1274.

Eränkö, O., and Teräväinen, H., 1967, Cholinesterases and eserine-resistant carboxylic esterases in degenerating and regenerating motor endplates of the rat, *J. Neurochem.* **14**:947.

Evans, R. H., 1974, The entry of calcium into the innervated region of the mouse diaphragm muscle, *J. Physiol. (London)* **240**:517.

Fambrough, D., and Rash, J. E., 1971, Development of acetylcholine sensitivity during myogenesis, *Dev. Biol.* **26**:55.

Feng, T. P., Wu, W. Y., and Yang, F. Y., 1965, Selective reinnervation of a "slow" or "fast" muscle by its original motor supply during regeneration of a mixed nerve, *Sci. Sinica* **14**:1717.

Fertuck, H. C., and Salpeter, M. M., 1974, Localization of acetylcholine receptor by ^{125}I-labeled α-bungarotoxin binding at mouse motor end-plates, *Proc. Natl. Acad. Sci. USA* **71**:1376.

Filogamo, G., and Gabella, G., 1967, Cholinsterase behavior in the denervated and reinnervated muscles, *Acta Anat.* **63:**199.

Fischbach, G. D., 1972, Synapse formation between dissociated nerve and muscle cells in low density cell cultures, *Dev. Biol.* **28:**407.

Fischman, D. A., 1970, The synthesis and assembly of myofibrils in embryonic muscle, *Curr. Top. Dev. Biol.* **5:**235.

Fluck, R. A., and Strohman, R. C., 1973, Acetylcholinesterase activity in developing skeletal muscle cells *in vitro, Dev. Biol.* **33:**417.

Frank, E., Jansen, J. K. S., Lomo, T., and Westgaard, R. H., 1974, Effect of foreign innervation on the reinnervation of muscle by its original nerve, *J. Physiol. (London)* **240:**24P.

Fukuda, J., 1974, Chloride spike: A third type of action potential in tissue-cultured skeletal muscle cells from chick, *Science* **185:**76.

Giller, E. L., Schrier, B. K., Shainberg, A., Fisk, H. R., and Nelson, P. G., 1973, Choline acetyltransferase activity is increased in combined cultures of spinal cord and muscle cells from mice, *Science* **182:**588.

Gonzenbach, H. R., and Waser, P. G., 1973, Electron microscopic studies of degeneration and regeneration of rat neuromuscular junctions, *Brain Res.* **63:**167.

Goodwin, B. C., and Sizer, I. W., 1965, Effects of spinal cord and substrate on acetylcholinesterase in chick embryonic skeletal muscle, *Dev. Biol.* **11:**136.

Gordon, T., and Vrbova, G., 1974, Synapse formation during development (abst), XXVI International Congress of Physiological Sciences, Jerusalem Satellite Symposium "Mechanisms of Synaptic Action," p. 21.

Grimm, L., 1971, An evaluation of myotopic respecification in axolotls, *J. Exp. Zool.* **178:**479.

Guth, L., 1956, Regeneration in the mammalian peripheral nervous system, *Physiol. Rev.* **36:**441.

Guth, L., 1968, "Trophic" influences of nerve on muscle, *Physiol. Rev.* **48:**645.

Guth, L., and Brown, W. C., 1965*a*, The sequence of changes in cholinesterase activity during reinnervation of muscle, *Exp. Neurol.* **12:**329.

Guth, L., and Brown, W. C., 1965*b*, Changes in cholinesterase activity following partial denervation, collateral reinnervation, and hyperneurotization of muscle, *Exp. Neurol.* **13:**198.

Guth, L., and Frank, K., 1959, Restoration of diaphragmatic function following vagophrenic anastomosis in the rat, *Exp. Neurol.* **1:**1.

Guth, L., and Samaha, F. J., 1969, Qualitative differences between actomyosin ATPase of slow and fast mammalian muscle, *Exp. Neurol.* **25:**138.

Guth, L., Zalewski, A. A., and Brown, W. C., 1966, Quantitative changes in cholinesterase activity of denervated sole plate following implantation of nerve into muscle , *Exp. Neurol.* **16:**136.

Gutmann, E., and Young, J. Z., 1944, The reinnervation of muscle after various periods of atrophy, *J. Anat.* **78:**15.

Gutmann, E., Hanzlikova, V., and Holeckova, E., 1969, Development of fast and slow muscles of the chicken *in vivo* and their latent period in tissue culture, *Exp. Cell Res.* **56:**33.

Hall, Z. W., and Kelly, R. B., 1971, Enzymatic detachment of endplate acetylcholinesterase from muscle, *Nature (London) New Biol.* **232:**62.

Hamburger, V., 1946, Isolation of the brachial segments of the spinal cord of the chick embryo by means of tantalum foil blocks, *J. Exp. Zool.* **103:**113.

Hamburger, V., 1952, Development of the nervous system, *Ann. N.Y. Acad. Sci.* **55:**117.

Harris, A. J., Heinemann, S., Schubert, D., and Tarakis, H., 1971, Trophic interaction between cloned tissue culture lines of nerve and muscle, *Nature (London)* **231**:296.
Harris, J. B., and Marshall, M. W., 1973, Tetrodotoxin-resistant action potentials in newborn rat muscle, *Nature (London) New Biol.* **243**:191
Harris, J. B., Marshall, M. W., and Wilson, P., 1973, A physiological study of chick myotubes grown in tissue culture, *J. Physiol. (London)* **229**:751.
Harrison, R. G., 1904, An experimental study of the relation of the nervous system to the development of musculature in the embryo of the frog, *Am. J. Anat.* **3**:197.
Harrison, R. G., 1910, The outgrowth of the nerve fibers as a mode of protoplasmic movement, *J. Exp. Zool.* **9**:787.
Hartzell, H. C., and Fambrough, D. M., 1972, Acetylcholine receptors: Distribution and extrajunctional density in rat diaphragm after denervation correlated with acetylcholine sensitivity, *J. Gen. Physiol.* **60**:248.
Hess, A., 1970, Vertebrate slow muscle fibers, *Physiol. Rev.* **50**:40.
Heuser, J. E., and Reese, T. S., 1973, Evidence for recycling of synaptic vesicle membrane during transmitter release at frog neuromuscular junction, *J. Cell Biol.* **47**:315.
Heuser, J. E., Reese, T. S., and Landis, D. M. D., 1974, Functional changes in frog neuromuscular junction studied with freeze-fracture *J. Neurocytol.* **3**:108.
Hirano, H., 1967, Ultrastructural study on the morphogenesis of the neuromuscular junction in the skeletal muscle of the chick, *Z. Zellforsch.* **79**:198.
Hnik, P., Jirmanova, I., Vyklicky, L., and Zelena, J., 1967, Fast and slow muscles of the chick after nerve cross-union, *J. Physiol. (London)* **193**:309.
Hoh, J. F. Y., 1971, Selective reinnervation of fast-twitch and slow-graded muscle fibers in the toad, *Exp. Neurol.* **30**:263.
Hubbard, J. I., 1973, Microphysiology of vertebrate neuromuscular transmission, *Physiol. Rev.* **53**:674.
Hughes, A. F. W., 1965, A quantitative study of the development of the nerves in the hindlimb of *Eleutherodactylus martinicensis*, *J. Embryol. Exp. Morphol.* **13**:9.
Hughes, A., 1968, *Aspects of Neural Ontogeny*, Academic Press, New York.
Hughes, A. F. W., and Prestige, M. C., 1967, Development of behavior in the hindlimb of *Xenopus laevis*, *J. Zool.* **152:347.**
Hunt, C. C., and Kuffler, S. W., 1954, Motor innervation of skeletal muscle: Multiple innervation of individual muscle fibers and motor unit function, *J. Physiol. (London)* **126**:293.
Iwayama, T., 1969, Relation of regenerating nerve terminals to original end-plates, *Nature (London)* **229**:81.
Jansen, J. K. S., Lomo, T., Nicolaysen, K., and Westgaard, R., 1973, Hyperinnervation of skeletal muscle fibers: Dependence of muscle activity, *Science* **181**:559.
Jirmanova, I., and Thesleff, S., 1972, Ultrastructural study of experimental muscle degeneration and regeneration in the adult rat, *Z. Zellforsch. Microsk. Anat.* **131**:77.
Jirmanova, I., Hnik, P., and Zelena, J., 1971, Implantation of "fast" nerve into slow muscle in young chickens, *Physiol. Bohemoslov.* **20**:199.
Katz, B., and Miledi, R., 1964, The development of acetylcholine sensitivity in nerve-free segments of skeletal muscle, *J. Physiol. (London)* **170**:389.
Kelly, A. M., and Zacks, S. I., 1969, The fine structure of motor endplate morphogenesis, *J. Cell Biol.* **42**:154.
Kidokoro, Y., 1973, Development of action potentials in a clonal rat skeletal muscle cell line, *Nature (London) New Biol.* **241**:158.
Koenig, J., 1963, Innervation motrice experimentale d'une portion de muscle strié normalement depouvrée de plaques motrices chez le rat, *C. R. Acad. Sci.* **256**:2918.

Koening, J., and Pecot-Dechavassine, M., 1971, Relation between appearance of miniature endplate potentials and ultrastructure of reinnervating or newly formed endplates in rats, *Brain Res.* **27**:43.

Kuno, M., Miyata, Y., and Munoz-martinez, E. J., 1974, Properties of fast and slow alpha motoneurons following motor reinnervation, *J. Physiol. (London)* **242**:273.

Landmesser, L., 1971, Contractile and electrical responses of vagus-innervated frog sartorius muscles, *J. Physiol. (London)* **213**:707.

Landmesser, L., 1972, Pharmacological properties, cholinesterase activity and anatomy of nerve–muscle junctions in vagus-innervated frog sartorius, *J. Physiol. (London)* **220**:243.

Langley, J. N., and Anderson, H. K., 1904, The union of different kinds of nerve fibers, *J. Physiol. (London)* **31**:365.

Lentz, T. L., 1969, Development of the neuromuscular junction. I. Cytological and cytochemical studies on the neuromuscular junction of differentiating muscle in the regenerating limb of the newt *Triturus*, *J. Cell Biol.* **42**:431.

Letinsky, M. S., 1974, The development of nerve–muscle junctions in *Rana catesbeiana* tadpoles, *Dev. Biol.* **40**:129.

Levi-Montalcini, R., 1950, The origin and development of the visceral system in the spinal cord of the chick embryo, *J. Morphol.* **86**:253.

Lewis, P. R., and Hughes, A. F. W., 1960, Patterns of myoneural junctions and cholinesterase activity in the muscles of tadpoles of *Xenopus laevis*, *Q. J. Microsc. Sci.* **101**:55.

Liu, H.-C., and Maneely, R. B., 1968, The development of motor endplates in the embryonic and regenerative tail of *Hemidactylus bowringi* (Gray), *Acta Anat.* **71**:249.

Luco, C. F., and Luco, J. V., 1971, Sympathetic effects on fibrillary activity of denervated striated muscles, *J. Neurophysiol.* **34**:1066.

Lüllmann-Rauch, R., 1971, The regeneration of neuromuscular junctions during spontaneous reinnervation of the rat diaphragm, *Z. Zellforsch.* **121**:593.

Mackay, B., and Harrop, T. J., 1969, An experimental study of the longitudinal growth of skeletal muscle in the rat, *Acta anat.* **72**:38.

Mark, R. B., 1969, Matching muscles and motoneurons: A review of some experiments on motor nerve regeneration, *Brain Res.* **14**:245.

Mark, R. F., 1970, Chemospecific synaptic repression as a possible memory store, *Nature (London)* **225**:178.

Mark, R., 1974, *Memory and Nerve Cell Connections*, Clarendon Press, Oxford.

Mark, R. F., and Marotte, L. R., 1972, The mechanism of selective reinnervation of fish eye muscles. III. Functional electrophysiological and anatomical analysis of recovery from section of the IIIrd and IVth nerves, *Brain Res.* **46**:131.

Mark, R., Marotte, L., and Johnstone, J., 1970, Reinnervated eye muscles do not respond to impulses in foreign nerves, *Science* **170**:193.

Mark, R. F., Marotte, L. R., and Mart, P. E., 1972, The mechanism of selective reinnervation of fish eye muscles. IV. Identification of repressed synapses, *Brain Res.* **46**:149.

Marotte, L. R., and Mark, F. F., 1970*a*, The mechanisms of selective reinnervation of fish eye muscle. I. Evidence from muscle function during recovery, *Brain Res.* **19**:41.

Marotte, L. R., and Mark, R. F., 1970*b*, The mechanism of selective reinnervation of fish eye muscle. II. Evidence from electron microscopy of nerve endings, *Brain Res.* **19**:53.

McLachlan, E. M., 1974, The formation of synapses in mammalian sympathetic ganglia reinnervated with pre-ganglionic or somatic nerves, *J. Physiol. (London)* **237**:217.

Mendez, J., Aranda, L. C., and Luco, J. V., 1970, Antifibrillary effect of adrenergic fibers on denervated striated muscles, *J. Neurophysiol.* **33**:882.

Miledi, R., 1960, Properties of regenerating neuromuscular synapses in the frog, *J. Physiol. (London)* **154**:190.
Miledi, R., 1962, Induced innervation of endplate free muscle segments, *Nature (London)* **193**:281.
Miledi, R., 1963, Formation of extra nerve–muscle junctions in innervated muscle, *Nature (London)* **199**:1191.
Miledi, R., and Slater, C. R., 1968, Electrophysiology and electron microscopy of rat neuromuscular junctions after nerve degeneration, *Proc. R. Soc. London Ser. B* **169**:289.
Miledi, R., and Slater, C. R., 1970, On the degeneration of rat neuromuscular junctions after nerve section, *J. Physiol. (London)* **207**:507.
Miledi, R., and Stefani, E., 1969, Non-selective reinnervation of slow and fast muscle fibers in the rat, *Nature (London)* **222**:569.
Mumenthaler, M., and Engel, W. K., 1961, Cytological localization of cholinesterase in developing chick embryo skeletal muscle, *Acta Anat.* **47**:274.
Nelson, P. G., 1975, Nerve and muscle cells in culture, *Physiol. Rev.* **55**:1.
Nickel, E., and Waser, P. G., 1968, Electronenmikroskopische Untersuchungen am Diaphragma der Maus nach einseitiger Phrenikotomie, *Z. Zellforsch.* **88**:278.
Obata, K., 1974, Transmitter sensitivities of some nerves and muscle cells in culture, *Brain Res.* **73**:71.
Oh, T. H., Johnson, D. D., and Kim, S. U., 1972, Neurotrophic effect on isolated chick embryo muscle in culture, *Science* **178**:1298.
Patterson, B., and Prives, J., 1973, Appearance of acetylcholine receptor in differentiating cultures of embryonic chick breast muscle, *J. Cell Biol.* **59**:241.
Pecot-Dechavassine, M., 1968, Course of the activity of cholinesterases and their functional capacity at the neuromuscular and musculotendinous junctions in the frog after motor nerve section, *Arch. Int. Pharmacodyn.* **176**:118.
Peper, K., Dreyer, F., Sandri, C., Akert, K., and Moor, H., 1974, Structure and ultrastructure of the frog motor endplate, *Cell Tissue Res.* **149**:437.
Peterson, E. R., and Crain, S. M., 1972, Regeneration and innervation in cultures of mammalian skeletal muscle coupled with fetal rodent spinal cord, *Exp. Neural.* **36**:136.
Piatt, J., 1940, Nerve–muscle specificity in *Amblystoma,* studied by means of heterotopic cord grafts, *j. Exp. Zool.* **85**:211.
Prestige, M. C., 1967, The control of cell number in the lumbar spinal ganglion during the development of *Xenopus laevis* tadpole, *J. Embryol. Exp. Morphol.* **17**:453.
Prestige, M.C., and Wilson, M. A., 1972, Loss of ventral roots during development, *Brain Res.* **41**:467.
Prives, J. M., and Patterson, B. M., 1974, Differentiation of cell membranes in cultures of embryonic chick breast muscle, *Proc. Natl. Acad. Sci.* **71**:3208.
Rash, J. E., and Ellisman, M. H., 1974, Studies on excitable membranes. I. Macromolecular specializations at the neuromuscular junction and the nonjunctional sarcoplasm, *J. Cell Biol.* **63**:567.
Rash, J. E., Ellisman, M. H., Staehelin, L. A., and Porter, K. R., 1975, Molecular specializations of excitable membranes in normal, chronically denervated and dystrophic muscle fibers, in: *Exploratory Concepts in Muscular Dystrophy II: Proceedings of an International Conference, Carefree, Arizona, October 15–19, 1973,* Exerpta Medica, Amsterdam.
Redfern, P. A., 1970, Neuromuscular transmission in new-born rats, *J. Physiol. (London)* **209**:701.

Redfern, P., and Thesleff, S., 1971, Action potential generation in denervated rat skeletal muscle. II. The action of tetrodotoxin, *Acta Physiol. Scand.* **82:**70.

Reger, J. F., 1959, Studies on the fine structure of normal and denervated neuromuscular junctions from mouse gastrocnemius, *J. Ultrastruct. Res.* **2:**269.

Reier, P. J., and Hughes, A., 1972, Evidence for spontaneous axon degeneration during peripheral nerve maturation, *Am. J. Anat.* **135:**147.

Robbins, N., and Yonezawa, T., 1971, Physiological studies during formation and development of rat neuromuscular junctions in tissue culture, *J. Gen. Physiol.* **58:**467.

Romanul, C. A., and Van der Meulen, J. P., 1967, Slow and fast muscles after cross-innervation: Enzymatic and physiological changes, *Neurology* **17:**387.

Rosenbluth, J., 1974, Structure of amphibian motor endplate: Evidence for a granular component projecting from the outer surface of the receptive membrane, *J. Cell Biol.* **62:**755.

Saito, A., and Zacks, S. I., 1969, Fine structure observations of denervation and reinnervation of neuromuscular junctions in mouse foot muscle, *J. Bone Jt. Surg.* **51A:**1163.

Salpeter, M. M., 1967, Electron microscope autoradiography as a quantitative tool in enzyme cytochemistry. I. The distribution of acetylcholinesterase at motor endplates of a vertebrate twitch muscle, *J. Cell Biol.* **32:**379.

Sandbank, U., and Bubis, J. J., 1974, *The Morphology of Motor Endplates,* Brain Information Service, University of California, Los Angeles.

Scott, S. A., 1975, Persistence of foreign innervation on reinnervated goldfish extraocular muscles, *Science* **189:**644.

Sonesson, B., and Thesleff, S., 1968, Cholinesterase activity after DFP application in botulinum poisoned, surgically denervated or normally innervated rat skeletal muscles, *Life Sci.* **7:**411.

Sperry, R. W., 1945, The problem of central nervous reorganization after nerve regeneration and muscle transposition, *Q. Rev. Biol.* **20:**311.

Steinbach, J. H., Harris, A. J., Patrick, J., Schubert, D., and Heinemann, S., 1973, Nerve–muscle interaction *in vitro*: Role of acetylcholine, *J. Gen. Physiol.* **62:**255.

Stefani, E., and Schmidt, H., 1972, Early stage of reinnervation of frog slow muscle fibers, *Pfluegers Arch.* **336:**271.

Sytkowski, A. J., Vogel, Z., and Nirenberg, M. W., 1973, Development of acetylcholine receptor clusters on cultured muscle cells, *Proc. Natl. Acad. Sci. USA* **70:**270.

Taylor, A. C., 1943, Development of the innervation pattern in the limb bud of the frog, *Anat. Rec.* **87:**379.

Tennyson, V. M., Brzin, M., and Kremzner, L. T., 1973, Acetylcholinesterase activity in the myotube and muscle satellite cell of the fetal rabbit: An electron microscopic-cytochemical and biochemical study, *J. Histochem. Cytochem.* **21:**634.

Teräväinen, H., 1968, Development of the myoneural junction in the rat, *Z. Zellforsch.* **87:**249.

Thesleff, S., 1960, Supersensitivity of skeletal muscle produced by botulinum toxin, *J. Physiol. (London)* **151:**598.

Tiegs, D. N., 1953, Innervation of voluntary muscle, *Physiol. Rev.* **33:**90.

Tonge, D. A., 1974*a*, Synaptic function in experimental innervated muscle in the mouse, J. Physiol. (London) **239:**96P.

Tonge, D. A., 1974*b*, Reinnervation of skeletal muscle in the mouse, *J. Physiol. (London)* **236:**22P.

Tuffery, A. R., 1971, Growth and degeneration of motor endplates in normal cat hind limb muscles, *J. Anat.* **110:**221.

Vera, C. L., and Luco, J. V., 1967, Reinnervation of smooth and striated muscle by sensory nerve fibers, *J. Neurophysiol.* **30**:620.
Vogel, Z., Sytkowski, A. J., and Nirenberg, M. W., 1972, Acetylcholine receptors of muscle grown *in vitro, Proc. Natl. Acad. Sci. USA* **69**:3180.
Wenger, B. S., 1951, Determination of structural patterns in the spinal cord of the chick embryo studied by transplantations between brachial and adjacent levels, *J. Exp. Zool.* **116**:123.
Weiss, P., and Edds, M. V., 1945, Sensory-motor nerve cross in the rat, *J. Neurophysiol.* **8**:173.
Weiss, P., and Hoag, A., 1946, Competitive reinnervation of rat muscles by their own and foreign nerves, *J. Neurophysiol.* **9**:413.
Weiss, P., and Taylor, A. C., 1944, Further experimental evidence against "neurotropism" in nerve regeneration, *J. Exp. Zool.* **95**:233.
Williams, P. E., and Goldspink, G., 1971, Longitudinal growth of striated muscle fibers, *J. Cell Sci.* **9**:751.
Wilson, B. W., Nieberg, P. S., Walker, C. R., Linkhart, T. A., and Fry, D. M., 1973, Production and release of acetylcholinesterase by cultured chick embryo muscle, *Dev. Biol.* **33**:285.
Yaffe, D., 1969, Cellular aspects of muscle differentiation *in vitro, Curr. Top. Dev. Biol.* **4**:37.
Yaffe, D. and Feldman, M., 1965, The formation of hybrid multinucleated muscle fibers from myoblasts of different genetic origin, *Dev. Biol.* **11**:300.
Yellin, H., 1967, Neural regulation of enzymes in muscle fibers of red and white muscle, *Exp. Neurol.* **19**:92.
Yonezawa, T., Saida, T., Robbins, N., and Ibata, Y., 1973, Electron microscopic studies on the neuromuscular junctions developed *in vitro:* Cholinesterase activity. (Japanese) *Advan. Neurol. Sci.* **17**:170.
Zacks, S. I., 1964, *The Motor Endplate,* Saunders, Philadelphia.
Zalewski, A. A., 1970, Reinnervation of denervated skeletal muscle by axons of motor, sensory and sympathetic neurons, *Physiologist* **12**:354.
Zelena, J., and Jirmanova, I., 1973, Ultrastructure of chicken slow muscle after nerve cross union, *Exp. Neurol.* **38**:272.
Zelena, J., and Szentagothai, J., 1957, Verlagerung der Lokalisation specifischer Cholinesterase Während der Entwicklung der Muskelinnervation, *Acta Histochem.* **3**:284.
Zelena, J., Vyklicky, L., and Jirmanova, I., 1967, Motor enplates in fast and slow muscles of the chick after cross-union of their nerves, *Nature (London)* **214**:1010.

3

Reactive Synaptogenesis in the Adult Nervous System

The Effects of Partial Deafferentation on New Synapse Formation

CARL W. COTMAN and GARY S. LYNCH

1. INTRODUCTION

In 1885 Exner suggested that the recovery of muscular contraction observed after partial transection of a motor nerve, but prior to regeneration of the damaged fibers, resulted from collateral growth of intact fibers and reinnervation of the muscle. Subsequently Edds (1950), Hoffman (1950), and others (Weddell *et al.*, 1946; Hones *et al.*, 1945; Weiss and Edds, 1946) demonstrated conclusively that the transection of few motor fibers could in fact result in axon collateral sprouting by the remaining undamaged fibers. This phenomenon was extended to connections between neurons when Murray and Thompson (1957) provided direct anatomical evidence for axon collateral sprouting in the partially denervated sympathetic ganglion and Liu and Chambers (1958) reported evidence for axon sprouting in the spinal cord. Over the last 20 years there has been an explosive growth of research on axon sprouting in the central nervous system. It is now clear that the phenomenon exists and can be highly selective in terms

---

CARL W. COTMAN and GARY S. LYNCH · Department of Psychobiology, University of California, Irvine, California.

of which fibers sprout, which neurons are reinnervated, and at what ages it can be demonstrated.

In this chapter, we will examine the properties of reinnervation after partial deafferentation of the sympathetic ganglion and of various regions in the CNS of adult animals and will discuss the possible underlying mechanisms. We will consider a number of questions relevant to the functional significance of this process. Are specific, topographically precise connections formed? Is the response general to all neurons in the circuit? What molecular and cellular processes control reinnervation?

The experimental paradigm which has been used in all the studies that we will discuss consists of removing part of the innervation to a given region of the nervous system and monitoring by various means the response of the remaining afferents. In the adult nervous system, at least three types of responses could result in the formation of anomalous synapses. The lesion may induce growth of axon collaterals which then traverse some distance before contacting the denervated target cell (collateral sprouting). In other cases, new boutons may bud off of existing axon terminals and form synapses adjacent to those already present (paraterminal sprouting). Since both these types of growth involve the branching of existing axons and since they are seldom distinguishable experimentally, Bernstein and Goodman (1973) have proposed the general term *axon sprouting* for cases in which either process may be operative. Their recommendation will be followed in this chapter. New synapses may also be formed after deafferentation if the deafferented neurons and intact afferent fibers shift position so that they contact one another at new locations. The formation of new synapses at these points of contact would not require axonal growth. We will refer to this process as *contact synaptogenesis*.

Because the strict application of these terms requires precise data on the nature of the effect, we shall use the general term reactive synaptogenesis or reinnervation to describe these responses to deafferentation in order to avoid inappropriate inferences. Reactive synaptogenesis should not be confused with regeneration, which will not be discussed in this chapter. In regeneration, damaged axons regenerate; in reactive synaptogenesis, deafferentation elicits the nearby undamaged fibers to grow and form new connections.

2. GENERAL PROPERTIES OF REACTIVE SYNAPTOGENESIS

There are several basic properties of the reinnervation process which have been established and illustrate the remarkable capacity of the adult nervous system to modify its circuitry in response to deafferentation.

1. New synaptic connections can be formed on deafferented neurons, thus restoring their synaptic input. Electron microscopic evidence for the formation of new synapses after denervation has been reported in the septum (Raisman, 1969; Raisman and Field, 1973), red nucleus (Nakamura *et al.*, 1974), dentate gyrus (Matthews *et al.*, 1976), spinal cord (Bernstein and Bernstein, 1973), and lateral geniculate (Lund and Lund, 1971). In the septum and dentate gyrus, quantitative studies suggest that the reinnervation process is nearly complete.
2. The new synapses can be electrophysiologically functional. In the sympathetic ganglion (Guth and Bernstein, 1961), red nucleus (Tsukahara *et al.*, 1974), and dentate gyrus (West *et al.*, 1975; Steward *et al.*, 1974), electrophysiological studies indicate that the new connections are functional. The finding that deafferentation induces the formation of new functional connections means that neurons in the adult brain have the capacity to carry out all essential processes in synaptogenesis: a synaptic junction can form; the apparatus to manufacture, store, and release the transmitter can be induced; and the secreted transmitter can successfully elicit a postsynaptic response.
3. Axon sprouting and the formation of new synapses begin rapidly after a lesion. Axon sprouting of noradrenergic fibers in the septum (Moore *et al.*, 1971) and lateral geniculate (Stenevi *et al.*, 1972) occurs within 1–2 weeks, and the initial signs of new functional connections are seen within 2 weeks in the red nucleus (Tsukahara *et al.*, 1974) and dentate gyrus (West *et al.*, 1975; Steward *et al.*, 1974). While the onset of reinnervation may be rapid, this process can continue for months. In the dentate gyrus and septum, while synapse formation begins quickly, electron microscopic analysis shows that there is a net addition of synapses well past 30 days after lesion (Raisman and Field, 1973; Matthews *et al.*, 1976). Similarly, in the lateral geniculate reactive synaptogenesis appears to progress very slowly (Wong-Riley, 1972).
4. The new synapses are similar in appearance to those normally present in the denervated region. Aberrations have occasionally been detected, but relatively infrequently. Most synapses display synaptic vesicles clustered near the typical synaptic junction. When the new connections are on dendritic spines,

the spines are stereotypical of that cell. In the dentate gyrus, for example, small spines and spines shaped in a "U" or "W" form are characteristic of both the normal input and the new connections (Matthews *et al.*, 1976). In the septum, boutons which form multiple synaptic junctions are much more common than in unoperated animals, but the junctions and spines appear normal (Raisman, 1969; Raisman and Field, 1973). In the lateral geniculate after ablation of the visual cortex, there is a loss of geniculate neurons; and after these neurons degenerate the retinal afferents which formerly terminated on them form an increased number of axoaxonic synapses (Ralston and Chow, 1972; Wong-Riley, 1972). Thus even in this unusual type of reactive synaptogenesis, the normal pre- and postsynaptic relationships are maintained. These data indicate that the reinnervation process is able to recreate synaptic morphology in a manner typical of the particular brain regions.

5. Intact axons can innervate parts of neurons normally devoid of their connections, if those parts are deprived of their normal input by a lesion. As seen in the septum (Raisman, 1969) and red nucleus (Nakamura *et al.*, 1974; Tsukahara *et al.*, 1974), fibers which normally contact the dendrites, and which account for little if any input to the soma, can grow and form synapses on the soma when it is denervated. Synapses can also be formed on parts of a denervated dendrite by axons which do not normally terminate there. In the dentate gyrus, for example, fibers which normally make synaptic connections on the proximal part of the dendrite can expand their termination field outward onto the more distal portion when this part of the dendrite is deafferented (Lynch *et al.*, 1973*c*; West *et al.*, 1975). Thus it appears as if highly specific laminar relationships on a neuron can be respecified following deafferentation which disrupts these relationships.
6. The reinnervation process is selective. Both positive and clearly negative cases of reactive synaptogenesis are known. Only particular afferents in particular regions react following particular lesions. For example, noradrenergic fibers sprout in the lateral geniculate after lesion of the visual cortex, but not after retinal lesions (Stenevi *et al.*, 1972). Similarly, certain types of deafferentation in the spinal cord result in reactive changes by certain inputs only in certain spinal centers (Liu and Chambers, 1958; Goldberger and Murray, 1974; Murray and Goldberger, 1974; Kerr, 1972).

No evidence exists at present to indicate that an afferent can reinnervate a new cell type. Thus reactive synaptogenesis probably results only in an increase in normal types of input to a cell. The highest order of selectivity would be achieved if homologous axonal systems specifically recapture the synaptic sites made available by the lesion. For example, unilateral retinal lesions should elicit reactive synaptogenesis by the contralateral optic tract, while other lesions should not. However, the reinnervation process is never this specific. In the superior colliculus, intrinsic connections appear to direct terminals to sites vacated by retinal lesions (Lund and Lund, 1971). In the lateral geniculate nucleus after removal of optic tract input, neither retinal nor noradrenergic inputs appear affected (Lund *et al.*, 1973; Guillery, 1972*b*; Stenevi *et al.*, 1972), whereas after visual cortex lesions noradrenergic fibers and uncrossed retinal fibers replace the degenerated boutons (Stenevi *et al.*, 1972; Goodman and Horel, 1966). Thus the optic tract apparently reacts to lesions of nonretinal input, and other inputs can reinnervate the synaptic sites of the optic tract. Similarly, in the dentate gyrus after a unilateral lesion of the entorhinal cortex, fibers from the remaining entorhinal cortex do not appear to have a selective advantage over other systems in recapturing available synaptic space (see Section 4). Thus, in these cases examined, homologous afferents do not have a selective advantage in reestablishing input to the denervated cell. The specificity of synaptic sites for particular afferents thus appears to be altered by lesions.

Little if any advantage is gained by homology to the degenerating input, distance of the reactive cell body from the denervated zone, or the nature of the transmitter. For example, in the dentate gyrus both commissural and associational inputs appear to sprout, even though the commissural cell bodies are located much farther away from the denervated field. Also, the cholinergic septal input and intrinsic inhibitory connections react after loss of excitatory noncholinergic input. In the septum, noradrenergic fibers sprout in response to loss of nonaminergic input. Thus it appears as if transmitter identity is not required to allow reactive synaptogenesis. Additional information appears required before the selection rules are understood.

As illustrated by the six conclusions described above, a growing body of evidence indicates that deafferented peripheral and central neurons, in many instances, are reinnervated by undamaged fibers. The reinnervation produces normal-appearing functional synapses and these are organized in a precise manner. However, the process(es) responsible for reinnervation, and even the afferents involved, are often uncertain. Conclusions that axon sprouting has occurred in the adult CNS are often confounded by atrophy and limitations inherent in the present methods of detection. As discussed in detail elsewhere (Lynch and Cotman, 1975), the degree of atrophy must be taken into account and it is also desirable to verify the findings by use of more than one technique since this compensates for the limitations of the others.

At present, in no case are the data such that sprouted branches can be traced from their point of origin and shown to form new synapses, and until this is accomplished there must be some reservation about the relation between reactive synaptogenesis as discussed here and the classical descriptions of collateral sprouting at the neuromuscular junction. As we have discussed previously, the processes of collateral sprouting, paraterminal sprouting, and contact synaptogenesis can produce the same final result and are distinctive only in the amount of axonal growth required. Thus, because of the difficulty of establishing distinctions between the possible means of forming new connections in most cases, we prefer the term *reactive synaptogenesis* or *reinnervation* as opposed to the more commonly but prematurely used mechanistic terms (*collateral sprouting, paraterminal sprouting, contact synaptogenesis*, etc.) to describe the recovery of synaptic input after deafferentation.

It is of interest, however, to distinguish insofar as possible the type of process responsible for the restoration of input. The type of process operative in a given situation may provide information about the selection rules governing reinnervation, such as the distance over which an initiation signal can evoke a response and over which growth can occur.

Axon collateral sprouting in the nervous system has been unequivocally demonstrated only in the reinnervation of the superior cervical ganglion. The studies on the superior cervical ganglion strongly suggest that the sprouts from remaining preganglionic axons form new connections which underlie the recovery of ganglionic function (Murray and Thompson, 1957; Williams *et al.*, 1973). Nevertheless, this conclusion can only be considered tentative until the formation of these anomalous connections has been followed with the electron microscope. In the CNS it appears that noradrenergic fibers to the lateral geniculate and septum sprout as do the commissural fibers to the dentate gyrus. In the septum, as indicated from electron microscopic analysis, axon sprouting (paraterminal) probably accounts for much of the reinnervation (Raisman, 1969), but contact synaptogenesis may contribute. In the rat lateral geniculate after retinal lesions, the increase in synapses displaying "flat" vesicle terminals on asymmetrical sites after eye enucleation (Lund and Lund, 1971) may be an instance of paraterminal sprouting or contact synaptogenesis. Similarly, in the rabbit or squirrel monkey lateral geniculate after ablation of the visual cortex, the formation of new axoaxonic synapses (Ralston and Chow, 1972; Wong-Riley, 1972) may be a case of axon sprouting, but contact synaptogenesis is as likely to be involved.

Thus all types of processes may be represented. Most likely the type of growth seen may be predicted on the basis of the distance between the reactive afferent and denervated synaptic field.

In the next section, we consider in detail the nature of the reinnervation process in the dentate gyrus. The dentate gyrus is one of the brain areas where reactive synaptogenesis has been most thoroughly studied and is best understood. In discussing the actual experimental data, we hope to provide an indication of the current level of analysis in the field. The conclusions on reactive synaptogenesis in the dentate gyrus will subsequently be used to evaluate the underlying cellular and molecular mechanisms.

3. CHARACTERISTICS OF REINNERVATION IN THE DENTATE GYRUS

It is necessary to examine axon sprouting in an area where the circuitry is well known and where there are only a few well-separated cell types. Over the past several years, we have found the dentate gyrus of the hippocampal formation of the rat particularly well suited for examining synaptic plasticity induced by lesions. The hippocampal formation contains two major cell types, pyramidal and granule cells, and these are found in distinctive and well-separated layers (Fig. 1). The granule cells are the major cell type in the dentate gyrus, and the pyramidal cells are the major type in the hippocampus proper.

The dentate gyrus appears to be an associative area where a few inputs converge on a single type of cell called the *granule cell.* Inputs from the cortex, septum, and contralateral and ipsilateral hippocampus converge on the granule cells, which integrate this information and relay it to hippocampal CA3 subfield.

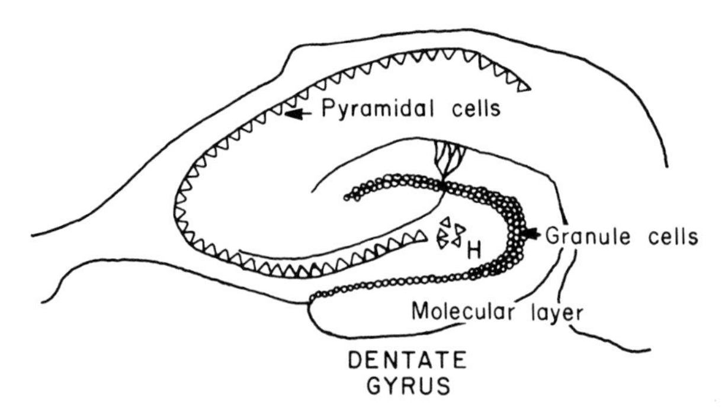

Fig. 1. The hippocampal formation consists of the hippocampus proper, where the pyramidal cells are the major cell population, and the dentate gyrus, where the granule cells are the major cell type. Granule cell dendrites ramify in the molecular layer.

In the dentate gyrus the layer of granule cells is arranged in a "U" or "V," and the dendrites of the granule cells extend outward into a zone called the *molecular layer*. The granule cell apical dendrites receive only a few inputs and these are precisely organized into discrete laminae. There are three major extrinsic synaptic inputs (commissural, associational, and entorhinal) and two minor ones (septohippocampal and noradrenergic). The commissural and associational fibers are exclusively localized in a narrow band above the granule cell bodies. The associational fibers originate from pyramidal cells in the ipsilateral hippocampus and commissural fibers arise from pyramidal cells of the contralateral hippocampus. Immediately above the commissural–associational zone is the zone occupied by entorhinal inputs. The commissural–associational zone occupies about 27% of the molecular layer, with the remaining 73% containing primarily entorhinal afferents (Fig. 2). The striking aspect of this lamination pattern is that there is little overlap between zones. The septal input is also laminated. A discrete zone of septal afferents is interspersed between the granule cell bodies and commissural–associational zone, and a rather diffuse septal input appears to exist in the same lamina as the entorhinal input. The adrenergic input, originating from the locus coeruleus, is diffusely spread over the molecular layer. A very few interneurons are scattered throughout the molecular layer and make up the intrinsic circuitry. As far as we know, these six synaptic populations account for the sum total of the synapses in the molecular layer; in fact, entorhinal, commissural, and associational account for the vast majority of the synapses (Lynch and Cotman, 1975).

This highly precise and stereotyped organization makes it possible to examine the effects of deafferentation on the remaining synaptic populations. Do all synaptic populations which remain intact in or near the denervation react to deafferentation? Since we know to a high degree all intrinsic and extrinsic connections, we can determine the selectivity of the growth process. What is the nature of the reaction? Does removal of one afferent, such as the entorhinal, stimulate the translaminar growth of commissural–associational afferents, or is any reaction restricted to afferents within the denervated zone? In the following paragraphs, we will describe the specificity of the response in terms of which neurons are reactive and in terms of the synaptic circuitry that emerges after deafferentation and reinnervation.

Our studies have centered primarily on removal of the entorhinal input. The entorhinal projection to the dentate gyrus in the rat is primarily unilateral (Zimmer and Hjorth-Simonsen, 1975; Goldowitz *et al.*, 1975) so that, after a unilateral lesion, the contralateral side can serve as a control. This allows a direct comparison of denervated and intact molecular layers in the same animal. The ipsilateral entorhinal input is massive. Unilateral removal of entorhinal cortex in adults results in degeneration of 85% of the total terminals in the entorhinal zone. This is about 60% of the total input to the granule cell.

After unilateral removal of the entorhinal cortex, the loss of input appears transient and over time there is a marked reinnervation of the

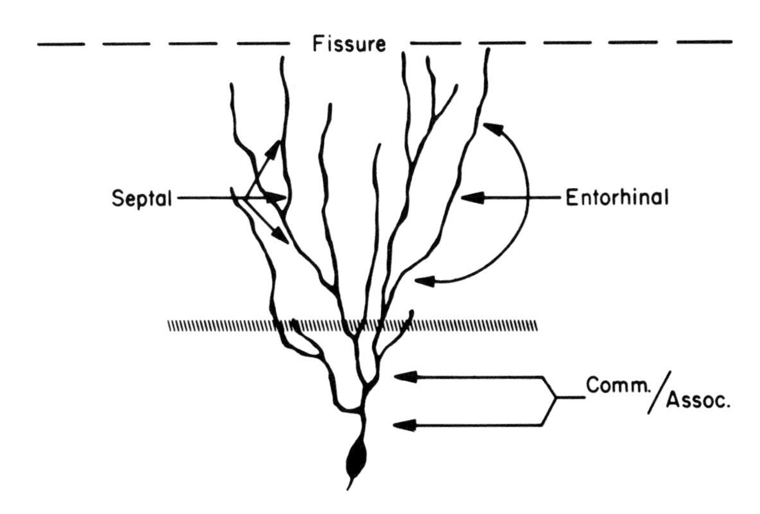

Fig. 2. Inputs are strictly ordered on granule cell dendrites. Septal and entorhinal afferents are found in the outer two-thirds of the molecular layer and commissural and associational afferents are found on the inner one-third. There is very little overlap between these synaptic fields.

granule cells by the remaining afferents. By 240 days after lesion, new synapses have grown and replaced most of those lost. We have analyzed the numbers of synapses in normal and in lesioned rats at various intervals after the lesion by electron microscopy (Matthews *et al.*, 1976). The density of terminals per unit area of 200 μm^2 in normal animals is 70 ± 4, and at the maximal period of degeneration (2 days) after a unilateral entorhinal lesion it drops precipitously to 10 ± 2. The density returns to about 70 ± 4 in rats which have survived 240 days after the operation.

As pointed out in the previous section, it is essential to quantify the contribution of tissue atrophy to changes in synaptic density. Shrinkage (or atrophy) can increase the density of the remaining intact boutons and give the impression of synaptic growth. The molecular layer becomes compressed by about 20% after an entorhinal lesion. This process is maximal by 10 days after operation; and remains relatively constant at longer survival times. Thus the restoration is not as large as implied by measurements of synaptic density. After correction for shrinkage, there appears to be about an 80% reinnervation of the granule cell.

The majority of contacts are qualitatively similar in structure to those of the normal molecular layer (Matthews *et al.*, 1976). In the normal molecular layer, complex spines stand out as one of its most distinctive features. These spines are most often "W" or "U" shaped. In view of the known susceptibility of spine morphology to the characteristics of the environment, spines might serve as a sensitive indicator of the restorative quality of the reinnervation. The types of spines in the reinnervated molecular layer are similar to those of normal animals and also are in about the same abundance. The major qualitative distinction is the presence of a few (less than 2%) vacant postsynaptic sites in the reinnervated layer. These can be apposed to glial processes but can also be open to the extracellular space of the brain.

The reorganization pattern which emerges after an entorhinal lesion is quite precise (Fig. 3). It involves both a reaction within the entorhinal terminal zone and growth of fibers beyond their normal lamina. Commissural fibers (Lynch *et al.*, 1973*c*) and associational fibers (Lynch *et al.*, 1975*b*) grow outward along the dendritic tree by 40 μm so as to increase their territory by about 40%. Septal (Lynch *et al.*, 1972) and crossed entorhinal (Steward *et al.*, 1974) projections appear to proliferate within the former entorhinal zone. The GABAnergic interneurons react to the lesion, but it is uncertain whether they form more synapses or hypertrophy (Nadler *et al.*, 1974). The noradrenergic input probably does not sprout (Moore *et al.*, unpublished observations).

The expanded commissural–associational zone in the lesioned rats, as in the normal, forms a very discrete lamina. In normal animals, the precise locus of the commissural–associational zone is one of the best examples of the precise organization of terminal fields in the adult mammalian brain (Cotman and Banker, 1974; Lynch and Cotman, 1975). No terminals have been found in the outer zone. Unilateral removal of the entorhinal cortex, however, allows the commissural–associational system to expand into part of the former zone of ipsilateral entorhinal input. It establishes a new boundary which is about 40 μm farther up the dendritic tree and which appears about as sharp as that in normal animals (Lynch *et al.*, 1973*c*, 1975*b*). Thus a highly specified system is respecified to a similar degree.

There are a number of independent lines of evidence which provide support for the sprouting of the commissural system after unilateral entorhinal lesions. In our original report (Lynch *et al.*, 1973*c*), we employed the paradigm used in most studies of collateral sprouting. After an initial entorhinal lesion, a secondary commissural

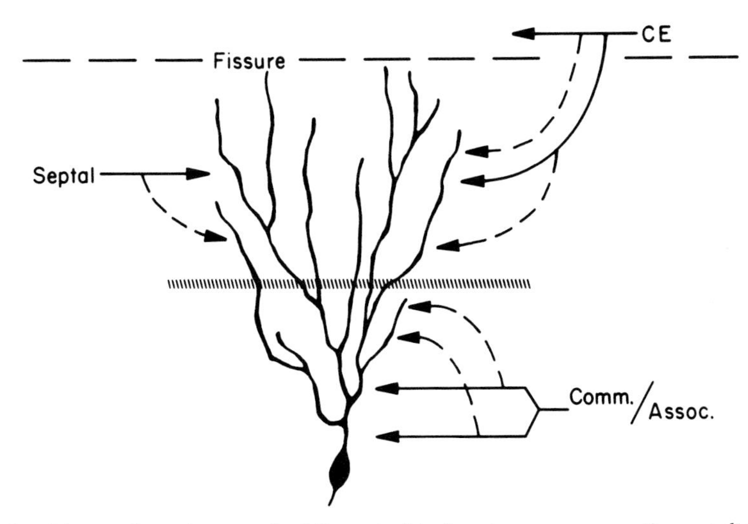

Fig. 3. After unilateral removal of the entorhinal cortex, axon sprouting results in a reordering of synaptic inputs on the granule cell dendrites. The commissural and associational systems appear to expand their terminal fields by about 40 μm into the denervated zone (dashed lines). In the remainder of the denervated zone, septal and residual entorhinal afferents appear to proliferate.

lesion was performed at a time when most degeneration from the original lesion had disappeared. Four days after the secondary commissural lesion, the animals were killed and the brains stained by the Fink–Heimer technique for terminal degeneration. On the side of the entorhinal lesion degeneration products were found about 40 μm beyond the normal commissural zone. These data indicated that the commissural fibers had expanded into the lower part of the entorhinal terminal zone and formed new synapses.

We have also carried out neurophysiological studies which indicate that the commissural system has grown and formed new synapses that are functional (West *et al.*, 1975). Neurophysiological analysis of the commissural system can be employed to provide an indication of its functional properties and, in addition, its organization. In normal animals, stimulation of the opposite CA3 field produces a short-latency response whose magnitude and polarity depends on the location of the electrode within the molecular layer. At the site of active synapses the extracellular field potentials are negative, whereas outside this zone the field potentials are positive. A plot of the field potentials as a function of the dendritic height (a laminar profile) reflects the organization of the activated synapses on the dendrites. In normal animals, the negativity never extends beyond 150–200 μm. After a unilateral entorhinal lesion, it appears that the commissural synapses activate a larger section of the granule cells' dendrites. In animals with a unilateral lesion, and a recovery time of more than 14 days, the negativity is recorded at least 100 μm farther out on the dendrites. This finding, together with the anatomical data, indicates that new functional synapses are formed.

Any proof of collateral sprouting requires the demonstration of new fiber growth. It would be expected that the expansion of the commissural–associational system should produce many new fibers. We have studied the fiber systems before and after an entorhinal lesion and have identified the appearance of new fibers in the expanded commissural–associational zone (Lynch *et al.*, in preparation). Degeneration disappears rapidly in this zone so that in a few days after an entorhinal lesion the zone is clear of degeneration and few fibers remain. Over time a new plexus of fibers appears in the expanded part of the commissural–associational zone. These fibers probably originated from parent fibers in the old commissural–associational zone. Finally, we have shown by electron microscopy that this new commissural–associational zone contains normal-appearing boutons and the synaptic density attains normal values. We have not identified the source of all terminals seen in the electron microscope or traced the origin of the new fibers, and these data are required to document the exact nature of the growth response. Taken together, however, we feel that all these data provide a solid case for lesion-induced axon sprouting and new synapse formation in the adult CNS. We have identified fiber invasion, identified new synapses by two anatomical methods, and tested them electrophysiologically. Commissural–associational systems normally occupy 27% of the granule cells' dendrites and come to occupy about 55% of the dendrites.

In the outer zone, there are changes in septal fibers and in fibers originating from the contralateral entorhinal cortex. A unilateral entorhinal lesion results in an intensification of acetylcholinesterase staining, which probably indicates a proliferation of cholinergic septohippocampal fibers (Lynch *et al.*, 1972). It is also possible the intensification signifies an increase of acetylcholinesterase activity in those fibers already present. Direct microchemical analyses have revealed an increase in the specific activity but not in the total enzymatic activity due to the extensive loss of protein encountered in these experiments (Storm-Mathisen, 1974). Thus the exact nature of the histochemical intensification is not presently established.

The fibers from the remaining entorhinal cortex seem to grow and form new synapses (Steward *et al.*, 1974). Normally the input from the contralateral entorhinal cortex is relatively sparse compared to the massive ipsilateral input (Zimmer and Hjorth-Simonsen, 1975; Goldowitz *et al.*, 1975). After a unilateral entorhinal lesion the projection from the contralateral entorhinal cortex appears to be more abundant, based on analysis by autoradiography (Steward *et al.*, 1974) or the silver degeneration method (Zimmer and Hjorth-Simonsen, 1975). In addition, electrophysiological analysis indicates an increase in the input from the contralateral entorhinal cortex (Steward *et al.*, 1974). Most likely the few fibers in this layer proliferate or fibers in the stratum moleculare of the CA1 field invade the molecular layer (Steward *et al.*, 1974; Goldowitz *et al.*, 1975). On the basis of present data, however, it is difficult to rigorously determine the nature of the effect or its magnitude. It may be that some presynaptic metabolic effect accounts for the increased autoradiographic signals or degeneration products; postsynaptic changes or changes in existing afferents may account for the augmented electrophysiological responses.

The intrinsic GABAnergic interneurons were found to undergo a small increase in the activity of the enzyme glutamic acid decarboxylase (Nadler *et al.*, 1974). Glutamic acid decarboxylase is highly concentrated in the boutons of GABAnergic neurons so that measurements of its activity provide an index of the state of these boutons.

Between 11 and 29 days after an ipsilateral entorhinal lesion, the specific activity of this enzyme increased by 63%, specifically in the outer part of the denervated zone. This effect was not accounted for by atrophy of the tissue. The increase in glutamic acid decarboxylase activity may reflect a biochemical adaptation of existing boutons or the growth of additional boutons. Its relatively slow onset may favor an adaptation within existing elements. Direct quantitative measurements on the number of GABAnergic terminals are required.

The extrinsic noradrenergic input does not appear to sprout. After a unilateral entorhinal lesion, the noradrenergic input into the molecular layer, as examined by the glyoxylic acid histofluorescence method, does not seem to sprout (Moore *et al.*, unpublished observations).

In summary, a unilateral entorhinal lesion removes up to 85% of the synapses within the outer zone of the granule cell dendrites. Most, but apparently not all, afferents in the molecular layer react morphologically and are potential contributors to the new synaptic input. The reinnervation process restores most of the granule cells' lost synaptic input, although in doing so it changes the lamination of afferent input. Prior to an entorhinal lesion, the synapses are arranged in discrete lamina, and after an entorhinal lesion a new laminar pattern emerges which appears similarily as discrete. Thus a highly selective synaptic organization is respecified in the adult brain.

4. MECHANISMS OF REACTIVE SYNAPTOGENESIS: THE DENTATE GYRUS AS A MODEL SYSTEM

Lesion-induced (reactive) synaptogenesis can be described by a temporal series of stages: initiation, growth, and synaptic junction formation. These stages correspond closely to critical events recognized in the early literature on sprouting in the peripheral nervous system (see Edds, 1953). The response is initiated by products of the degeneration, or by a consequence of deafferentation on other events which in turn initiate growth. Growth begins and axons sprout. Synapse formation is complete when sprouts locate their site of termination and a synaptic junction forms. The outcome of these events is the formation of new circuitry.

As seen in the previous sections, lesion-induced synaptogenesis in the hippocampus and other CNS systems can assume a number of forms, and thus possibly slightly different mechanisms. Initiation, growth, and the recognition of synaptic sites may operate along similar but not necessarily identical lines in all forms of lesion-induced synaptogenesis. For example, in axon collateral sprouting, growth occurs over longer distances as compared to paraterminal sprouting and contact synaptogenesis, where the formation of the synaptic complex is the major transformation. For this discussion, however, in view of the lack of clear identification at present among the types and/or mechanisms of reactive synaptogenesis in the CNS, we shall consider the phenomena as one, keeping in mind possible qualifications.

In this section, we describe the events as we currently understand them, from the initiation process to the cellular and molecular events in the formation of a synaptic junction. We begin with a discussion of the initiation process. We then proceed to an analysis of the control of growth rate. What factors set the pace in the sequence from damage to synaptogenesis? Having analyzed the events prior to synapse formation, we consider the process of synaptic junction formation. Finally we consider perhaps the most difficult question: what specifies the site of synapse formation in axon sprouting? In the hippocampus, growth begins and ends in a way that produces a highly specified reordering of connections. In this discussion our goal is to formulate the central issues and assess the current status of knowledge on these issues in terms of the underlying cellular and molecular events.

4.1. *Initiation*

What initiates the reactive response? Initiation is critical because it is the first level of control and determines whether or not the response will occur at all. Growth may be actively initiated or released after deafferentation. Deafferentation itself, however, does not provide a sufficient stimulus for all reactive growth. Afferents can remain unreactive, even when a stimulus is present and other systems are sprouting. For example, after an entorhinal lesion the noradrenergic fibers do not sprout whereas others do. It is likely that the noradrenergic input possesses the capacity to sprout since sprouting of CNS noradrenergic fibers is seen in other brain areas (e.g., septum). It may be that an entorhinal lesion does not deafferentate the same cell contacted by the noradrenergic fibers, in which case growth may either fail to be initiated or be initiated and reversed because the synapses are incompatible with the granule cell. Similarly, the cholinergic septal system responds to an entorhinal lesion but does not show a measurable response after a commissural lesion (Nadler *et al.*, manuscripts in preparation). In this case, the same postsynaptic cells are probably involved. Thus at present the initiation process(es) seems highly selective rather than nonspecific; however, beyond this we have little information on the initiation of reactive synaptogenesis in the CNS.

There is much more known about the processes which initiate axon sprouting in the peripheral nervous system. In the early literature on axon collateral sprouting in the peripheral nervous system, diffusible substances were presumed to cause initiation. Hoffman and Springell (1951), for example, isolated a substance called *neurocletin* which appeared to stimulate collateral sprouting at the neuromuscular junction. (As far as we know, this result has not been repeated and the analysis refined with more modern methods now available.) Hoffman and others (see Edds, 1953; also Lynch and Cotman, 1975) suspected that the initiating factors came from the degeneration. In the sympathetic ganglia, sprouts are found in all parts of the sympathetic trunk, far removed from sites of terminal degeneration (Williams *et al.*, 1973). This also suggests that the axonal degeneration reaction and perhaps the associated glial accompaniment initiate collateral growth. In the CNS, however, there is no direct evidence for either the possible involvement of factors from degenerating tissue or the release of growth-promoting substances from other sources. Although nerve growth factor, for example, appears to stimulate growth of central adrenergic collaterals (Bjorklund and Stenevi, 1972), as it does in the peripheral adrenergic system, we do not know if nerve growth factor is involved naturally in reactive processes. Active suppression, as opposed to induction, may also regulate the growth of collaterals. In this model, growth is tonically restrained but released by deafferentation because the damaged input can no longer release suppressive factors (Aguilar *et al.*, 1973). Experiments are needed to determine if the factors or processes which initiate growth in the peripheral nervous system are similar in the CNS.

4.2. *Control of Growth Rate*

As described in an earlier section, synaptogenesis in the hippocampus after an entorhinal lesion has a time course which begins within a few days and continues as long as 140 days. What accounts for the delay of a few days prior to the appearance of new synapses? What controls the rate of growth once the process begins?

During the 7–14-day postlesion period prior to new synapse formation, a series of processes must occur. We suggest that in the first few days there occur preparatory processes managed by the glia as well as time for initiation, while the later times are used for growth and synapse formation. Each event consumes a time block; and these together, unless independently manipulated, require the critical 7–14 days. It is not yet possible to establish rigid temporal relationships because the times required for individual events are not precisely known. An unknown but a particularly significant aspect is where to place the initiation phase in relationship to other events. We need to know when neurons first begin to grow in order to set other events in their appropriate relationships.

It is unlikely the 7–14 days is required solely to synthesize and transport materials for axonal growth. Some reactive cells have their cell bodies located within 2–4 mm of the site of new synapse formation, and, on the basis of the expected synthetic and transport rates, materials should reach the reacting axon in 1–2 days. In addition, axonal growth over a few micrometers probably consumes little time.

It appears as if glia are involved in the initial period preceding sprouting. Glial cells remove degeneration products, and, as previously hypothesized (Lynch and Cotman, 1975; Lynch *et al.*, 1975*a*), may also play a role in initiating or regulating afferent growth. As we have observed in the molecular layer, and many investigators have observed in other brain regions, astrocytes separate the dying terminals from dendritic spines and engulf them (or, in some cases, the terminal and its spine). We have found that astrocytic processes in the molecular layer rapidly hypertrophy in response to deafferentation (Lynch *et al.*, 1975*a*). The astrocytic reaction is well developed at about 1–2 days after lesion, when the extensive bouton degeneration is being rapidly removed. Astrocytic reactions are seen as early as 12 h after lesion, possibly the time required for those boutons, separated by the lesion from their cell bodies, to enter into a degenerating state. Clear signs of degeneration are seen at these times, and, once astrocytes hypertrophy, degeneration is rapidly removed. Microglia also increase in number at the early phase of induced neuronal reactivity. The number of microglia in the molecular layer, as described elsewhere (Lynch *et al.*, 1975*a*), approximately triples in the first 4 days after an entorhinal lesion. Thus the first couple of days after lesion may be a period of degeneration and the cellular transformation of the glia.

Glial cells may be important in providing a favorable milieu for the subsequent growth reaction. In view of the reactivity of glia prior to synaptogenesis, it would appear that astrocytes, microglia, or both may aid in sprouting and synapse formation. For example, glia may guide afferents and direct growth, stabilize the extracellular environment, give off signals which facilitate neuronal growth, or otherwise prepare the neuropil for a rapid transition into the next stage. Thus in a sense glia catalyze the reinnervation process. It may be that the first few days are used in activating the glial cells. Once glia are activated, the reinnervation process proceeds in full force: growth is initiated, axons grow, and new synapses form.

Descriptions of the underlying events and their temporal relationships must eventually incorporate the fact that synapse formation can continue at reduced rates for 1–2 months (Matthews *et al.*, 1976; Raisman and Field, 1973). What processes regulate the rate of synapse formation once it has begun?

One striking correlation emerges: the rate of reinnervation is directly related to the rate of removal of degenerating terminals. At present, we do not know how precise this relationship is, but this close correspondence between the loss of degenerating terminals and their replacement by new synapses suggests that the rate of degeneration can limit, at least in part, the reinnervation rate. In the septal nucleus, Raisman and Field (1973) also observed that the rate of loss of degeneration was closely related to the rate of synapse reappearance.

The removal of terminal degeneration may increase the capacity of the granule cell to receive new afferent input. That is, the loss of an old terminal increases the probability of gaining a new one. As suggested by Raisman (1969), this reciprocal relationship is easily explained at a mechanistic level if new synapses are formed on old sites. As we argue in the next section, however, this does not appear to be the sole mechanism operative in the dentate gyrus. Yet it seems the granule cell can accept a new synapse only when an old one is lost. Because astrocytes remove nonfunctional terminals, these cells may indirectly set the reinnervation rate.

4.3. *Synaptic Junction Formation*

After deafferentation, when a reactive afferent reaches its site of termination, synaptogenesis *per se* begins through the formation of a synaptic junction. In this stage, the synaptic cleft is formed and joins specialized pre- and postsynaptic membranes. The formation of a synaptic junction marks the beginning of the critical commitment to engage in synaptic contact. All types of reactive synaptogenesis, while distinctive in several ways, share this common end point—the formation of a synaptic junction. Thus a central issue is understanding the principles and constraints in synaptic junction formation.

a. Morphological Correlates. Synaptic junction formation after a lesion in adults can proceed either by reoccupation of old sites or by the generation of new ones. As suggested by Raisman (1969), afferents may reoccupy sites vacated by degenerated boutons and form a synaptic junction using an old postsynaptic site. Alternatively, a synaptic junction can form *de novo* at a new site (Fig. 4).

The most widely held model at present is the reoccupation model. Reoccupation of previous sites has been suggested to account for the appearance of "flat" vesicle terminals within asymmetrical synaptic junctions in the lateral geniculate (Lund and Lund, 1971), the reappearance of input in the septum (Raisman, 1969; Raisman and Field, 1973), and the presence of synapses in the spinal cord which apparently display postsynaptic membrane from the previous connection (Bernstein and Bernstein, 1973). In the spinal cord, a few of the new synapses near the site of transection display a number of small synaptic junctions with a vacant postsynaptic site interposed between the intact junctions (Fig. 5). Normally most synapses display a continuous junction, and never is a vacant postsynaptic site interposed between normal junctions. It was suggested (Bernstein and Bernstein, 1973) that old postsynaptic sites survive and influence the site of formation of new synaptic junctions. Although the reoccupation model is still speculative, increasing amounts of supportive (albeit indirect) data are generating a widespread acceptance of this model.

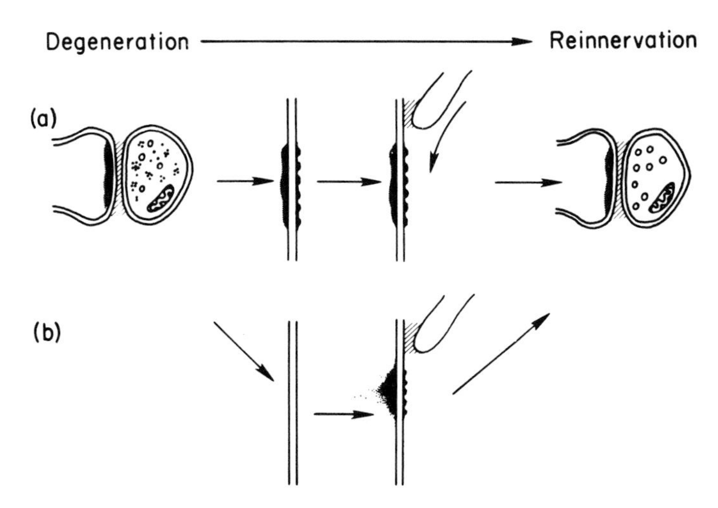

Fig. 4. After degeneration of terminals, reactive synaptogenesis can follow two courses. The postsynaptic sites may be preserved or reoccupied by a sprouted afferent (a) or the postsynaptic site may be removed and reassembled (b). There is evidence that either course can be followed.

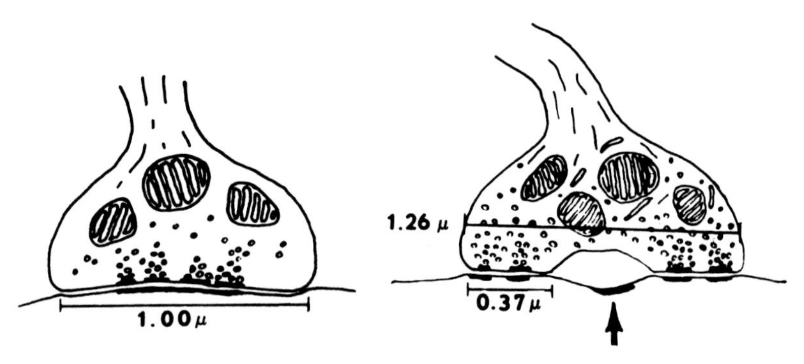

Fig. 5. Comparison of the normal synaptic types in the spinal cord (left) and an abnormal type seen after transection and apparent sprouting (right). The abnormal synapses are larger and have smaller synaptic contacts. In some cases a vacant postsynaptic density (arrow) is seen in the cleft. This has been interpreted as evidence that the new terminal has formed at a site of a previous one. From Bernstein and Bernstein (1973), courtesy of *International Journal of Neuroscience*.

Nonetheless, definitive evidence for this as the major mechanism for lesion-induced synaptogenesis is lacking.

In the reoccupation model postsynaptic sites are conserved so that the number of these should be constant throughout the degeneration–reinnervation process, whereas in the reassembly model postsynaptic sites are lost and restored over time. In order to determine which mechanism is used in the dentate gyrus, and the extent of its involvement, we measured two parameters. At various times after a unilateral entorhinal lesion, we counted the number of postsynaptic sites (as determined by the presence of a postsynaptic density) per given length of dendritic surface, and we analyzed the total number of postsynaptic sites within the denervated zone over time. Both measures showed a loss and subsequent return of postsynaptic sites (Matthews *et al.*, 1976). For example, we found that in unoperated adult rats there were 36 sites per 100 μm of dendritic surface. This decreased to about 10 per 100 μm within 4 days after lesion. Over time the synapses along the dendrites were restored so that by 240 days after lesion there were about 25 per 100 μm. Therefore, it seems that a portion of new postsynaptic sites are generated *de novo*, by synthesis of new postsynaptic densities and specialized membrane. As far as we know, our results are the first to show that the adult CNS has the capacity to carry out all the essential processes in synaptogenesis.

It is more difficult to determine whether reoccupation of vacated synaptic sites also contributes to the formation of the remaining new connections. This process should contribute less than half of the new terminal population, based on our estimates of the change in postsynaptic sites over time. A very few synapses display a small "slip" of membrane in the cleft. This is never seen in normal connections, and, as suggested by Bernstein and Bernstein (1973), this may originate from a small segment of membrane which remains from the previous terminal. Also, some boutons appear to be "off the center" of postsynaptic sites, as if in reoccupation the match were imperfect. Vacant postsynaptic sites appear maximally between 2 and 10 days and decrease at longer times, but never totally disappear. As others have argued (Bernstein and Bernstein, 1973; Raisman, 1969; Raisman and Field, 1973), the presence of the largest number of these at early postlesion times may indicate that the vacated sites are subsequently reoccupied by new terminals as synaptogenesis occurs. Taken together, these data suggest that a few of the synapses arise by reoccupation.

Thus in the dentate gyrus new synapses form in part by the assembly of a new synaptic junction (Fig. 4b) and probably in part by the reoccupation of vacated sites (Fig. 4a).

b. A Molecular Model for Reactive Synaptogenesis. What molecular events are involved in the assembly of a synaptic junction? The construction of a synaptic junction, whether the junction is formed from preexisting postsynaptic components or is newly created, can be considered in terms of a simple model. In brief, we propose that specialized proteins on the internal surface of the postsynaptic membrane such as the postsynaptic density (PSD), in conjunction with surface receptors*, serve to specialize a zone of the dendritic membrane and provide a fixed adhesive area to bind reactive afferents. In this model, the process of synapse formation centers about a nascent or preexistent postsynaptic site because this area has a fixed set of surface receptors with a strong affinity for afferents and because the postsynaptic site has the appropriate machinery to secure the connection. Surface receptors on the dendrite bind impinging afferents, but if the interaction is weak (incompatible), an afferent does not become fixed in position. In a sense, sprouts "test" dendritic surfaces. When an afferent with a compatible set of receptors locates a preexisting synaptic site, it binds tightly and becomes secured by synaptic junction formation.

Biochemical data have provided some support for this model and have allowed the designation of some molecules as candidates for participation in the hypothesized processes. We have possibly identified a class of molecules that represents the postulated postsynaptic

* The term *surface receptors* implies only that these molecules are on the outer surface of the membrane and not that they interact with transmitters.

surface receptor, and we have recently been able to characterize the composition of PSDs. In addition, we have established some of the chemical bonding responsible for maintaining synaptic connectivity.

c. Localization of Lectin Receptors at Synaptic Junctions. Are there specialized surface receptors at the synaptic junction? We have approached this question through the use of plant lectins conjugated to ferritin. Lectins are proteins which bind to specific carbohydrate molecules. When they are coupled to ferritin, the location of ferritin molecules allows the cytochemical localization of specific carbohydrates to a resolution of about 50 Å. In order to study the binding of ferritin-conjugated lectins to the granule cell postsynaptic junctional membrane, the molecular layer was dissected from 1-mm-thick slabs of fresh dentate gyrus and homogenates of the tissue were prepared. In homogenates a few clefts split open, so that ferritin–lectin conjugates are accessible to the postsynaptic membrane and so that binding to postsynaptic membrane can be distinguished from presynaptic binding. Since the granule cell dendrite is the major source of postsynaptic membrane in the molecular layer, we assume that the vast majority of open postsynaptic sites are derived from these dendritic membranes. After incubation of the tissue with lectin conjugate, the sample is prepared for electron microscopy and analyzed. We have examined the binding of concanavalin A–ferritin and *Ricinis communis* agglutinin–ferritin conjugates (Kelly *et al.*, manuscript in preparation). Concanavalin A (Con A) binds to carbohydrates which contain mannopyranosyl or related groups, and *Ricinis communis* agglutinin (RCA) binds to carbohydrates which contain galactosyl or related residues.

Studies on the binding of Con A and RCA conjugates indicate that the postsynaptic membrane of the granule cells is rich in carbohydrate residues at the synaptic junction. Clusters of ferritin molecules are densely distributed on the postsynaptic membrane surface overlying the PSD (Fig. 6). RCA and Con A conjugates show similar degrees of binding, although usually fewer ferritin molecules bind with RCA than with Con A conjugates. Most synapses in the molecular layer bind conjugate. (In Section 4.4, we will discuss the possible relation of surface codes to afferent placement.)

Synaptic junctions can be isolated so that the lectin receptor can be directly identified. In synaptic junctions (isolated from rat forebrain), the lectin receptors appear to be primarily glycoproteins (Cotman and Taylor, 1974) so that it is probable that the surface lectin receptors of the granule cell are glycoproteins. These glycoproteins may be associated with the bristles seen in high-resolution electron

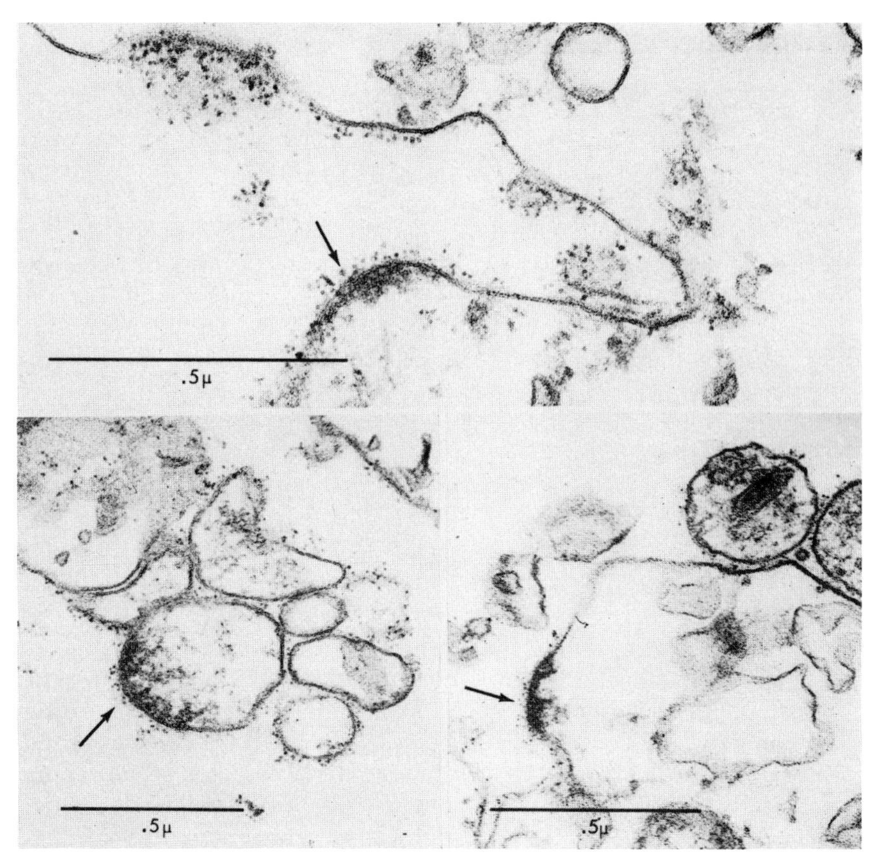

Fig. 6. Binding of ferritin–*Ricinis communis* agglutinin conjugates to the postsynaptic surface of the granule cell. Ferritin molecules are present on the surface of the postsynaptic membrane overlying the postsynaptic density (arrows), indicating the presence of lectin receptors at this specialized membrane area.

microscopy protruding from the surface of the postsynaptic junctional membrane. Such protrusions appear at vacant synaptic sites and seem to be present also at sites of abnormal attachment to glia or myelin in the deafferented molecular layer (authors' unpublished observations). A similar situation seems to exist in the lateral geniculate after deafferentation and restoration (see Lund and Lund, 1971; Raisman and Field, 1973).

It may be that these lectin receptors are the surface component first contacted by an afferent. Lectinlike molecules may also be included in the presynaptic membrane and be involved in the initial stabilization and/or recognition between afferent and receptive site.

In our model, a postsynaptic site provides a stable as well as an adhesive surface for afferent attachment. It is known that lectin receptors on dendrites outside the postsynaptic junction appear to be mobile (Matus *et al.*, 1973). Is there any evidence that membrane surface receptors at the postsynaptic membrane are stable? As suggested elsewhere (Cotman and Taylor, 1972; Banker *et al.*, 1974), the PSD may serve as an anchor to restrict the molecules in the membrane overlying the PSD from freely diffusing over the plane of the membrane. Furthermore, PSD proteins positioned on the internal surface of the membrane may provide transmembrane control over the surface receptors in a manner similar to other proteins on the internal surface of the membrane described for a variety of cell types (e.g., Nicolson and Painter, 1973; Edelman *et al.*, 1973; Berlin *et al.*, 1974). Surface receptors and the "bristles" seem to be positioned in reference to the PSD, and it is most often the case that ferritin–lectin conjugates cluster over the surface of membrane delineated by the PSD. Thus it appears reasonable to postulate the membrane overlying the PSD has a stable potentially adhesive surface.

d. Postsynaptic Density Components. As described above, in some cases postsynaptic sites are lost and later replaced. Thus the PSD must be assembled and repositioned on the dendritic membrane. In order to understand the reassembly processes, we need to know the composition of PSDs. PSDs can be isolated by the technique of subcellular fractionation in 80–90% purity (Cotman *et al.*, 1974). Isolated PSDs are primarily protein and 1–2% carbohydrate which is protein bound (Churchill *et al.*, manuscript in preparation). On polyacrylamide gels, PSD proteins resolve into three prominent bands. The major component has a molecular weight of 54,000, and we have estimated that it could account for seven out of every ten polypeptides in the PSD (Banker *et al.*, 1974). Because of its abundance, it is probably the molecular counterpart to the structure seen in electron micrographs after heavy metal staining. The two other polypeptides are of higher molecular weight and, in the purest preparations, account for most of the remaining protein. Thus the PSD consists of relatively few polypeptides.

The production of a PSD during lesion-induced synaptogenesis would then involve the synthesis and assembly of a few polypeptides at particular positions along the cytoplasmic surface of the dendrite. At present, little is known about the placement process except that PSD components may insert into the dendrite as prepacked units (small vesicles coated with PSD material; Altman, 1971). It is not known whether PSDs are regularly or irregularly spaced along the surface. They cover about 10% of the surface of the granule cell dendrite, and, in view of the extensive deafferentation which decreases the population by one-half to two-thirds, their restoration would necessarily involve an active synthetic and coordinated reassembly effort by the granule cell. The spacing of postsynaptic sites, their morphology, and the configuration of dendritic spines appear very normal after reinnervation.

Thus in adding new sites we suggest that the constituents of postsynaptic membrane, possibly only a few proteins or glycoproteins, are assembled and positioned along the dendritic membrane in a normal-looking array. These structures may provide a durable and adhesive template for the formation of new synapses.

e. The Nature of Synaptic Connectivity. The critical stage in synaptogenesis is the actual connection of the pre- and postsynaptic membranes. What is the nature of the linkage between membranes? We are beginning to obtain information on the nature of the bonds joining synapses. Our approach, which is similar to that previously used by Pfenninger (1971), has been to determine what treatments dissociate synapses in the molecular layer (Kelly *et al.*, 1975). In this way it may be possible to infer what holds membranes together. Thin slices of the molecular layer (200 μm) were prepared and incubated in various media which contained chemicals known to dissociate ionic, coordination, and weak hydrophobic bonds. We found that synaptic connections resist most treatments. Increased ionic strength (4 M NaCl), alkaline pH (10–11), or a chelating agent (10 mM EDTA) has little dissociative power. Therefore, ionic bonds or coordination complexes were insufficient in themselves to join synaptic membranes. In addition, urea (4 M), guanidine hydrochloride (2 M), Triton X100 (1% v/v), and SDS (0.2% v/v) did not produce marked disruption of synapses. Since more than 50% of the synapses remain intact, it seems that weak hydrophobic bonds were not responsible for junctional integrity. Incubation in the presence of a variety of sugars expected to break lectinlike interactions was also without effect. However, it is uncertain which treatments can dissociate this type of interaction between cells so that we cannot completely exclude lectinlike interactions. Dissociation has been achieved only when the membrane was dissolved or broken, except when enzymes were used. Trypsin under conditions which did not appear to create intracellular damage split the junction. We conclude that the junction is very durable except to enzymatic action. Thus proteins are required in junctional integrity, but their exact role is unclear.

We are not yet certain how synaptic membranes are connected, but the evidence favors covalent binding, possibly in the form of peptide bonds. Thus we might presume that the initial securing of the junction involves an enzymatic process.

Once a synaptic junction has formed, it may not be fully functional. The capacity to store and release transmitter may not be intact, and the appropriate receptors for the postsynaptic response may be absent. Clearly, the complete process can take place since functional synapses are formed. However, it is not clear that the transformation to a functional synapse is inevitable since abnormal junctions form, and those devoid of synaptic vesicles are sometimes present (Matthews *et al.*, 1976).

After lesions, the process of synaptic junction formation probably occurs very rapidly. Perhaps, as is the case in development, the spectrum of events in synaptic junction formation may require only a few hours (see Cotman and Banker, 1974). As we envision the assembly of a new site, the process appears quite similar to that in ontogenic development, although the possible presence of a preexisting site may pose even less stringent time demands. Thus the time for the production and assembly of components and the linkage of the membranes may be brief relative to the days required for the appearance of new synapses after lesions.

4.4. *Specification of Synaptic Fields*

As growth is initiated or released, synapses form in highly precise topographical order. Commissural–associational synapses form a lamina at a new position on the dendrites and other afferents are excluded from this zone; there is laminar specificity in the same way as in normal animals. What processes restrict sprouting of commissural–associational systems to a 40-μm portion of the dendrites after entorhinal deafferentation in an adult? There are a number of possibilities. Other afferents may restrict growth by competing for synaptic sites. There may exist chemical or physical barriers in the neuropil which prevent further growth. Some alteration in dendritic properties may be incompatible with growth beyond a certain point. Growth may be arrested simply on the basis of processes totally autochthonous to the cell itself. Although the factors responsible for the reordering of lamination patterns are largely unknown at present, some clues are emerging.

The complementarity of afferent ordering on the dendrites suggests that interactions between growing systems can influence the growth patterns. Perhaps growing commissural–associational fibers expand into the entorhinal zone until they encounter another afferent in the zone. Processes from different afferent systems may compete for synaptic sites: commissural–associational fibers would advance until other systems within the zone outcompete for the now limiting synaptic sites. For example, the fibers from the crossed perforant path may arrest competitively the growth of commissural–associational fibers. Two arguments indicate that competition is unlikely to account for the highly specific commissural–associational relamination pattern. First, both commissural and associational systems show the same degree of invasion after bilateral removal of the entorhinal cortex as they do after unilateral lesions of the entorhinal cortex (Lynch *et al.*, 1975*b*). Thus the normal or reactive crossed entorhinal input does not suppress or facilitate fiber growth into the outer molecular layer. Second, strict competition on a one-to-one basis should produce a mixed field (Cotman and Banker, 1974) so the intermediate zone would be expected to contain a mixture of septal, crossed, entorhinal, and commissural–associational inputs. Although far fewer in number, septal and crossed entorhinal inputs should gain some extra connections. However, septal projections actually appear displaced from the expanded commissural–associational zone (Storm-Mathisen, 1974), as are perhaps the crossed entorhinal afferents (authors' unpublished observations). On the basis of the effectiveness of the commissural and associational systems to gain extra space after removal of entorhinal input in adults, it seems as if these systems should overwhelm other inputs and capture the majority of the available dendritic space in the denervated molecular layer. Thus, probably factors other than competition restrict the spread of commissural and associational systems in adults.

We have previously suggested (Lynch and Cotman, 1975) that the reordered lamination pattern after entorhinal deafferentation in adults may result from an interaction between the residual, slowly degenerating, entorhinal axons and the invading commissural–associational fibers. Fibers grow into the entorhinal zone until they encounter a plexus of degenerating axons. We noted that in the normal molecular layer there are two bundles of axons in the molecular layer which course the length of the molecular layer: one is found in the normal commissural–associational zone, and the other is midway into the entorhinal zone. Between the two axon groupings, there is a zone relatively free of axons containing fine-caliber axons and terminals from the entorhinal cortex, but few thick axons. After an entorhinal lesion, fibers and terminals are rapidly removed from this fiber clear zone while degenerating entorhinal axons persist in the more superficial fiber plexus. The striking feature is that the commissural–associational expansion stops at about the position where there is persistent entorhinal axon degeneration. The invasion may in some way, then, be retarded by the common degenerating fascicles of myelinated entorhinal axons. Alternatively, the correspondence may also be a mere coincidence and not reflect any fundamental relationships. However, further experiments are necessary.

There may also be some alteration in dendritic properties which is incompatible with the formation of commissural–associational connections beyond a certain point. This feature would necessarily have to shift from a zone of about 75 μm above the granule cell bodies to a zone 115 μm above the cells. Changes in the specificity of the dendritic surface are a possibility. We have examined the molecular layer for discrete differences in lectin-binding sites and as yet have found no differences which seem to account for such selectivity: postsynaptic junctional receptors in the commissural–associational zone and entorhinal zone show little difference in the binding of RCA–ferritin or Con A–ferritin conjugates. This obviously does not exclude the possibility that different molecular sequences or different carbohydrates not revealed by Con A or RCA conjugates are the critical molecules in synaptic specificity. However, it is not obvious how a dendrite or synaptic site can be recoded with such sharp boundaries. Some type of intracellular control over the surfaces which can move the boundary a set distance would seem necessary. It has been suggested that the dendrites grow from the base the required amount, and this extension of the dendrite may account for the commissural–associational spread (Storm–Mathisen, 1974). This model seems unlikely because existing afferents would rise along with the extending dendrite. It appears that old degeneration products coexist with new terminals (authors' unpublished observations) and dendritic extension should result in the movement of both populations together.

A simple explanation for the limited commissural–associational expansion seen after adult lesions is that the boundary may reflect a viability state of the dendrite. Commissural–associational fibers grow rapidly over that receptive surface which is rapidly cleaned of degeneration and is viable and most receptive to the new input. The outer dendrites are shrinking and lose terminals somewhat slower and may be less receptive to afferents. By the time the outer dendrites recover from trauma, if at all, the trigger which had elicited commissural–associational growth has declined and the system is in a new steady state where further growth of commissural–associational synaptogenesis is suppressed. Another possibility is that the commissural and associational systems respond only to degeneration of the most medial entorhinal fibers.

At present, the principles which allow the strict reordering of lamination patterns are not known. More definitive data on the underlying phenomenon itself are required since many possible processes are consistent with the present information.

4.5. *Summary*

Reactive synaptogenesis constitutes a series of closely associated events which have as their common end point the formation of a synaptic junction.

The rate of reinnervation during reactive synaptogenesis is determined by the temporal course of initiation and growth. The processes controlling the duration of initiation, onset of growth, and rate of growth, however, remain unestablished and are clearly variable from system to system. The direct relationship between the disappearance of degeneration and appearance of terminals in the septum and hippocampus suggests at least one possible process. Glial cells are responsible for the removal of the degeneration, and the reactive glial proliferation precedes synapse formation, allowing for a possibly direct role for glial elements in determining reinnervation rate.

In the hippocampus, as perhaps in other brain areas, reactive synaptic junction formation probably requires either synthesis of new postsynaptic sites or reutilization of old sites vacated by degenerated boutons. That is, all afferents form connections at postsynaptic membrane specializations, whether previously extant or derived *de novo*. The production of a new site involves the assembly and insertion of new proteins into the membrane and the formation of the postsynaptic density (PSD), whereas the reutilization of a previously vacated site may require little if any site modification. If site properties determine selectivity of reinnervation, there must be either a turnover of highly specific receptors or specificity on the basis of a class of afferent types.

We propose a general model for the mechanism of synaptic junction formation which can be accommodated within either the reoccupation or the *de novo* concepts for the origin of the postsynaptic specialization. Within the proposed model, the postsynaptic site provides a stable, adhesive surface necessary to immobilize the arriving, reactive afferent. Carbohydrate moieties on the external surface overlying the PSD demonstrated for postsynaptic junctions in the dentate gyrus constitute recognition sites for the presynaptic element. A successful recognition interaction between pre- and postsynaptic surfaces is then secured by linking the membranes, possibly via covalent bonding, thus forming the synaptic junction.

The selectivity of reactive growth and synapse formation probably occurs during both the initiation and junction formation phases. Systems that possess reactive potential do not respond to all types of denervation. Such failures to respond could reflect selective initiation at the level of activation or depression, but could as well reflect selective failures of the initiated system to induce growth. The reactive process is apparently further specified by selective synaptogenesis in that successful innervation is restricted to only certain regions of certain cell types. This specificity of reinnervation may arise from selective surface carbohydrate receptor binding characteristics native to the denervated postsynaptic site.

Many mysteries remain within the phenomena of growth plasticity of central nervous tissue. Only as the phenomena themselves are clarified will the underlying cellular and molecular events emerge.

5. DEVELOPMENTAL DIFFERENCES IN REACTIVE SYNAPTOGENESIS

As opposed to the relatively restricted changes produced in the remaining systems after lesions in adults, lesions in developing rats produce a striking alteration in terminal fields. In general, all loci which show sprouting after lesions in adults also display altered growth after lesions during development. The reciprocal is not true. Alterations in growth patterns by lesions in developing rats commonly are not seen after adult lesions. For example, anomalous pathways are created after lesions in neonates, or very young subjects, which are never seen in adults (e.g., Lund *et al.*, 1973; Schneider, 1970). Thus, unlike the situation in the adult, new connections can be formed to cells not normally innervated.

In the dentate gyrus, axon sprouting in response to lesions can be demonstrated both during early postnatal development and in adulthood. In this brain area, those afferents which respond to an entorhinal lesion in developing animals also react to the loss of entorhinal input when the lesion is performed in adults. However, the area occupied by the anomalous synapses differs significantly with age, and this produces different postlesion laminar patterns (Fig. 7).

When the ipsilateral entorhinal cortex is removed in developing rats, the commissural fibers form synapses throughout the inner 80% of the molecular layer (Lynch *et al.*, 1973*a*,*b*), and the associational projec-

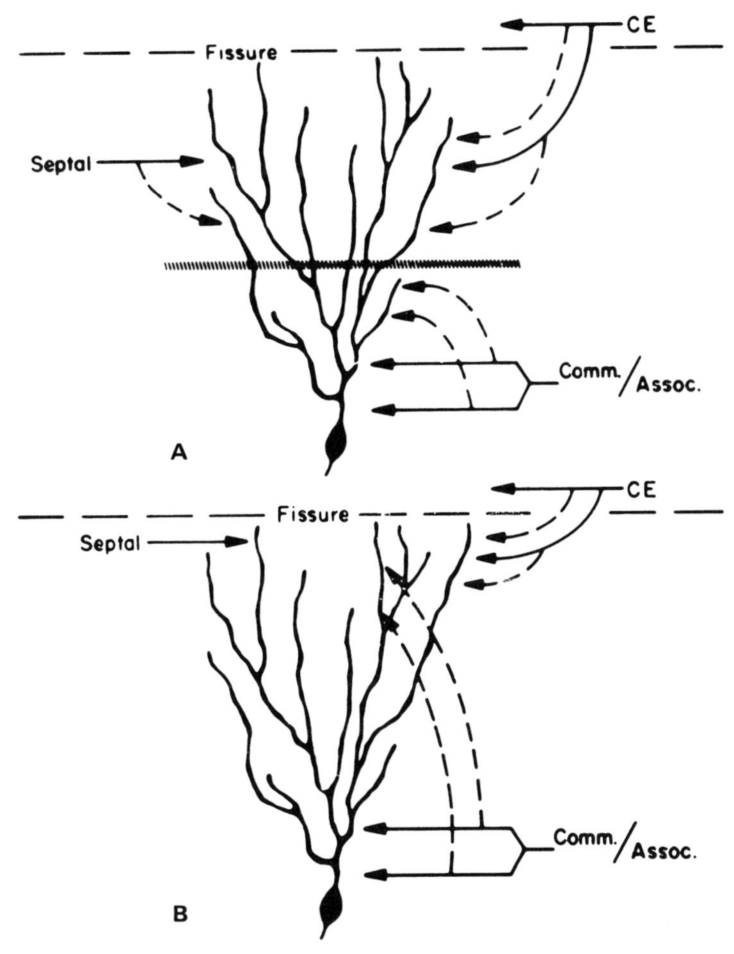

Fig. 7. Comparison between reinnervation patterns of the granule cell after a unilateral lesion in an adult rat (A) and in one lesioned at 11 days of age (B). The lamination patterns are different. After lesions in adult rats, commissural (Comm.) and associational (Assoc.) systems expand part way up the dendrites while the remainder is occupied by septal and contralateral entorhinal (CE) inputs. After lesions in young rats, commissural and associational afferents fill almost the entire dendrite. Septal afferents are concentrated in the remaining zone.

tion may even innervate the entire molecular layer (Zimmer, 1973). The same lesion in adults leads to an outward spread of these projections by only about 40 μm, so that they merely occupy an additional 15–20% of the molecular layer (Lynch *et al.*, 1973*c*). On the other hand, the reaction of the septohippocampal projection varies inversely with age. After early entorhinal deafferentation, these fibers proliferate only along the superficial edge of the molecular layer, whereas after the same lesion in adults they appear to proliferate in the outer 50–60% of the molecular layer (Cotman *et al.*, 1973). The crossed perforant path forms anomalous synapses within the outer half of the molecular layer in both developing and adult rats (Steward *et al.*, 1973, 1974; Goldowitz *et al.*, 1975; Zimmer and Hjorth-Simonsen, 1975). These synapses appear to be located more superficially in the molecular layer after lesions during development, but this apparent difference with age has not been analyzed in any detail. In both developing animals and adults, anomalous septohippocampal and crossed perforant path connections are made only at septotemporal levels where these projections normally terminate and only within their normal lamina of termination. Commissural and associational afferents differ in response to the entorhinal lesion, in that these fibers extend beyond their normal laminar boundaries. Thus, lesion studies clearly demonstrate that normal development is an interdependent process where the final synaptic organization depends on the proper growth of each element.

Results of a recent study in our laboratory may shed some light on the factors which determine the altered laminar pattern. Recently we analyzed the response of the septohippocampal input to entorhinal lesion at a number of points in development by acetylcholinesterase histochemistry (Nadler *et al.*, manuscripts in preparation). The earliest response after lesion (less than 3 days) was an intensification of staining throughout the denervated part of the molecular layer regardless of the age of the animal at the time of operation (5–30 days of age). Within 3–5 days after lesion, in animals operated on at 5 or 11 days of age, the histochemical intensification became concentrated along the superficial edge of the molecular layer. However, when the entorhinal lesion was performed at 26 or 30 days of age, the staining simply continued to intensify in the outer 50–60% of the molecular layer. Most significantly, a lesion made during the period of synaptogenesis (16–21 days of age) also caused the staining to intensify in the outer 50–60% of the molecular layer, but it was noticeably darker in the outer one-fifth to one-third. The basic difference in the response of very immature and older rats seems to lie in the ability of the septohippocampal axons to become concentrated in more superficial parts of the molecular layer. These results suggest that, in the immature dentate gyrus, some of the septohippocampal fibers probably relocate themselves. It has been suggested (Cotman *et al.*, 1973) that the pattern of septohippocampal synaptic reorganization after entorhinal lesion is a result of competition between afferents for dendritic space. Thus the expanded commissural–associational systems might outcompete the septal fibers and result in their presence in the outermost dendritic field largely unoccupied by commissural fibers. We have found, however, that after removal of commissural input and bilateral input from entorhinal cortices, septohippocampal fibers still redistribute in the same pattern seen after a unilateral entorhinal lesion alone. Although a possible competitive role of the associational system cannot be excluded, these results suggest a role for other factors. The ability of septohippocampal axons to move through the molecular layer may be restricted during development (1) by septohippocampal synaptogenesis, (2) by the maturation of the granule cell, or (3) by the larger amount and longer persistence of degenerating elements. Similar developmental changes also could restrict the extension of commissural and associational fibers into the denervated area.

In addition to possible differences in mobility of axons, there was a striking developmental difference in the response to partial entorhinal deafferentation. In adults, intensification of acetylcholinesterase-dependent histochemical staining was obtained in any part of the molecular layer which was denervated by a partial lesion. In contrast, distinct intensification was observed after lesions at 11 days when the most lateral part of the lateral entorhinal cortex was removed. This area projects to the most superficial part of the molecular layer (Hjorth-Simonsen, 1972). Removal of the remainder of the entorhinal cortex produced no intensification in any part of the molecular layer, if this lateral portion was spared. The reasons for this highly lesion-selective response of the septohippocampal fibers are presently unknown. It is important to determine whether the adjustments of other afferents are similarly lesion selective and whether the rules of selectivity change during development.

Although competition between afferents does not appear to account for the reorganization of synaptic fields in the dentate gyrus, competition may account for the reorganization pattern of retinal projections after removal of a portion of their normal target. In a series of experiments, Schneider (1970, 1973) has analyzed the development of retinal inputs if the superior colliculus is damaged. In animals with the superior colliculus removed, the most pronounced effect was a rerouting of optic fibers to the intact contralateral colliculus. This crossed projection is normally very sparse and not readily apparent. However, after a lesion it was so prevalent that it could be seen in a fresh dissected brain as a large branch to the remaining colliculus (see Fig. 16 in Schneider, 1973). This anomalous crossed retinal projection and the normal one terminated in the same lamina and formed two discrete nonoverlapping zones. Thus some mechanism restricted the proportion of the zone each could occupy.

The projection pattern in the remaining colliculus appeared to result from competitive interaction between the normal and abnormal retinal fibers. It was found that the simultaneous ablation of the damaged colliculus and the opposite eye resulted in the growth of the remaining retinal projection across the entire colliculus (Schneider, 1973). This result supports the view that a competitive interaction between retinal projections establishes the abnormal pattern of input to the colliculus after deafferentation (Schneider, 1973). Guillery (1972*a*) has also provided evidence that competition may regulate the establishment of retinal fields in normal animals. The competition model is unique in that it is the only model established to this degree in a vertebrate species for the development of a synaptic field. It is not unexpected in this case, however, since retinal fibers from either eye are likely to be homologous and replace each other when they possess the growth potential.

Perhaps the most remarkable example of a lesion-induced growth change is the creation of a retinal projection to the medial geniculate, which normally receives auditory input but never receives, as far as is known, retinal fibers. In animals where a fiber tract which projects to the medial geniculate body was damaged, retinal axons were redirected into the medial geniculate (Schneider, 1973). The growth of visual system fibers into an auditory center amounts to an unprecedented fiber rerouting. Thus it appears as if rerouted fibers can connect to a cell type not normally innervated.

As indicated by the induced innervation of the medial geniculate, the pursuit for synaptic sites is a powerful driving force where specifications are either overridden or miscoded as one element of the system is disturbed during development. This is true, however, only if one element is disturbed. Thus although retinal axons are capable of innervating a number of structures, a loss of actual or prospective input is necessary (Schneider, 1973).

The extensive circuit changes produced after lesions in developing animals seem to result in aberrant function in some cases (Lynch *et al.*, 1973*a*; Schneider, 1970, 1973). However, anomalous connections may be functionally suppressed. For example, the anomalous septohippocampal input formed after a unilateral entorhinal lesion in 11-day-old rats showed a permanent increase in the number of cholinergic synapses, but the increase in choline acetyltransferase activity relative to the corresponding zone on the control side was transient. This surprising result suggests that the anomalous cholinergic boutons eventually became functionally suppressed (Nadler *et al.*, 1973). It appears as if the crossed perforant path fibers are required for the suppression of choline acetyltransferase activity (Nadler *et al.*, in preparation). Similarly, the anomalous uncrossed retinal synapses formed after unilateral enucleation appear nonfunctional in an adult despite indications that abnormal synapses form (Chow *et al.*, 1973). These data may indicate some additional active plasticity controls the abnormal synapses (Chow *et al.*, 1973; Nadler *et al.*, 1973). Thus it appears as if the CNS can correct for violations in the formation of abnormal connections during development.

6. SIGNIFICANCE OF REINNERVATION IN THE CENTRAL NERVOUS SYSTEM

Reactive synaptogenesis is of direct significance to understanding the nature of the CNS response to injury and cell loss. Axon sprouting may underlie the partial recovery of function which generally follows injury, as it probably does in the partially deafferented superior cervical ganglion. There is now some indication from studies on denervated spinal cord that this could be the case (Goldberger and Murray, 1974; Murray and Goldberger, 1974). On the other hand, the abnormal connections may irreversibly alter the operation of physiological circuits and further handicap an already disturbed system (Raisman, 1969; Schneider, 1973; Cragg, 1974). Changes in connectivity often are even more extensive after damage to the developing CNS than in adults, and the functional consequences favorable or unfavorable are expected to be more pronounced. However, the ability to compensate for the adverse effects of abnormal circuitry may be greater in developing animals. These developmental differences may be relevant to the clinical observation that children generally recover more completely from brain damage than adults. However, we do not yet have a set of rules which allows us to predict when axon sprouting can occur and what consequences it will have.

Axon sprouting may be a normal event in the adult brain. Sotelo and Palay (1971) have detected degenerating synapses in the normal brain, and they hypothesized that terminals are constantly degenerating and being replaced. Fibers may retract their synapses and form new connections. Alternatively, other fibers may replace the lost connections. Remodeling of circuitry in the brain may therefore be a lifelong process, rather than simply a developmental event. At present, this hypothesis is largely speculative, but it is gaining support from detailed electron microscopic studies on certain brain regions (Sotelo and Palay, 1971; Chan-Palay, 1973; Angaut and Sotelo, 1973) and from studies on reactive synaptogenesis after denervation as described here. As indicated from studies on reactive synaptogenesis, it would be expected that remodeling would be selective and would produce subtle, but progressive, changes in brain circuitry and function. These changes may be especially significant in explaining the neurological consequences of aging and disease, conditions which can lead to loss of neurons and their connections.

As we and others have noted, there are important similarities between the formation of anomalous synapses after deafferentation and normal synaptogenesis. The synapses formed are similar in appearance and both processes follow highly restrictive rules which govern the localization of connections. Further studies of the reinnervation process should reveal the factors which influence the formation of new connections, and these will likely be applicable also to the complex problem of synaptogenesis in normal brain function.

Acknowledgments

We wish to thank Mrs. Pat Lemestre for secretarial aid and Dr. J. Victor Nadler, Dr. John Haycock, and W. Frost White for their aid and helpful discussions in the preparation of the manuscript.

7. REFERENCES

Angaut, P., and Sotelo, C., 1973, The fine structure of the cerebellar central nuclei in the cat. II. Synaptic organization, *Exp. Brain Res.* **16**:431.

Aguilar, C. E., Bisby, M. A., Cooper, E., and Diamond, J., 1973, Evidence that axoplasmic transport of trophic factors is involved in the regulation of peripheral nerve fields in salamanders, *J. Physiol. (London)* **234**:449.

Altman, J., 1971, Coated vesicles and synaptogenesis: A developmental study in the cerebellar cortex of the rat, *Brain Res.* **30**:311.

Banker, G., Churchill, L., and Cotman, C. W., 1974, Proteins of the postsynaptic density, *J. Cell Biol.* **63**:456.

Berlin, R. D., Oliver, J. M., Ukena, T. E., and Yin, H. H., 1974, Control of cell surface topography, *Nature (London)* **247**:45.

Bernstein, J. J., and Bernstein, M. E., 1973, Regeneration of axons and synaptic complex formation rostral to the site of hemisection in the spinal cord of the monkey, *Int. J. Neurosci.* **5**:15.

Bernstein, J. J., and Goodman, D. C., 1973, Overview, *Brain Behav. Evol.* **8**:162.

Bjorklund, A., and Stenevi, U., 1972, Nerve growth factor: Stimulation of regenerative growth of central noradrenergic neurons, *Science* **175**:1251.
Chan-Palay, V., 1973, Neuronal plasticity in the cerebellar cortex and lateral nucleus, *Z. Anat. Entwicklungsgesch*. **142**:23.
Chow, K. L., Mathers, L. H., and Spear, P. D., 1973, Spreading of uncrossed retinal projections in superior colliculus of neonatally enucleated rabbits, *J. Comp. Neurol*. **151**:307.
Churchill, L., Banker, G., and Cotman, C. W., 1976, Carbohydrate composition of central nervous system synapses: Analysis of isolated synaptic junctional complexes and postsynaptic densities, in preparation.
Cotman, C. W., and Banker, G., 1974, The making of a synapse, in: *Reviews of Neuroscience*, Vol. 1 (S. Ehrenpreis and I. Kopin, eds.), pp. 1–62, Raven Press, New York.
Cotman, C. W., and Taylor, D., 1972, Isolation and structural studies on synaptic complexes from rat brain, *J. Cell Biol*. **55**:696.
Cotman, C. W., and Taylor, D., 1974, Localization and characterization of concanavalin A receptors in the synaptic cleft, *J. Cell Biol*. **62**:236.
Cotman, C. W., Matthews, D. A., Taylor, D., and Lynch, G., 1973, Synaptic rearrangement in the dentate gyrus: Histochemical evidence of adjustments after lesions in immature and adult rats, *Proc. Natl. Acad. Sci. USA* **70**:3473.
Cotman, C. W., Banker, G., Churchill, L., and Taylor, D., 1974, Isolation of postsynaptic densities from rat brain, *J. Cell Biol*. **63**:441.
Cragg, B. G., 1974, Plasticity of synapses, *Br. Med. Bull*. **30**:141.
Edds, M. V., Jr., 1950, Collateral regeneration of residual motor axons in partially denervated muscles, *J. Exp. Zool*. **113**:517.
Edds, M. V., Jr., 1953, Collateral nerve regeneration, *Q. Rev. Biol*. **28**:260.
Edelman, G. M., Yahara, I., and Wang, J. L., 1973, Receptor mobility and receptor–cytoplasmic interactions in lymphocytes, *Proc. Natl. Acad. Sci. USA* **70**:1442.
Exner, S., 1885, Notiz zur der Frage von der Faservertheilung mehrerer Nerven in einem Muskel, *Arch. Gen. Physiol*. **36**:572.
Goldberger, M. E., and Murray, M., 1974, Restitution of function and collateral sprouting in the cat spinal cord: The deafferented animal, *J. Comp. Neurol*. **158**:37.
Goldowitz, D., White, W., Steward, O., Cotman, C., and Lynch, G., 1975, Anatomical evidence for a projection from the entorhinal cortex to the contralateral dentate gyrus of the rat, *Exp. Neurol*. **47**:433.
Goodman, D. C., and Horel, J. A., 1966, Sprouting of optic tract projections in the brain stem of the rat, *J. Comp. Neurol*. **127**:71.
Guillery, R. W., 1972*a*, Binocular competition in the control of geniculate cell growth, *J. Comp. Neurol*. **144**:117.
Guillery, R. W., 1972*b*, Experiments to determine whether retinogeniculate axons can form translaminar collateral sprouts in the dorsal lateral geniculate nucleus in the cat, *J. Comp. Neurol*. **146**:407.
Guth, L., and Bernstein, J. J., 1961, Selectivity in the re-establishment of synapses in the superior cervical sympathetic ganglion of the cat, *Exp. Neurol*. **4**:59.
Hines, H. M., Wehrmacher, W. H., and Tomson, J. D., 1945, Functional changes in nerve and muscle after partial denervation, *Am. J. Physiol*. **145**:48.
Hjorth-Simonsen, A., 1972, Projection of the lateral part of the entorhinal area to the hippocampus and fascia dentata, *J. Comp. Neurol*. **146**:219.
Hoffman, H., 1950, Local reinnervation in partially denervated muscle: A histophysiological study, *Aust. J. Exp. Biol. Med. Sci*. **28**:383.

Hoffman, H., and Springell, P. H., 1951, An attempt at the chemical identification of "neurocletin" (The substance evoking axon sprouting), *Aust. J. Exp. Biol. Med. Sci.* **29**:417.

Kelly, P., Kaups, P., Lynch, G., and Cotman, C., 1975, Studies on the nature of the chemical bonds underlying synaptic connectivity, *5th Ann. Mtg. Soc. Neurosci.*, p. 607.

Kelly, P. T., Cotman, C. W., Gentry, C., and Nicolson, G. L., 1976, The distribution of concanavalin A and *Ricinis cummunis* agglutinin receptors on specific brain neurons, manuscript in preparation.

Kerr, F. W., 1972, The potential of cervical primary afferents to sprout in the spinal nucleus of V following long term trigeminal denervation, *Brain Res.* **43**:547.

Liu, C. N., and Chambers, W. W., 1958, Intraspinal sprouting of dorsal root axons: Development of new collaterals and preterminals following partial denervation of the spinal cord in the cat, *Arch. Neurol. Psychiat. (Chicago)* **79**:46.

Lund, R. D., and Lund, J. S., 1971, Synaptic adjustment after deafferentation of the superior colliculus of the rat, *Science* **171**:804.

Lund, R. D., Cunningham, T. J., and Lund, J. S., 1973, Modified optic projections after unilateral eye removal in young rats, *Brain Behav. Evol.* **8**:51.

Lynch, G., and Cotman, C. W., 1975, The hippocampus as a model for studying anatomical plasticity in the adult brain, in: *The Hippocampus, Vol. 1* (R. L. Isaacson and K. H. Pribram, eds.), pp. 123–155, Plenum Press, New York.

Lynch, G., Matthews, D., Mosko, S., Parks, T., and Cotman, C., 1972, Induced acetylcholinesterase-rich layer in rat dentate gyrus following entorhinal lesions, *Brain Res.* **42**:311.

Lynch, G. S., Deadwyler, S. A., and Cotman, C. W., 1973*a*, Postlesion axonal growth produces permanent functional connections, *Science* **180**:1364.

Lynch, G. S., Mosko, S., Parks, T., and Cotman, C. W., 1973*b*, Relocation and hyperdevelopment of the dentate gyrus commissural system after entorhinal lesions in immature rats, *Brain Res.* **50**:174.

Lynch, G. S., Stanfield, B., and Cotman, C. W., 1973*c*, Developmental differences in postlesion axonal growth in the hippocampus, *Brain Res.* **59**:155.

Lynch, G., Rose, G., Gall, C., and Cotman, C. W., 1975*a*, The response of the dentate gyrus to partial deafferentation, in: *Golgi Centennial Symposium, Proceedings* (M. Santini, ed.), pp. 305–317, Raven Press, New York.

Lynch, G., Gall, C., Rose, G., and Cotman, C. W., 1975*b*, Changes in the distribution of the dentate gyrus associational system following unilateral or bilateral entorhinal lesions in the adult rat, *Brain Res.* (in press).

Lynch, G., Gall, C., and Cotman, C. W., 1976, Evidence on the temporal parameters of axon "sprouting" in the brain of the adult rat, *Nature*, in preparation.

Matthews, D. A., Cotman, C., and Lynch, G., 1976, An electron microscopic study of lesion-induced synaptogenesis in the dentate gyrus of the adult rat, *Brain Res.*, in press.

Matus, A., de Petris, S., and Raff, M. C., 1973, Mobility of concanavalin A receptors in myelin and synaptic membranes, *Nature* **244**:278.

Moore, R. Y., Bjorklund, A., and Stenevi, U., 1971, Plastic changes in the adrenergic innervation of the rat septal area in response to denervation, *Brain Res.* **33**:13.

Murray, J. G., and Thompson, J. W., 1957, The occurrence and function of collateral sprouting in the sympathetic nervous system of the cat, *J. Physiol. (London)* **135**:133.

Murray, M., and Goldberger, M. E., 1974, Restitution of function and collateral sprouting in the cat spinal cord: The partially hemisected animal, *J. Comp. Neurol.* **158**:19.

Nadler, J. V., Cotman, C. W., and Lynch, G. S., 1973, Altered distribution of choline acetyltransferase and acetylcholinesterase activities in the developing rat dentate gyrus following entorhinal lesion, *Brain Res.* **63**:215.

Nadler, J. V., Cotman, C. W., and Lynch, G. S., 1974, Biochemical plasticity of short-axon interneurons: Increased glutamate decarboyxlase activity in the denervated area of rat dentate gyrus following entorhinal lesion, *Exp. Neurol.* **45**:403.

Nadler, J. V., Cotman, C. W., and Lynch, G. S., 1976, Histochemical evidence of altered development of cholinergic fibers in the rat dentate gyrus following lesions. I. Time course after complete unilateral entorhinal lesion at various ages, *J. Comp. Neurol.*, manuscript in preparation.

Nadler, J. V., Paoletti, C., Cotman, C. W., and Lynch, G. S., 1976, Histochemical evidence of altered development of cholinergic fibers in the rat dentate gyrus following lesions. II. Effects of partial entorhinal and simultaneous multiple lesions, *J. Comp. Neurol.*, manuscript in preparation.

Nakamura, Y., Mizuno, N., Konishi, A., and Sato, M., 1974, Synaptic reorganization of the red nucleus after chronic deafferentation from cerebellorubral fibers: An electron microscope study in the cat, *Brain Res.* **82**:298.

Nicolson, G. L., and Painter, R. G., 1973, Anionic sites of human erythrocyte membranes, *J. Cell Biol.* **59**:395.

Pfenninger, K. H., 1971, The cytochemistry of synaptic densities. II. Proteinaceous components and mechanism of synaptic connectivity, *J. Ultrastruct. Res.* **35**:451.

Raisman, G., 1969, Neuronal plasticity in the septal nuclei of the adult rat, *Brain Res.* **14**:25.

Raisman, G., and Field, P. M., 1973, A quantitative investigation of the development of collateral reinnervation after partial deafferentation of the septal nuclei, *Brain Res.* **50**:241.

Ralston, H. J., and Chow, K. L., 1972, Synaptic reorganization in the degenerating lateral geniculate nucleus of the rabbit, *J. Comp. Neurol.* **147**:321.

Schneider, G. E., 1970, Mechanisms of functional recovery following lesions of visual cortex or superior colliculus in neonate and adult hamsters, *Brain Behav. Evol.* **3**:295.

Schneider, G. E., 1973, Early lesions of superior colliculus: Factors affecting the formation of abnormal retinal projections, *Brain Behav. Evol.* **8**:73.

Sotelo, C., and Palay, S. L., 1971, Altered axons and axon terminals in the lateral vestibular nucleus of the rat, *Lab. Invest.* **25**:653.

Stenevi, U., Bjorklund, A., and Moore, R. Y., 1972, Growth of intact central adrenergic axons in the denervated lateral geniculate body, *Exp. Neurol.* **35**:290.

Steward, O., Cotman, C. W., and Lynch, G. S., 1973, Re-establishment of electrophysiologically functional entorhinal cortical input to the dentate gyrus deafferented by ipsilateral entorhinal lesions: Innervation by the contralateral entorhinal cortex, *Exp. Brain Res.* **18**:396.

Steward, O., Cotman, C. W., and Lynch, G. S., 1974, Growth of a new fiber projection in the brain of adult rats: Re-innervation of the dentate gyrus by the contralateral entorhinal cortex following unilateral entorhinal lesions, *Exp. Brain Res.* **30**:45.

Storm-Mathisen, J., 1974, Choline acetyltransferase and acetylcholinesterase in fascia dentata following lesions of the entorhinal afferents, *Brain Res.* **80**:181.

Tsukahara, N., Hultborn, H., and Murakami, F., 1974, Sprouting of corticorubral synapses in red nucleus neurones after destruction of the nucleus interpositus of the cerebellum, *Experientia* **30**:57.

Weddell, G., Guttmann, L., and Guttmann, E., 1946, Local extension of nerve fiber into denervated areas of skin, *J. Neurol. Psychiat. (London)* **4**:206.

Weiss, P. A., and Edds, M. V., 1946, Spontaneous recovery of muscle following partial denervation, *Am. J. Physiol.* **145**:587.

West, J. R., Deadwyler, S. A., Cotman, C. W., and Lynch, G. S., 1975, Time dependent changes in commissural field potentials in the dentate gyrus following lesions of the entorhinal cortex in adult rats. *Brain Res.* **97**:215.

Williams, T. H., Jew, J., and Palay, S. L., 1973, Morphological plasticity in the sympathetic chain, *Exp. Neurol.* **39**:181.

Wong-Riley, M. T. T., 1972, Changes in the dorsal lateral geniculate nucleus of the squirrel monkey after unilateral ablation of the visual cortex, *J. Comp. Neurol.* **146**:519.

Zimmer, J., 1973, Extended commissural and ipsilateral projections in postnatally deentorhinated hippocampus and fascia dentata demonstrated in rats by silver impregnation, *Brain Res.* **64**:293.

Zimmer, J., and Hjorth-Simonsen, A., 1975, Crossed pathways from the entorhinal area to the fascia dentate. II. Provokable in rats, *J. Comp. Neurol.* **161**:71.

4

The Expression of Neuronal Specificity in Tissue Culture

RICHARD P. BUNGE

1. INTRODUCTION*

The earliest work with nerve tissue in culture was undertaken by experimental embryologists and was designed to address several pressing questions that could not be definitively studied in the developing embryo. Of this early work, the contribution of Harrison (1910) to the understanding of the mechanisms involved in axonal elongation and that of Weiss (reviewed in 1955) to the study of the mechanisms of axon guidance are particularly well known. More recently, the use of tissue culture made a further noteworthy contribution to neuroembryology when employed to demonstrate the humoral nature of the nerve growth factor of the mouse submandibular gland (Levi-Montalcini *et al.*, 1954). Considering the substantial contributions made by these relatively few efforts, it is surprising that culture techniques did not find more frequent use in neuroembryological research.

This course of events may be explained (at least in part) by the fact that after World War II much of nerve tissue culture work began to be addressed primarily to the problems of neuropathology. The differing

* Abbreviations: CNS, central nervous system; PNS, peripheral nervous system; SC, spinal cord; SCG, superior cervical ganglion; DRG, dorsal root ganglion; NE, norepinephrine; ACh, acetylcholine.

RICHARD P. BUNGE · Department of Anatomy, Washington University School of Medicine, St. Louis, Missouri.

viewpoints of the pathologist and the experimental embryologist may substantially influence the premises from which experimental design originates. The view derived from embryological experimentation is exemplified, for example, in Detwiler's (1920) experiments with heterotopic limb transplantation in *Ambystoma* (showing the early determination and rigidity of the behavioral repertoire of the developing spinal cord) (see also Narayanan and Hamburger, 1971) and Jacobson's (1968) demonstration that the cessation of DNA synthesis in retinal ganglion cells in *Xenopus laevis* embryos correlated with the time of specification of their central connections. Against this background, tissue culture experiments may be expected to begin with the premise that the characteristics which identify the neuron type are determined very early in development, and therefore it is to be expected that the postmitotic neurons removed from their embryonic environment will express their basic neuronal characteristics (if they survive in the culture milieu). Expression of their distinguishing differentiated characteristics might well be the expected result. Within the conceptual framework of the pathologist, however, the expectations for tissue taken from its natural environment and placed in the nutrient care of an ill-defined culture medium might be quite different. The pathologist would seem naturally to view the neuron a few minutes removed from crisis, critically dependent on its environment, and often diseased because of environmental changes. The immediate environment of the neuron thus becomes the critical factor in the expression of functional characteristics. Certainly many postwar practitioners of the tissue culture art (including this writer) tended to view the expression of differentiation by neurons in culture as quite remarkable. They thus tended to announce detection of the expression of a characteristic neuronal function in culture with the zeal appropriate for the discovery of the unexpected.

The point of view taken in this brief chapter will be that it may be beneficial in the use of tissue culture to take the view of the neuroembryologist, to expect full expression of neuronal function and improve culture conditions until this is obtained. In regard to the synapse, which will be the particular subject of this review, we might then expect that neurons under proper care in culture will manufacture and release transmitters and mount in their membranes receptors known to characterized their type *in vivo:* furthermore, they will form specific synaptic connections to allow the appropriate interaction between these components.

It will be my purpose here to review the considerable number of recent experiments which indicate the remarkable degree of synaptic function and specificity which may be expressed in culture, and also to discuss several dissenting reports. I consider this exercise important now, for if this view is generally supportable then tissue culture preparations may be suitable for the study of one of the current central issues in developmental neurobiology, the issue of how it is determined which transmitters and receptors neurons will make, and which synapses they will provide and which they will receive—in other words, what kind of neurons they will be.

2. OBSERVATIONS ON CULTURES OF THE CENTRAL NERVOUS SYSTEM

The first physiological as well as anatomical evidence for the presence of synapses in tissue in culture was obtained from explant cultures of CNS tissue and in cultures containing combinations of spinal cord explants and sensory ganglia (reviewed by Crain, 1966; Crain *et al.*, 1968; Crain and Peterson, 1974; see also Bunge *et al.*, 1967*b*). These studies, as summarized by Crain and Peterson (1974), provided valuable discoveries bearing on several aspects of neural tissue development. Examples are (1) the development of synaptic networks in tissues placed in culture prior to the period of *in vivo* synapse development, (2) the development of synaptic interactions between separately placed explants of various CNS regions (cerebral cortex, brains stem, and spinal cord), (3) the development of synaptic networks in tissues in which all bioelectrical discharges within CNS explants were prevented by high levels of xylocaine or Mg^{2+}, (4) the occurrence of complex bioelectrical discharges undoubtedly derived from substantial synaptic interaction from several types of nervous system tissue completely disaggregated prior to culture, and (5) the formation of synaptic networks *in vitro* with bioelectrical activities indicative of both excitatory and inhibitory synapses, each with characteristic pharmacological sensitivities.

From these now extensive studies of CNS explants, one gains a general impression of the *in vitro* development of histotypic synaptic organization within various tissue types, but critical evaluation of synaptic specificity depends on the ability to undertake intracellular recordings from or electron microscopic analysis of identifiable types of neurons. An example may be cited from the work of Peacock *et al.* (1973) in which certain general neuronal types could be recognized after *in vitro* maturation of cells cultured from a dissociate culture of mixed spinal cord (SC) and dorsal root ganglion (DRG) cells prepared from midterm mouse embryos. Observations pertinent to the question of the expression of neoronal specificity included delineation of characteristic electrophysiological activities for SC and DRG neurons, particularly the observation that cells that could unambiguously be identified as DRG neurons in no case showed clearly identified postsynaptic potentials, although a high percentage of SC neurons showed this definitive evidence of synaptic activity.

Less directly related to the question of the site of synapse formation but rather to the question of transmitter metabolism are the observations of Lasher (1974) (as well as earlier observations by Hösli *et al.*, 1972, and England and Goldstein, 1969) showing that neurons apparently retain their specific transmitter uptake mechanisms in long-term tissue culture. Lasher's work with cultures of 2-day rat cerebellum

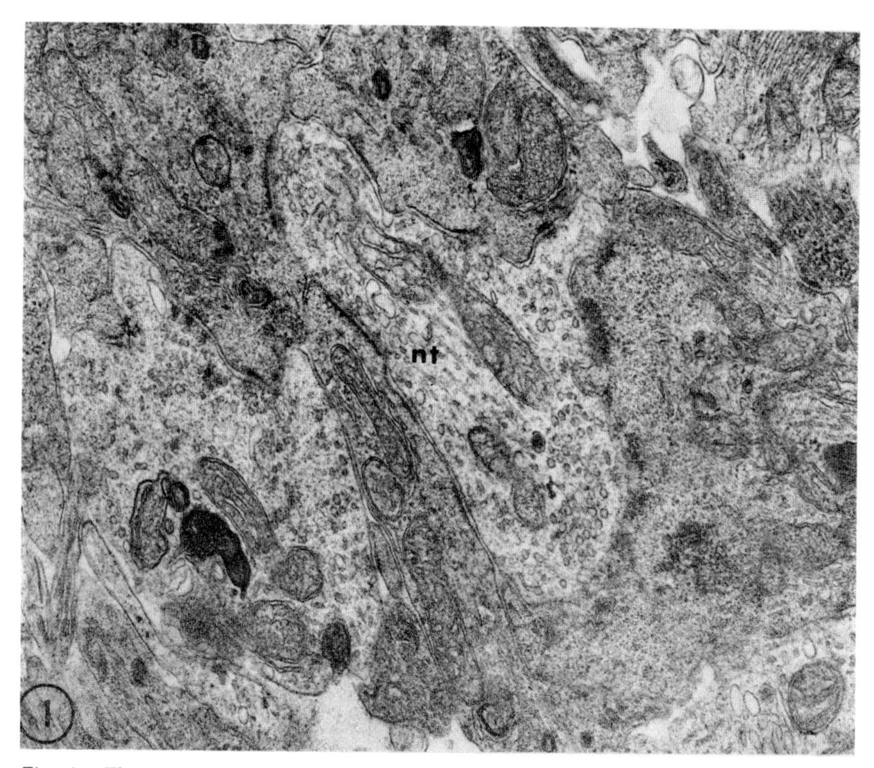

Fig. 1. Electron micrograph showing a nerve terminal (nt) in a combination culture prepared with tissue from both the cerebellum and the superior vestibular nucleus taken from a newborn rat and allowed to develop for several weeks in culture. The ending provides several points of apparent synaptic interaction with adjacent processes considered to be granule cell dendrites. The morphology of this ending is similar to that of mossy fiber–granule cell synapses *in vivo* and demonstrates the expression of complex synaptic interaction that may occur in tissue culture preparations. The photo was kindly provided by Dr. A. Privat (see Privat and Drian, 1975). $\times 25{,}000$.

demonstrated that high-affinity, Na^+-dependent, rapid GABA uptake can be demonstrated specifically by cells identified as stellate neurons, Purkinje neurons, and interstitial neurons, the neuron types thought, from *in vivo* experiments, to utilize this transmitter. Lasher concludes that GABA uptake is a function of early cell differentiation, not of membrane changes depending on subsequent neuronal interaction, and that this step in differentiation appears during or soon after the terminal division in these neuron types. It should also be mentioned that studies of the expression of neuronal differentiation in terms of histotypic organization and electrophysiological properties, including excitatory and inhibitory postsynaptic potentials, make the cultured cerebellum one of the most completely studied of the CNS systems *in vitro* (see review and data in Nelson and Peacock, 1973).

Privat and Drian (1975) have obtained evidence that a highly organized synaptic complex, the cerebellar glomerulus, is expressed if cerebellum explant cultures (Fig. 1) are prepared with a portion of the superior vestibular nucleus. Neurons from this nucleus apparently provide the mossy fibers necessary for the organization of the glomerulus, and it does not form in cultures from which vestibular components are excluded. These observations have, however, been questioned by Hendelman (1975) who found glomerular complexes in cerebellum cortex cultures not containing brain stem neurons. (See also discussion by Nelson, 1975.)

3. OBSERVATIONS ON NEUROMUSCULAR JUNCTIONS AND DISSOCIATED SENSORY NEURONS

Three short reviews (Crain and Peterson, 1974; Fischbach *et al.*, 1974; Fambrough *et al.*, 1974) provide an opportunity for an overview of the use of tissue culture in the study of muscle development, as well as in the study of the neuromuscular junction, particularly in regard to its development and its role in trophic interactions between nerve and muscle. These, as well as Shimada and Fischman's more extensive earlier review (1973), cite numerous examples of the tissue culture systems in which neuromuscular junctions have been shown to form. Either spinal cord explants or dissociated spinal cord neurons may be utilized as a source of innervating cells, and explants of embryonic or adult muscle (or dissociated embryonic muscle) may be employed as target tissue. In general, it has been observed that the more "organotypic" culture systems provide neuromuscular junctions with tested functions most closely correlated with *in vivo* observations; the simpler dissociated systems may more closely resemble immature neuromuscular junctions in, for example, the failure to develop demonstrable concentrations of acetylcholinesterase.

Whereas these observations repeatedly document the fact that neuron-to-muscle synapses form in culture, and in several instances establish that these are cholinergic, most do not bear directly on the question of whether only spinal cord "motor" neurons are involved. This point must be made because in all instances it has not been possible to make a clear identification of the neuronal type providing the fiber which synapses with the muscle. Four cases may be cited, however, in which the question of specificity appears worthy of some discussion. Working with combination cultures of SC and striated muscle, Crain and Peterson (1974) have reported that if sympathetic ganglion nerve fibers are present among cultured muscle fibers the muscle atrophy which regularly ensues after SC extirpation in their culture system is prevented. They interpret this as an indication that autonomic ganglion neurons can substitute for SC neurons in providing the necessary trophic influence to cultured muscle in their preparations. Furthermore, in some cases, stimulation of these autonomic neurons could evoke muscle contractions. This stimulus-evoked contraction was blocked by *d*-tubocurarine (1–10 μg/ml). It was emphasized that these signs of innervation were inconstant, and often could not be evoked in muscle fibers whose cytology clearly indicated that they were under a beneficial trophic influence from the autonomic neurons. Does this visceral innervation of somatic muscle indicate an aberration of synaptic specificity in this tissue culture system? Almost certainly not, for as the authors point out there is evidence (e.g., Landmesser, 1971) that various neurons of the autonomic nervous system make functional synaptic connections and provide trophic maintenance for striated muscle *in vivo*.

A report by Hooisma *et al.* (1975) of the innervation of striated muscle by the chick ciliary ganglion in tissue culture may, on superficial study, suggest an abnormality of neuromuscular relationships *in vitro*. As the authors point out, however, certain neurons of this ganglion normally innervate (via cholinergic endings) the ciliary body and irides, which in the chick are composed of fast striated muscle fibers. It does not seem surprising, then, that neurons of this ganglion are able to provide cholinergic innervation to cultured striated muscle from the chick leg.

A third case in which a tissue culture observation may be interpreted as indicating abnormal expression of synaptic potentialities in culture is that of apparent axon-to-neuroglial synapses seen in the outgrowth region of chick spinal cord explants by James and Tresman (1969). These indisputably "synaptic" profiles were clearly on cells of glial type. With the similar observations by Henrikson and Vaughn (1974) during normal development in the intact mouse spinal cord, the *in vitro* observation is no longer unique. Henrikson and Vaughn (1974) observed these synapses between neurites and radial glial processes between days 11 and 14 of embryonic development, but not after day 15. They interpret this as a transient phase during normal development when the specificity of synaptic relationship is not yet strictly expressed.

A fourth example, in which culture work with spinal cord tissues provides evidence for the expression of specificity in tissue culture, comes from the work of Fischbach (1972) with cultures in which both the SC tissues and the muscle are dissociated prior to culturing. His exploration of many neurons and neuronal types spread over the surface of the muscle cells indicated that only 5–10% of apparent endings on muscle could provide detectable postsynaptic potentials, suggesting that only 10% of the surviving neurons were motor neurons. Fischbach points out that "the great majority of [cells] destined to form neurons are probably not multipotent, and one explanation for the low incidence of functional nerve–muscle contacts is simply that only a few cholinergic motor neuroblasts were added to the muscle cultures." Fischbach also notes that he sought but did not find evidence for synaptic transmission from dorsal root ganglion neurons added to these cultures.

I and my colleagues have been observing rat sensory ganglia of various ages in long-term culture, in a variety of normal and experimental conditions (e.g., Bunge *et al.*, 1967*a*), for over a decade, and we have never observed a synapse in these cultures. Similarly, Okun (1972) found no clear evidence of synaptic interaction in extensive electrophysiological studies of dissociated sensory neurons grown in tissue cultures designed to attain substantial dispersion of the neuronal somas. Also, in less strictly neuronal sensory ganglion preparations (in that they contained many more supporting cells) Varon and Raiborn (1971) found no electrophysiological evidence of synaptic interaction. Against this background, the brief report of Miller *et al.* (1970) (see also Crain, 1971) of synaptic profiles in dissociated chick sensory ganglion cultures was particularly surprising. These authors noted a low incidence of unquestionable synaptic profiles in electron micrographs of these monolayer preparations. This observation remains unexplained except as an aberration of the culture situation, and perhaps of the dispersion and lack of ensheathment of the neurons. An additional instance will be presented later (Section 5) in which dispersion and lack of ensheathment lead to an accentuation of numbers of a synaptic type normally present only rarely. As of this writing, these several synapses found by Miller *et al.* (1970) represent the only example of synapse formation in culture for which an *in vivo* counterpart cannot be cited.

4. OBSERVATIONS ON SYNAPSES BETWEEN SPINAL CORD AND AUTONOMIC NEURONS IN CULTURE

4.1. *Explant Cultures*

Olson and Bunge (1973) have used combinations of explants from fetal rat spinal cord, cerebral cortex, and autonomic ganglia to test the

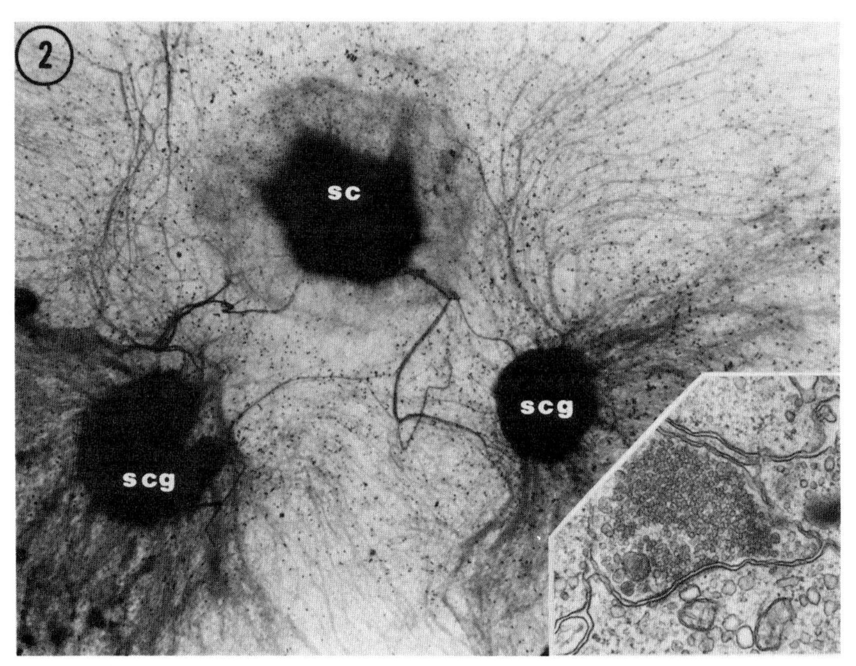

Fig. 2. Light micrograph showing an organotypic culture prepared from rat spinal cord (sc) and superior cervical ganglion (scg) fragments and allowed to mature for about 2 months in culture. Ventral root fibers have grown from the ventrolateral region of the spinal cord, and some of these have invaded the superior cervical ganglion to provide synapses of the type shown in the electron micrograph in the inset. These endings, distinguished by their substantial size and large number of synaptic vesicles, are not found in the superior cervical ganglion explants from which the spinal cord explant has been excised. From Olson and Bunge (1973). Photo $\times 12.5$; inset $\times 20{,}500$.

ability of neurons normally unrelated *in vivo* to form synaptic connections in culture. As may be anticipated, SC explants will innervate explants of the superior cervical ganglion (SCG) grown in juxtaposition (Fig. 2). It is interesting to note that when the number of synapses formed is quantitated per neuron cell body domain and compared with the number of synapses seen in the SCG *in vivo* the numbers are remarkably similar (Raisman *et al.*, 1974). When the SC explant is removed from these cultures the number of synapses seen in the SCG explant falls (by about 20 times) to a very low number. This is in contrast to the substantial number of "intrinsic" synapses seen among SCG neurons is dissociated cell culture (as discussed below). Evidence for the expression of synaptic specificity in culture was obtained from the observations that (1) juxtaposed cerebral cortex explants did not provide synapses among the SCG neurons and (2) when $KMnO_4$ fixation was used to identify noradrenergic endings the SCG neurites did not invade and synapse within either SC or cerebral cortex tissues. These observations on synaptology, derived from electron microscopic scrutiny of the cultured explants, were supported by light microscopic observations on their general organization (Fig. 2). After several weeks in culture, ventral root fibers (sometimes myelinated) can be seen leaving the appropriate SC area and coursing into the outgrowth. Several of these fibers can be seen to penetrate the SCG explant; others pass by and are dispersed in the general outgrowth. In contrast, the multitude of neurites originating from the SCG explant begin their course in a radial pattern, but those directed toward the SC are seen to turn and veer away before reaching its immediate vicinity. These light and electron microscopic data were presented as "evidence that certain basic properties of the axonal surface necessary for synaptic specifications are expressed in the tissue culture system."

4.2. *Synapses Formed Between Spinal Cord Explants and Dissociated Neurons of the Rat Superior Cervical Ganglion*

In the system described in a preliminary report by Bunge *et al.* (1974), cultures are prepared by initially seeding dissociated SCG neurons from perinatal rats into small culture dishes and 2 or 3 days later adding short segments of meninges-free 15-day fetal rat thoracic SC (Fig. 3). The small interneuron known to be present in the rat SCG does not survive under the culture conditions used. By use of antimitotic agents early in their development, these cultures are made to mature with a minimal development of nonneuronal cells. The SCG target neuron is then available for penetration by intracellular electrodes

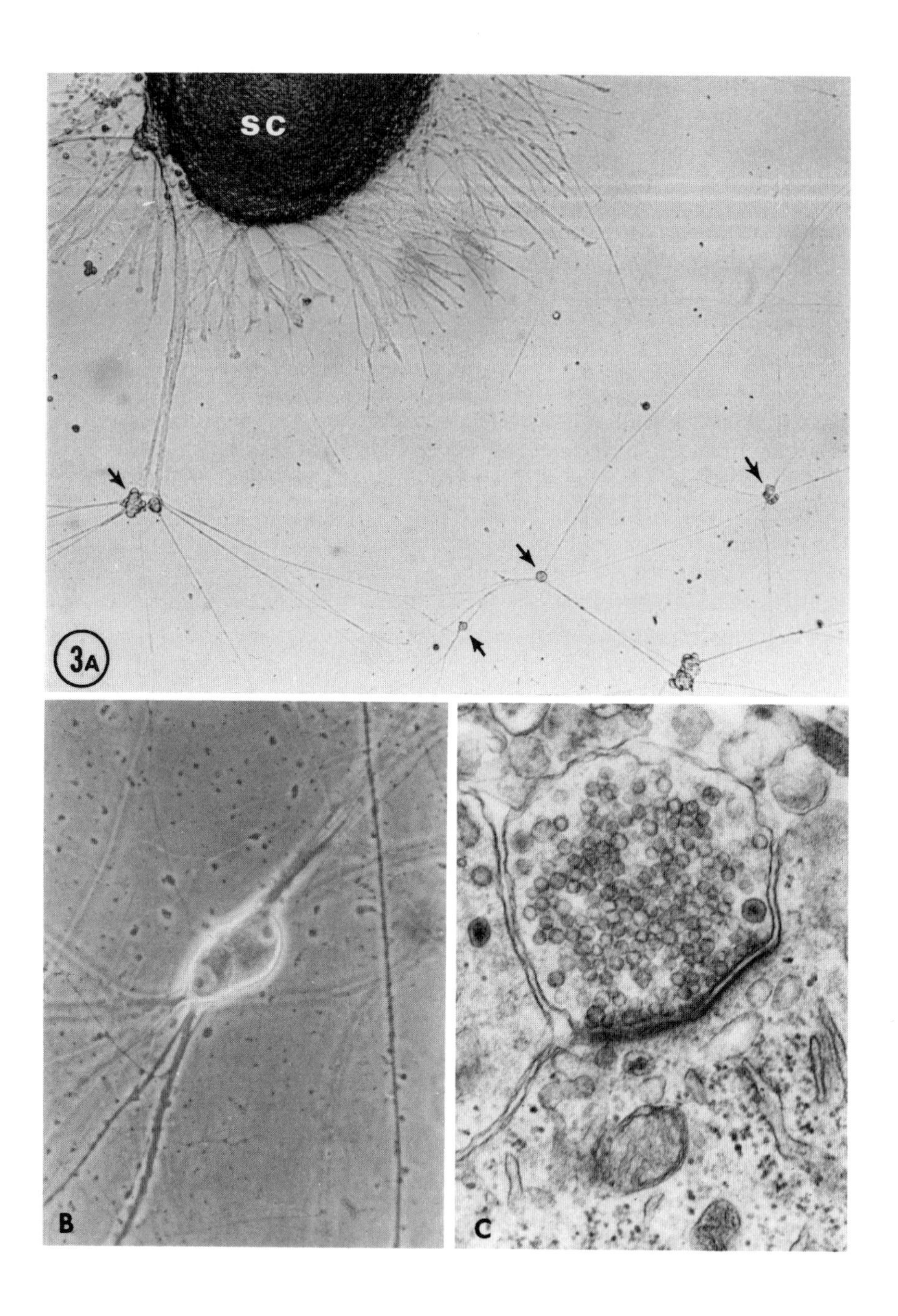
SC
3A
B
C

under direct observation, as well as being conveniently disposed for iontophoretic application of neurotransmitters and drugs. It should be noted that in this culture system the type of neuron within the SC explant that forms the synapses with the outlying SCG neurons has not been established. Also, as discussed in detail below, the synaptology is to some extent complicated by the presence of numerous intrinsic synapses between the SCG neurons themselves. It is possible, however, to distinguish this intrinsic synapse on morphological criteria (Rees and Bunge, 1974) and to limit electrophysiological studies to the synapses between SC and SCG neurons by stimulation of the SC directly while recording intracellularly from the dissociated SCG neurons. Using this approach, it has been established that the morphology of the SC to SCG neuron synapse is similar to that of its *in vivo* counterpart (Bunge *et al.*, 1974), providing a typical synaptic junction with pre- and postsynaptic membrane densities and quite uniformly round 50–60 nm synaptic vesicles (Fig. 3). This synapse has now been studied in some detail using physiological and pharmacological tools and is cholinergic nicotinic, as we report in detail elsewhere (Ko *et al.*, 1976*a*).

Thus the CNS to autonomic sympathetic neuron synapse formed *de novo* in tissue culture of rat fetal tissues is similar in its anatomical as well as in its pharmacological and physiological properties to its *in vivo* counterpart. The availability of dissociated "target" neurons in this culture system allows for detailed study of the physiology of any synapse impinging on these neurons. This system would appear to be particularly useful in testing the ability of a variety of neuron cell types to provide effective synaptic interaction, and we are currently utilizing this approach.

←

Fig. 3. Low-magnification view (A) of a tissue culture prepared from newborn rat dissociated superior cervical ganglion neurons (arrows) (seeded either as single cells or as small groups of cells) and with an explant of fetal thoracic spinal cord (sc). The neurites growing from the spinal cord explant have, after several days, contacted some of the superior cervical ganglion neurons and will in time make contact with others. Part B shows two superior cervical ganglion neurons after several days in culture; part C shows the type of synapse formed between the spinal cord neurites and the superior cervical ganglion neuron. For additional discussion of this experimental system, see the text. Details of the early events occurring when spinal cord neurites first contact superior cervical ganglion neurons are now available (Rees *et al.*, 1976), as well as data on the physiological and pharmacological properties of this cholinergic synapse (Ko *et al.*, 1976*a*). A, ×65; B, ×230; C, ×36,000. A, Sudan black stain after OsO_4 fixation.

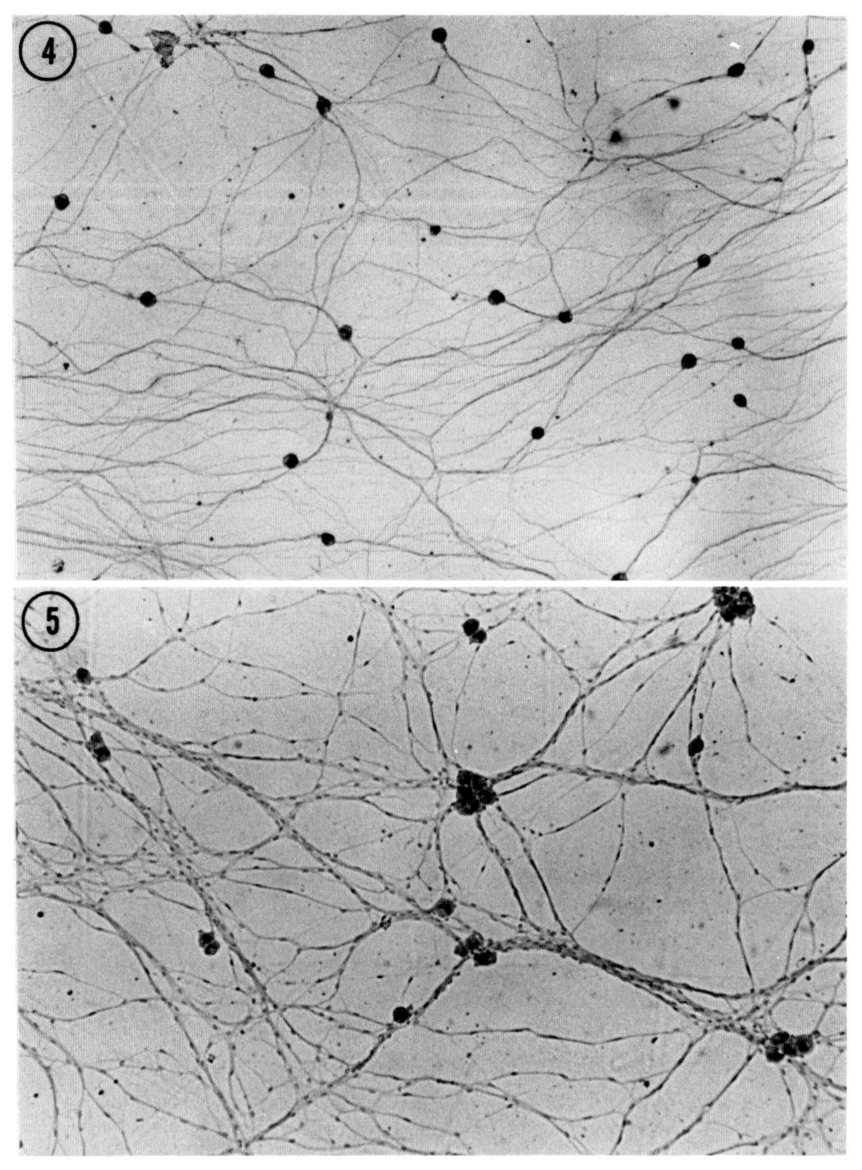

Figs. 4 and 5. Photomicrographs showing networks formed by dissociated rat superior cervical ganglion neurons grown without (Fig. 4) and with (Fig. 5) Schwann cells. The only supporting cell present in the preparation shown in Fig. 5 is the Schwann cell; the fibroblast population has been eliminated. Details of methods used in obtaining this type of preparation are given in the text and in Wood and Bunge (1975). Fig. 4, ×60; Fig. 5, ×60.

5. OBSERVATIONS ON CONNECTIVITY BETWEEN ISOLATED AUTONOMIC NEURONS

As discussed in the foregoing section, isolated neurons prepared from the SCG of perinatal rats form extensive networks as they extend their processes over the base of the culture dish. By adjustment of culture conditions, it is now possible to obtain either (1) pure principal neuron preparations without any supporting cells, (2) neuronal preparations with added Schwann cells but without fibroblasts, and (3) neuronal preparations containing both Schwann cells and fibroblasts (Wood and Bunge, 1975) (Figs. 4, 5, and 6). As the neuronal processes in these cultures contact neighboring neurons, they form numerous synapses. The cytochemical characteristics of these "intrinsic" synapses have been described in some detail (Claude 1973; Rees and Bunge, 1974). As studied in relatively young cultures (about 3 weeks *in vitro*), these show (Fig. 6) many of the cytochemical characteristics of adrenergic synapses *in vivo* (discussion in Rees and Bunge, 1974; see also O'Lague *et al.*, 1974). Also, this type of culture has been shown to contain the enzymes necessary for norepinephrine (NE) synthesis, and many of the metabolic aspects of this synthesis are known from the reports of Mains and Patterson (1973*a,b,c*). As discussed in some detail by Rees and Bunge (1974), this "intrinsic" synapse among what appeared to be a uniform population of adrenergic neurons can be considered a quantitative accentuation of a small number of intrinsic adrenergic synapses probably present in the rat SCG *in vivo* and certainly present in substantial numbers in some autonomic ganglia in other species. As such, it would represent an *in vitro* aberration in synapse number but not in kind. This increased number may relate to the lack of targets (other than neurons of similar type) in this culture system.

Against this background, the report by O'Lague *et al.* (1974) that electrophysiological signaling between these neurons is via cholinergic synapses is quite surprising. These workers report that one-fourth to one-half of the neuron pairs tested in these complex networks (after 4–6 weeks of *in vitro* maturation) showed cholinergic interaction. These interactions were both excitatory postsynaptic potentials (EPSPS) and action potentials, they were reduced by increasing Mg^{2+} and decreasing Ca^{2+} concentrations, and they were reversibly blocked by *d*-tubocurare, hexamethonium, and mecamylamine. These sensitivities were comparable to those for nicotinic cholinergic transmission from preganglionic to principal cell in the intact ganglion *in vivo*. Rare instances of electrical interaction between neurons were also observed.

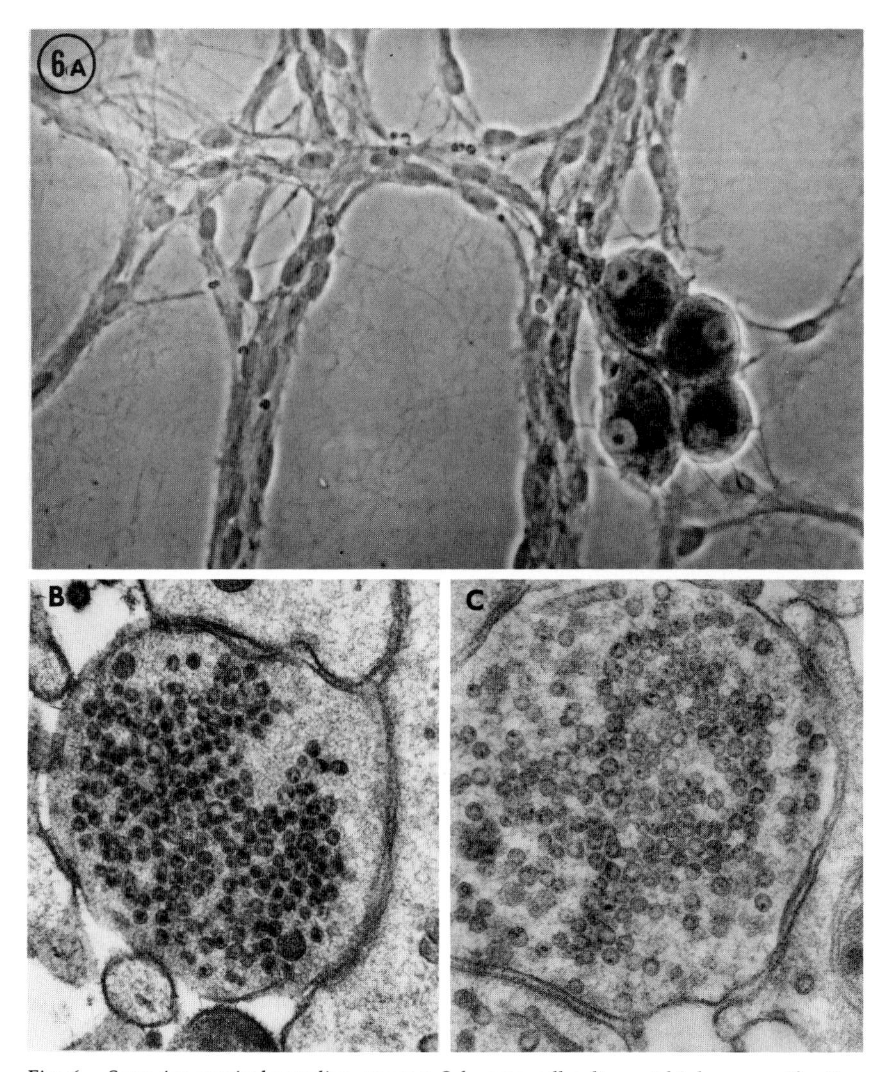

Fig. 6. Superior cervical ganglion neuron–Schwann cell culture at higher magnification (A) and the type of synapse formed in dissociated superior cervical ganglion neuron cultures early in development (B) and after 7 weeks in culture (C). The $KMnO_4$ fixation technique was employed (in B and C) after norepinephrine loading (for details, see Rees and Bunge, 1974). The dense cores within the synaptic vesicles are typical of adrenergic nerve endings; whether the differences in the core incidence and density between B and C represent a significant shift in transmitter metabolism is not known. For additional discussion of this point, see the text. A, ×350; B, ×34,000; C, ×45,000.

It is also worthy of note that a somewhat similar culture system has been employed by Obata (1974) to test the transmitter sensitivities of nerve cells in culture. Autonomic neurons from the rat SCG were found to be depolarized by acetylcholine (ACh) and (surprisingly) GABA. These cells were found not to respond to NE or to dopamine. These results are of considerable interest in the attempt to understand the complex responses of the intact autonomic ganglion and suggest a particularly promising use of dissociated neuron cultures. It may by this means be possible to quite precisely determine which transmitter receptors are present on specific neuron types, if these can be grown in dispersed cultures and their identity clearly established.

Working primarily with similar SCG neuronal networks grown with added segments of thoracic spinal cord as described above, we have confirmed the observation that many of the neurons of the SCG networks interact via cholinergic mechanisms (Ko *et al.*, 1976*b*). In addition, we have observed that the proportion of SCG neurons interacting in this way (30–40%) in our preparations is not influenced by the presence of several SC explants (providing SC to SCG neuronal synapses) placed within the culture dish. Thus the presence of "normal" input to these neurons in culture apparently does not influence their tendency to interact synaptically with one another.

The observation that neurons from an adrenergic population interact with one another via cholinergic mechanisms after several weeks in culture would appear to seriously challenge the hypothesis that neurons can be expected to express their *in vivo* characteristics *in vitro*. An explanation that comes immediately to mind is that the source tissue of these cultures (the SCG of the rat) contains a substantial number of cholinergic neurons among the dominant adrenergic population and that these express themselves in culture by providing cholinergic endings on their adrenergic siblings. The presence of cholinergic neurons in sympathetic ganglia is known in the cat, but the rat appears not to have been systematically studied in this regard (see discussion in O'Lague *et al.*, 1974). The number of cholinergic interactions that may appear in certain of these cultures (up to 50%; O'Lague *et al.*, 1974) would suggest that this is an unlikely explanation. An alternate explanation (suggested by the much debated hypothesis of Burn and Rand, 1965 regarding adrenergic neuron transmitter mechanisms) is that individual neurons are capable of synthesizing both NE and ACh and that certain culture conditions foster ACh synthesis and thus cholinergic interaction. Deciding definitively between these possibilities would appear to depend on the use of techniques capable of simultaneously demonstrating enzymes of the NE- and ACh-synthesizing pathways within the same cell.

The report of O'Lague *et al.* (1974) was accompanied by a study from Patterson and Chun (1974) demonstrating a relationship between the amount of ACh synthesis in dissociated SCG neuron cultures and the type of nonneuronal cells cocultured with these neurons. Whereas NE synthesis seemed little affected by the presence or absence of nonneuronal cells or by their type, ACh synthesis was substantially affected by these variables. ACh synthesis was increased 100–1000 times in the presence of nonneuronal cells derived from sympathetic ganglia; synthesis was also increased by C6 rat glioma cells but not by 3T3 mouse fibroblasts. If, in fact, the influence on ACh synthesis described occurs uniformly among individual neurons of these cultures (and not selectively, affecting a cholinergic subpopulation), then it is clearly suggested that the differentiation of these autonomic neurons is not occurring according to the concept discussed above (that the critical steps in neuron differentiation occur very early, and are related to the neuron's last mitotic experience). Instead, the expression of differentiation in this neuronal type is markedly influenced by its immediate cellular environment.

There is now evidence that similar mechanisms may be at work during autonomic neuron differentiation *in vivo.* A noteworthy work on neural crest differentiation by Le Douarin and Teillet (1974) describes experiments on chick–quail chimeras. These workers grafted quail neural crest anlagen into chick embryos and were able to recognize the derivatives of these grafts within the chick tissues because the quail cell interphase nucleus is distinguished by large heterochromatic nucleolar condensations. Their methods allow determinations of nuclear type and the presence of catecholamine-derived fluorescence within the same cell. Their conclusions are based on transplantation of presumptive quail neural crest material into various regions of the developing chick neuraxis, as well as *in vitro* maintenance of quail neural tube in direct contact with chick embryo gut tissue.

When tissues from the lumbosacral regions of quail embryos (which normally form the brightly fluorescent adrenomedullary cells) were placed in the "vagal" region of developing chick embryos, the quail cells followed a migratory path to the anterior gut region of the chick and there differentiated into enteric ganglia exhibiting no specific fluorescence. When tissues from the quail cephalic neural crest (which normally colonizes the gut to provide neurons of the enteric plexus) were placed in the lumbosacral region of the chick embryo, the quail cells migrated into the suprarenal gland and there differentiated into adrenomedullary cells. These authors conclude that "the differentiation of the autonomic neuroblasts is controlled by the environment in which crest cells are localized at the end of their migration." Whereas during their migration these cells are mitotic, they apparently become postmitotic as they take up their final positions in the various ganglionic, plexus, and glandular positions of neurons of the autonomic nervous system. This evidence from chick–quail chimera development suggests a similar conclusion to that derived from the tissue culture work described above; autonomic neurons express their transmitter synthetic capacities in response to environmental cues and this expression may occur after the neuron has become postmitotic. In this light, the *in vitro* results of Patterson and Chun (1974), as discussed above, may reflect influences on autonomic neuron differentiation which occur *in vivo*. For this neuronal type, then, results in the culture system call into question not the ability of the *in vitro* system to usefully mimic *in vivo* development but some of the most fundamental tenets of our present understanding of neuronal function. The tenets include the concept that neuronal differentiation is essentially fixed at the time of last neuroblast mitosis (e.g., see Jacobson, 1968) and the concept known as Dale's principle which states that only one transmitter synthetic mechanism is present in each neuron type. In fact, as Patterson and Chun (1974) have pointed out, tissue culture studies would appear to offer a most promising method of searching for the specific factors influencing these aspects of autonomic neuronal differentiation during development.

It would seem propitious, however, to end by urging caution in the making of extensive extrapolations from observations on autonomic neuron development (which occurs exclusively outside the CNS and within the mesodermal tissues of the body) to developmental mechanisms within the CNS. It would seem more judicious to extend our observations in tissue culture and use these results as a basis for designing *in vivo* experiments.

6. SUMMARY AND CONCLUSIONS

A review of observations on central and peripheral neural tissue in culture indicates that, in general, cultured neurons express cytological specificity (in their anatomical and electrical properties and in their capacities for transmitter synthesis, uptake and release) as well as synaptic specificity (in selecting cells with which they can be demonstrated to interact). Certain puzzling observations, such as the formation of cholinergic synapses between cultured neurons derived from (largely) adrenergic autonomic neurons, need further investigation before their implications can be fully understood. When viewed in the light of recent observations from experimental embryology, these tissue culture results suggest the following: If novel neuronal interactions are observed in tissue culture, these may reflect important developmental and physiological mechanisms operative *in vivo* and suggest experiments worthy of test in *in vivo* systems. If cultured neurons are found speaking a language we cannot at first comprehend, we may be well advised not to dismiss these signals as nonsense but rather to consider them as clues which we must learn to understand if we are to gain added insight into the mechanisms of development and functioning of neural systems.

7. REFERENCES

Bunge, M. B., Bunge, R. P., Peterson, E. R., and Murray, M. R., 1967*a*, A light and electron microscope study of long-term organized cultures of rat dorsal root ganglia, *J. Cell Biol.* **32**:439.

Bunge, M. B., Bunge, R. P., and Peterson, E. R., 1967*b*, The onset of synapse formation in spinal cord cultures as studied by electron microscopy, *Brain Res.* **6**:728.

Bunge, R. P., Bunge, M. B., and Peterson, E. R., 1965, An electron microscope study of cultured rat spinal cord, *J. Cell Biol.* **24**:163.

Bunge, R. P., Rees, R., Wood, P., Burton, H., and Ko, C.-P., 1974, Anatomical observations on synapses formed on isolated autonomic neurons in tissue culture, *Brain, Res.* **66**:401.

Burn, J. H., and Rand, M. J., 1965, Acetylcholine in adrenergic transmission, *Ann. Rev. Pharmacol.* **5**:163.

Claude, P., 1973, Electron microscopy of dissociated rat sympathetic neurons in culture, *J. Cell Biol.* **59**:57a.

Crain, S. M., 1966, Development of "organotypic" bioelectric activities in central nervous tissues during maturation in culture, *Int. Rev. Neurobiol.* **9**:1.

Crain, S. M., 1971, Intracellular recordings suggesting synaptic functions in chick embryo spinal sensory ganglion cells isolated *in vitro*, *Brain Res.* **26**:188.

Crain, S. M., and Peterson, E. R., 1974, Development of neural connections in culture, *Ann. N.Y. Acad. Sci.* **228**:6.

Crain, S. M., Peterson, E. R., and Bornstein, M. B., 1968, Formation of functional interneuronal connections between explants of various mammalian central nervous tissues during development *in vitro*, in: *Ciba Foundation Symposium: Growth of the Nervous System* G. E. W. Wolstenholme and M. O'Connor, eds.), pp. 13–31, Churchill, London.

Detwiler, S. R., 1920, Experiments on the transplantation of limbs in *Ambystoma*, *J. Exp. Zool.* **31**:118.

England, J. M., and Goldstein, M. N., 1969, The uptake and localization of catecholamine in chick embryo sympathetic neurons in tissue culture, *J. Cell Sci.* **4**:677.

Fambrough, D., Hartzell, H. C., Rash, J. E., and Ritchie, A. K., 1974, Receptor properties of developing muscle, *Ann. N. Y. Acad. Sci.* **228**:47.

Fischbach, G. D., 1972, Synapse formation between dissociated nerve and muscle cells in low density cell cultures, *Dev. Biol.* **28**:407.
Fischbach, G. D., Cohen, S. A., and Henkart, M. P., 1974, Some observations on trophic interaction between neurons and muscle fibers in cell culture, *Ann. N.Y. Acad. Sci.* **228**:35.
Harrison, R., 1910, The outgrowth of the nerve fiber as a mode of protoplasmic movement, *J. Exp. Zool.* **9**:787.
Hendelman, W., 1975, The synaptology of cultures of cerebellar cortex, *Neurosc. Abst.* **1**:812.
Henrikson, C. K., and Vaughn, J. E., 1974, Fine structural relationships between neurites and radial glial processes in developing mouse spinal cord, *J. Neurocytol.* **3**:659.
Hooisma, J., Slaaf, D. W., Meeter, E., and Stevens, W. F., 1975, The innervation of chick striated muscle fibers by the chick ciliary ganglion in tissue culture, *Brain Res.* **85**:79.
Hösli, E., Ljungdahl, A., Hökfelt, T., and Hösli, L., 1972, Spinal cord tissue cultures—A model for autoradiographic studies on uptake of putative neurotransmitters such as glycine and GABA, *Experientia* **28**:1342.
Jacobson, M., 1968, Cession of DNA synthesis in retinal ganglion cells correlated with the time of specification of their central connections, *Dev. Biol.* **17**:219.
James, D. W., and Tresman, R. L., 1969, Synaptic profiles in the outgrowth from chick spinal cord *in vitro, Z. Zellforsch.* **101**:598.
Johnson, D. G., Silberstein, S. D., Hanbauer, I., and Kopin, I. J., 1972, The role of nerve growth factor in the ramification of sympathetic nerve fibers into the rat iris in organ culture, *J. Neurochem.* **19**:2025.
Ko, C.-P., Burton, H., and Bunge, R. P., 1976*a*, Synaptic transmission between rat spinal cord explants and dissociated superior cervical ganglion neurons in tissue culture. Submitted for publication.
Ko, C.-P., Burton, H., Johnson, M. I., and Bunge, R. P., 1976*b*, Synaptic transmission between rat superior cervical ganglion neurons in dissociated cell cultures. Submitted for publication.
Landmesser, L., 1971, Contractile and electrical responses of vagus-innervated frog sartorius muscles, *J. Physiol. (London)* **213**:707.
Lasher, R. S., 1974, The uptake of (^{3}H) GABA and differentiation of stellate neurons in cultures of dissociated postnatal rat cerebellum, *Brain Res,* **69**:235.
Le Douarin, N. M., and Teillet, M. M., 1974, Experimental analysis of the migration and differentiation of neuroblasts of the autonomic nervous system and of neurectodermal mesenchymal derivatives using a biological cell marking technique, *Dev. Biol.* **41**:162.
Levi-Montalcini, R., Meyer, H., and Hamburger, V., 1954, *In vitro* experiments on the effects of mouse sarcoma 180 and 37 on the spinal and sympathetic ganglia of the chick embryo, *Cancer Res.* **14**:49.
Mains, R. E., and Patterson, P. H., 1973*a*, Primary cultures of dissociated sympathetic neurons. I. Establishment of long-term growth in culture and studies of differentiated properties, *J. Cell Bio.* **59**:329.
Mains, R. E., and Patterson, P. H., 1973*b*, Primary cultures of dissociated sympathetic neurons. II. Initial studies of catecholamine metabolism, *J. Cell Biol.* **59**:346.
Mains, R. E., and Patterson, P. H., 1973*c*, Primary cultures of dissociated sympathetic neurons. III. Changes in metabolism with age in culture, *J. Cell Biol.* **59**:361.
Miller, R., Varon, S., Kruger, L., Coates, P. W., and Orkand, P. M., 1970, Formation of synaptic contacts on dissociated chick embryo sensory ganglion cells *in vitro, Brain Res.* **24**:356.

Narayanan, C. H., and Hamburger, V., 1971, Motility in chick embryos with substitution of lumbrosacral by brachial and brachial by lumbrosacral spinal cord segments, *J. Exp. Zool.* **178:**415.

Nelson, P. G., 1975, Nerve and muscle cells in culture, *Physiol. Rev.* **55:**1.

Nelson, P. G., and Peacock, J. H., 1973, Electrical activity in dissociated cell cultures from fetal mouse cerebellum, *Brain Res.* **61:**163.

Obata, K., 1974, Transmitter sensitivities of some nerve and muscle cells in culture, *Brain Res.* **73:**71.

Okun, L. M., 1972, Isolated dorsal root ganglion neurons in culture: Cytological maturation and extension of electrically active processes, *J. Neurobiol.* **3:**111.

O'Lague, P. H., Obata, K., Claude, P., Furshpan, E. J., and Potter, D. D., 1974, Evidence for cholinergic synapses between dissociated rat sympathetic neurons in cell culture, *Proc. Natl. Acad. Sci.* **71(9):**3602.

Olson, M. I., and Bunge, R. P., 1973, Anatomical observations on the specificity of synapse formation in tissue culture, *Brain Res.* **59:**19.

Patterson, P. H., and Chun, L. L., 1974, The influence of non-neuronal cells on catecholamine and acetylcholine synthesis and accumulation in cultures of dissociated sympathetic neurons, *Proc. Natl. Acad. Sci.* **71:**3607.

Peacock, J. H., Nelson, P. G., and Goldstone, M. W., 1973, Electrophysiologic study of cultured neurons dissociated from spinal cords and dorsal root ganglia of fetal mice, *Dev. Biol.* **30:**137.

Privat, A., and Drian, M. J., 1975, Specificity of the formation of the mossy fibre-granule cell synapse in the rat cerebellum. An *in vitro* study, *Brain Res.* **88:**518.

Raisman, G., Field, P. M., Ostberg, A. J., Iverson, L. L., and Zigmond, R. E., 1974, A quantitative ultrastructural and biochemical analysis of the process of reinnervation of the superior cervical ganglia in the adult rat, *Brain Res.* **71:**1.

Rees, R., and Bunge, R. P., 1974, Morphological and cytochemical studies of synapses formed in culture between isolated rat superior cervical ganglion neurons, *J. Comp. Neurol.* **157:**1.

Rees, R. P., Bunge, M. B., and Bunge, R. P., 1976, Morphological changes in the neuritic growth cone and target neuron during synaptic junction development in culture. Accepted for publication in the *Journal of Cell Biology*.

Shimada, Y., and Fischman, D. A., 1973, Morphological and physiological evidence for the development of functional neuromuscular junctions *in vitro*, *Dev. Biol.* **31:**200.

Varon, S., and Raiborn, C., 1971, Excitability and conduction in neurons of dissociated ganglionic cell cultures, *Brain Res.* **30:**83.

Weiss, P., 1941, Self-differentiation of the basic patterns of coordination, *Comp. Psychol. Monogr.* **17:**1.

Weiss, P., 1955, Nervous system (neurogenesis), in: *Analysis of Development* (B. Willier, P. Weiss, and V. Hamburger, eds.), pp. 346–401.

Wood, P. M., and Bunge, R. P., 1975, Evidence that sensory axons are mitogenic for Schwann cells, *Nature*, **256:**662.

II

Morphological and Biochemical Studies of Synapses

5

From the Growth Cone to the Synapse

Properties of Membranes Involved in Synapse Formation

KARL H. PFENNINGER and ROSEMARY P. REES

1. INTRODUCTION

The development of specific neuronal connections must be based on a set of intricate control mechanisms. The complexity of this fascinating problem results from (1) the enormous number of elements involved (in the human CNS, an estimated 10^{14} synapses are formed), (2) the long pathway of certain nerve fibers, and (3) the high selectivity of their synaptic connections, not only between and within groups of neurons, but even with regard to the localization of the synapse on an individual postsynaptic cell. Numerous papers (e.g., Sperry, 1944; Attardi and Sperry, 1963; Crossland *et al.*, 1974*a*; for review, see Gaze, 1970; Jacobson, 1970; Jacobson, this volume) have suggested predetermined, high specificity of synaptic connections. Other experimental work with developing and regenerating parts of the central nervous system (for

KARL H. PFENNINGER · Section of Cell Biology, Yale University School of Medicine, New Haven, Connecticut. ROSEMARY P. REES · Department of Anatomy, Washington University School of Medicine, St. Louis, Missouri.

review, see Jacobson, 1970; Crossland *et al.*, 1974*a*; Raisman and Field, 1973) as well as studies on synapse formation *in vitro* (Rees and Bunge, 1974) clearly indicate that the specificity of synaptic connections is not absolute and that it is strictly effective only during a certain period of embryonic development. This suggests that the constellation of neurons—i.e., the availability of certain postsynaptic candidates and the timing of their appearance—in a developing brain area may greatly influence the resulting selectivity in synapse formation. These factors and the early conformation of the brain could contribute to a simpler pattern of recognition than is suggested by the immense complexity of the adult brain. However, the requirements of a basic mechanism of guidance (to bring the nerve fiber to the right area) and a mechanism of recognition (to initiate synapse formation with the right cell) remain.

For a long time, chemotropism seemed to be a good candidate for the guidance mechanism (Forssmann, 1900; Ramón y Cajal, 1913; reviewed in Jacobson, 1970). Generally, however, the experiments carried out *in vivo* (Weiss and Taylor, 1944; Hibbard, 1967) as well as those *in vitro* (e.g., see Nakai and Kawasaki, 1959; Olson and Bunge, 1973) failed to yield any evidence supporting the chemotropic hypothesis. In the latter case, it could be argued that the necessary concentration gradient of an "alluring substance" would not be maintained in the culture systems usually employed for these studies. In a paper on cultured sympathetic ganglia combined with the different autonomic effector organs, Chamley *et al.* (1973) presented data which led them to reconsider the chemotropism theory. These authors speculate, however, that a gradient of nerve growth factor might suffice to produce selective fiber growth.

Observations on the behavior of outgrowing nerve fibers *in vitro* suggest that, instead of directed advancement toward an appropriate target, the moving growth cones and filopodia probe their environment nonselectively (Nakai and Kawasaki, 1959; Rees *et al.*, 1976). They attach to whichever particles are in their way but detach within minutes from all but the appropriate target cells. On the other hand, they may grow past and miss (by several micrometers only) a postsynaptic candidate not precisely in their path of advancement (Rees *et al.*, 1976); steps of such an event are shown in Fig. 1. The nerve tips seem preferentially to grow and advance on supporting cells in culture (Grainger and James, 1970; Pfenninger and Bunge, 1974); they also show mutual contact reactions which lead either to fasciculation or to retraction of the elements involved (Nakai, 1960; Nakajima, 1965; Dunn, 1971). *In vivo*, proper migration of postmitotic neurons in the developing mammalian cerebrum and cerebellum seems to depend on

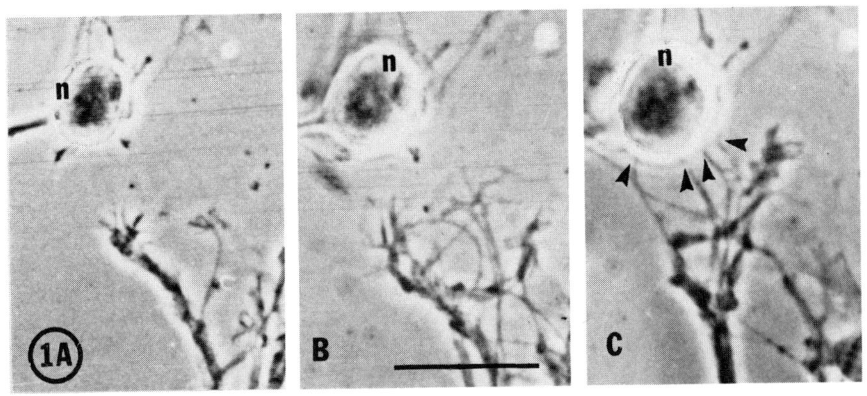

Fig. 1. Series of light micrographs of a living nerve tissue culture. In A, a bundle of spinal cord neurites is seen approaching an isolated superior cervical ganglion neuron (n). Fifteen minutes later (B), the neuritic bundle has broken up into several smaller groups of fibers and growth cones with widely spread filopodia. Forty-five minutes later (C), contacts have been established at four separate points (arrowheads) on the neuronal surface. However, the majority of fibers and growth cones at the right of the micrograph did not make contact with the neuron and continue their growth just past the putative target cell. Subsequent electron microscopic examination of this cell, fixed 48 h after initial contact, showed maturing synaptic profiles at the contact points. Magnification ×400, calibration 50 μm. From Rees *et al.* (1976).

their extensive, direct interaction with radial glia fibers (Sidman and Rakic, 1973; Sidman, 1974). Such observations on neuronal behavior both *in vivo* and *in vitro* suggest that *physical contact* may be required for the recognition process as well as for guidance of the migrating neuron and the growing nerve fiber to their targets. The existence of selective affinities between cell surfaces is further supported by the sorting out of specific cell types from mixed cell suspensions (for review, see Moscona, 1974, and this volume). In recent years, numerous publications have dealt with various components of the cell surface and some workers have tried to link alterations of these components with changes in cellular behavior (for review, see Burger, 1971; Edelman *et al.*, 1974). As a logical sequence of these considerations, we can suspect that clues for the guidance and recognition mechanisms exist in cell surface molecules.

It is therefore important to examine the properties of plasma membranes engaged in the initiation and formation of synaptic junctions with special reference to the composition of the external protein- and carbohydrate-rich surface layer. This chapter reviews relevant data on the structure and cytochemistry of mature synaptic membranes and

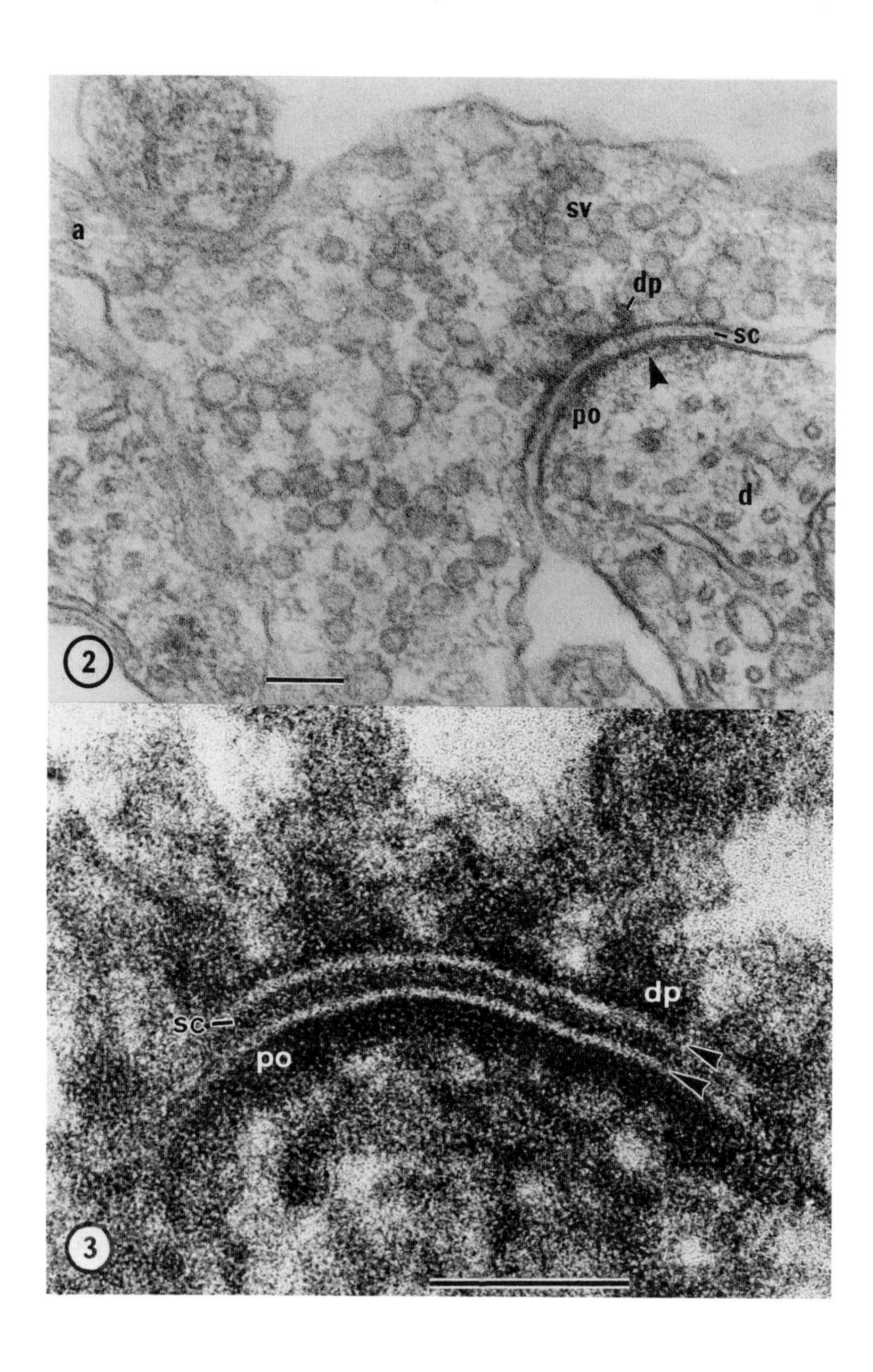
a
sv
dp
sc
po
d
2
dp
sc
po
3

presents more recent data on the properties of nerve growth cone plasmalemma. A third section on synapse formation explores the period during which the growth cone is transformed into a synapsing nerve terminal. As of now, probably the most important conclusion which can be drawn from these various data is that the recognizing structure, the growth cone, exhibits a membrane composition which is morphologically and cytochemically distinct from all other parts of the neuron, and especially from the presynaptic membrane. It becomes evident that synaptic cleft components (and other synaptic membrane specializations) are the result (rather than the substrates) of a recognition process which has taken place previously. The key molecules for this complex function will have to be sought in the growth cone plasmalemma.

2. THE MEMBRANE SURFACE AT THE SYNAPTIC SITE

The chemical synapse is a structurally distinct interneuronal contact where electrical signals are transmitted via the release of a chemical intermediate. This contact is characterized by an asymmetrical junction exhibiting (in vertebrates) an hexagonal array of presynaptic dense projections with associated synaptic vesicles, a wide cleft, and usually a postsynaptic (subjunctional) density of varying thickness.

The structural specializations forming the intra- and extracellular components of the synaptic junctional complex can be viewed as derivatives of the plasma membrane (Akert and Pfenninger, 1969; Pfenninger, 1973). Our current concept of the molecular organization of the

←

Fig. 2. Synapse formed in a 7-day-old rat spinal cord culture; aldehyde–osmium fixation, uranyl-lead staining (OsUL). Although this synapse appears young with its incompletely formed membrane densities, the characteristic synaptic structural features can be recognized: presynaptic dense projections (dp), wide-open cleft (sc), postsynaptic density (po), synaptic vesicles (sv). Note some fuzzy material in the junctional cleft, which in one area (arrowhead) seems to interconnect the opposing membranes in cross-striations. a, Axon giving rise to the nerve terminal; d, dendritic element. Magnification ×110,000, calibration 0.1 μm.

Fig. 3. Synapse in the cat subfornical organ, aldehyde fixed, bismuth iodide impregnated, and uranyl-lead stained (BIUL). Note the absence of unit membrane "tramlines" and the prominent staining of paramembranous densities (dp, presynaptic dense projections; po, postsynaptic density; sc, synaptic cleft material). The arrowheads point at the interspaces between the cytoplasmic and external membrane "coats" which contain the membrane lipid bilayer. Near the center of the junction the synaptic cleft material can be seen to consist of two separate layers of external membrane surface material. Magnification ×290,000, calibration 0.1 μm. From Pfenninger *et al.* (1972).

membrane (for review, see Singer and Nicolson, 1972; Bretscher, 1973; Steck, 1974) depicts it as a bimolecular, intrinsically fluid layer of lipids permeated partially or fully by proteins which stick polar portions out into the hydrophilic environment, the extracellular space or the intracellular cytosol, or both. These polar portions adjacent to the lipid bilayer, earlier termed "membrane coat material" (*cf.* "the greater membrane": Revel and Ito, 1967; Lehninger, 1968), are of protein and carbohydrate nature. The carbohydrates are linked to the glycoproteins or the glycolipids anchored in the membrane. In contrast to its intracellular counterpart, the extracellular portion—i.e., the outer surface layer—is particularly rich in carbohydrate moieties.

The most superficial layer of the plasma membrane is of particular interest in this chapter since it represents the substrate of mutual interaction between cells contacting each other (*cf.* James and Tresman, 1972). In fact, with the exception of tight and a few other junctions, plasmalemmal unit membranes are never seen to touch each other but instead maintain an "intercellular space" (defined as a gap between the apposed unit membrane "tramlines") of at least 100 Å.

2.1. *Synaptic Cleft Morphology*

In a typical synapse (Gray I or S type: Gray, 1959, 1969; Akert *et al.*, 1972; Pfenninger, 1973), the intercellular space is widened to about 160–

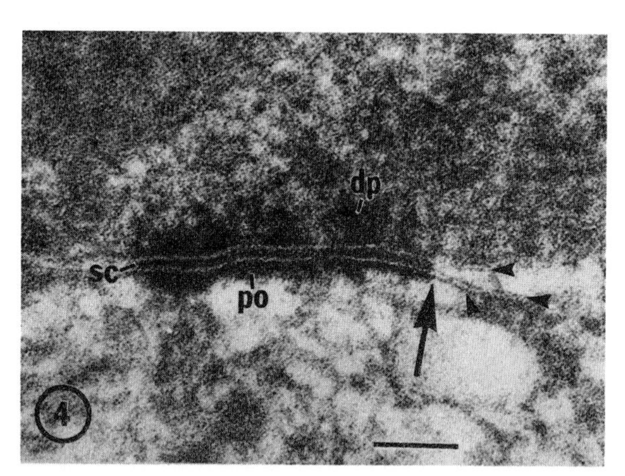

Fig. 4. BIUL-treated synapse in the rat spinal cord. Note the thin membrane "coats" (small arrowheads) of the nonspecialized plasmalemma at the right of the synapse which are continuous with (arrow) external and cytoplasmic junctional densities. sc, Synaptic cleft material; dp, presynaptic dense projections; po, postsynaptic density. Magnification ×110,000, calibration 0.1 μm. From Pfenninger (1973).

250 Å (depending on the type of preparation) as compared to about 120 Å for nonsynaptic areas. The synaptic cleft contains material that varies in appearance with the method of tissue processing. In osmicated, uranyl-lead stained (OsUL) specimens (Fig. 2), the cleft contains some fuzzy, slightly electron-dense material, and sometimes a dense line running in the middle between pre- and postsynaptic membranes can be detected (Pappas and Waxman, 1972; Pfenninger, 1973). By contrast, ethanolic phosphotungstic acid (E-PTA) reveals the cleft material as a more or less solid band of electron-dense substance (Bloom and Aghajanian, 1968) (see Fig. 5). Similarly, the bismuth iodide–uranyl-lead technique (BIUL), as shown in Figs. 3, 4, and 5, produces two cleft-filling electron-dense layers separated by a narrow gap of about 20 Å (Akert and Pfenninger, 1969; Akert *et al.*, 1969; Pfenninger *et al.*, 1969; Pfenninger, 1971*a*). In these preparations, the unit membranes cannot be seen. The two synaptic cleft layers are continuous with the thinner layers of nonsynaptic external membrane material (Pfenninger, 1971*a*, 1973), as shown in Fig. 4. The integrity of the pre- and postsynaptic external membrane "coats" in the cleft can be established by treatment of unfixed synapses with solutions of high ionic strength (Pfenninger, 1971*b*): in a high percentage of cases, the synaptic contacts are opened (see below), and the two BIUL-positive layers are separated from each other (Figs. 6 and 7). These membrane surface layers as well as the osmiophilic middle dense line (which is more pronounced in trypsinized tissue; Fig. 8) indicate a structural cleft organization in parallel to the synaptic membranes. Early workers have suggested, though, that there is also a perpendicular pattern crossing the synaptic cleft. A variety of filamentous structures across the gap have been described in osmicated material (De Robertis, 1964; Gray, 1966; Van der Loos, 1963) but could never be clearly defined (*cf.* Fig. 2). In BIUL-treated, partially opened synaptic contacts, pre- and postsynaptic membranes are sometimes found to remain interconnected by thin threads or bundles of filaments, a finding which suggests that the two layers of cleft material are actually composed of long molecules which may form filamentous units. These would then stick out from the plasma membranes like the bristles of a brush. In intact synapses, electron optical superimposition in the depth of the section would fuse their image to a compact layer (Pfenninger, 1971*b*, 1973).

2.2 *Synaptic Cleft Cytochemistry*

Chemical analysis of synaptosome fractions and synaptic junctional complexes has shown the presence of large amounts of polar

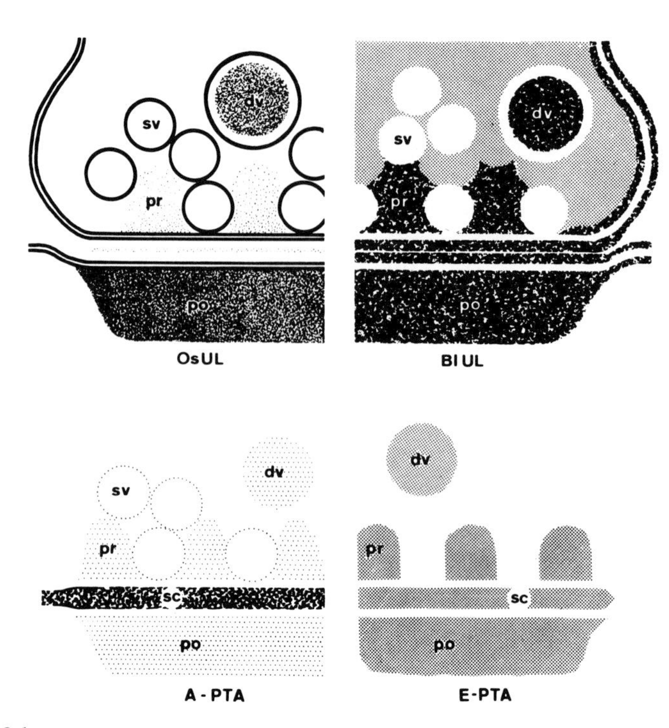

Fig. 5. Schematic comparison between differently impregnated and stained synaptic junctions. In OsO_4-impregnated and uranyl-lead stained specimens (OsUL), unit membrane structure is visualized (for the sake of simplicity the vesicular membranes have been drawn as single lines), and a postsynaptic density (po) is usually clearly recognizable. However, presynaptic dense projections (pr) and cleft material are only faintly stained. By contrast, bismuth iodide impregnation followed by uranyl-lead staining (BIUL) and ethanolic phosphotungstic acid impregnation (E-PTA) reveal only the surface material on the cytoplasmic and external sides of the synaptic membranes (and the core of the large dense-core vesicles, dv). BIUL clearly shows the continuity between nonspecialized and synaptic membrane "coats," i.e., dense projection network, postsynaptic thickening, and two layers of cleft material (sc). Acidic phosphotungstic acid impregnation (A-PTA), believed to be specific for carbohydrates, almost exclusively visualizes synaptic cleft material. sv, Synaptic vesicles. For further description, see text. Modified after Pfenninger (1971*a*).

protein (Cotman *et al.*, 1968), glycoprotein (Brunngraber *et al.*, 1967), and glycolipid (Wolfe, 1961); the latter two components are suspected to extend into the synaptic cleft (for review, see also Barondes, 1974). The exact localization of these molecular constituents by fractionation of synaptic junctional components has not yet been possible (however, postsynaptic densities have successfully been isolated: Cotman *et al.*, 1974). Thus our biochemical knowledge of synaptic cleft material is entirely based on relatively crude cytochemical studies. This type of work has been carried out on synapses from such varied tissues as rat hypothalamus (Bloom and Aghajanian, 1968), cat and rat subfornical organ (Pfenninger, 1971*a*, *b*), and rat forebrain (synaptosome fractions: Cotman and Taylor, 1972) so that the findings do not appear to be confined to particular brain areas. However, it seems that, within these regions, only S-type synapses (putative excitatory synapses with spherical vesicles, wide cleft, broad postsynaptic density: for review, see Akert *et al.*, 1972; Pfenninger, 1973) have been investigated with the more complex cytochemical experiments involving selective tissue digestion. The validity beyond S-type synapses of the experimental data presented below remains to be assessed.

Figure 5 schematically indicates the synaptic staining patterns obtained by OsUL, E-PTA, BIUL, and A-PTA (acidic phosphotungstic acid: Pease, 1966; Meyer, 1969, 1970; Pease, 1970) treatment of nervous tissue. In contrast to the OsUL image, BIUL, E-PTA, and A-PTA do not reveal unit membrane structure but, as we have previously reported, cause intense staining of synaptic cleft material and, in the case of E-PTA and BIUL, of cytoplasmic densities. From the staining specificity of E-PTA and BIUL (Bloom and Aghajanian, 1968; Pfenninger, 1971*a*), we can conclude that synaptic densities are rich in basic amino groups. The presence of large numbers of anionic groups in synaptic membrane coats, especially in the cleft substance, can be inferred from their high electron opacity after staining with ruthenium red (Bondareff, 1967; Tani and Ametani, 1971) or with uranyl-lead (UL) without prior osmication (Pfenninger, 1971*a*, 1973). These anionic groups of synaptic cleft components seem largely to belong to carboxyl residues, as indicated by the fact that their UL stainability can be suppressed almost completely by preceding carboxymethylation of the tissue (Pfenninger, 1971*a*, 1973). Additional information on cleft cytochemistry comes from staining with A-PTA, believed to be specific for carbohydrate (or polyhydroxyl) substrates (Pease, 1970). This reagent stains the synaptic cleft intensely, in contrast to the almost invisible cytoplasmic densities (Meyer, 1969, 1970), a finding which is confirmed by the application of other carbohydrate-contrasting methods, e.g., periodic acid–silver methenamine (Rambourg and Leblond, 1967). More specific data are obtained by studying the binding to synaptosomes of the ferritin-conjugated plant lectins concanavalin A (Cotman and Taylor, 1974) and ricin (Bittiger and Schnebli, 1974), which demonstrate the occurrence in the synaptic cleft of mannoside and/or glucoside and of galactoside, respectively.

To what larger molecules do these carbohydrate moieties and the

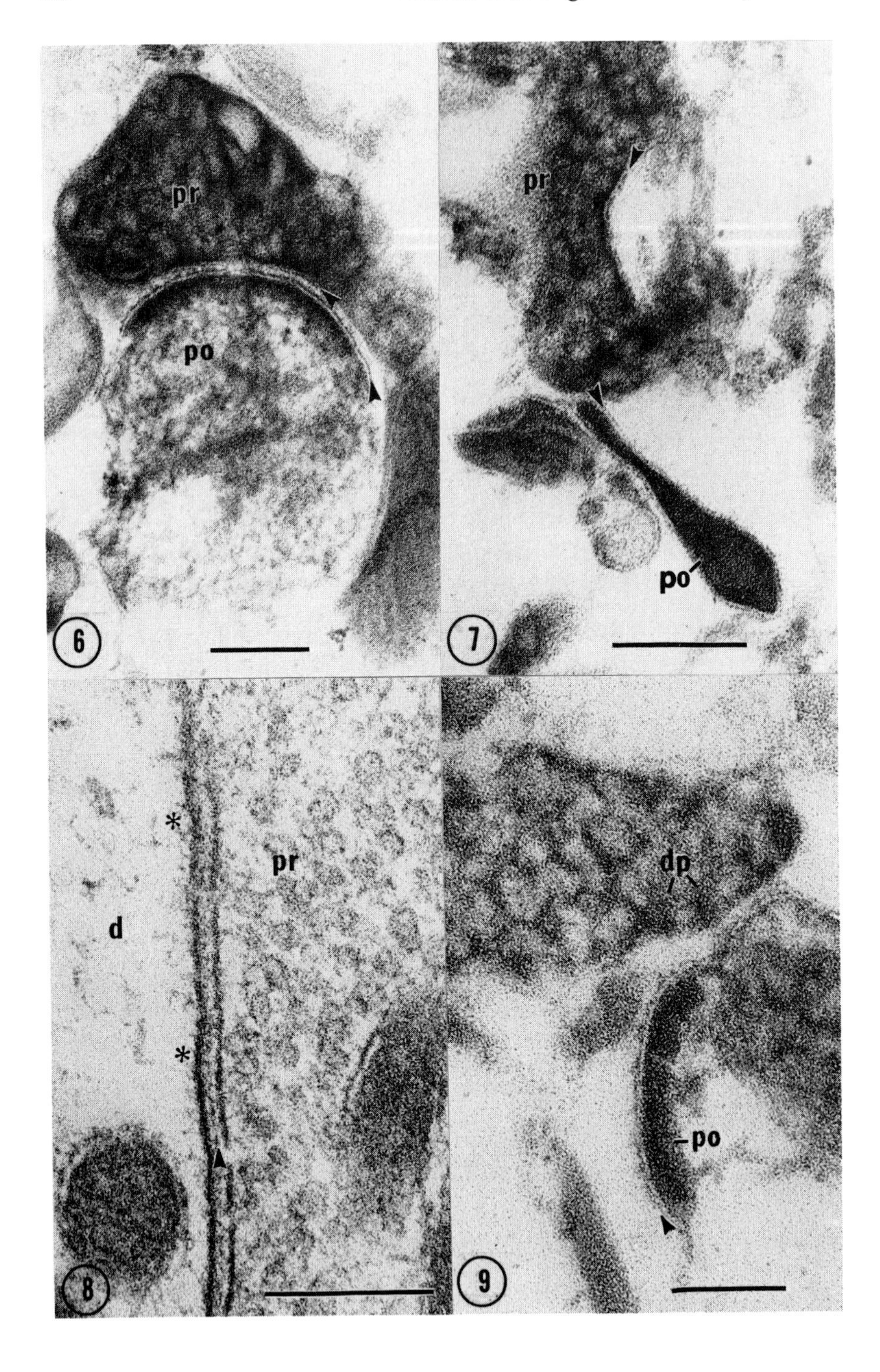
pr
po
6
pr
po
7
*
pr
d
*
8
dp
po
9

anionic and cationic groups in the synaptic cleft belong? The comparison of synaptic staining properties, before and after digestion with specific enzymes, has yielded a partial answer. Neuraminidase treatment causes only a minor decrease of staining with cationic reagents (Bondareff and Sjöstrand, 1969; Pfenninger, 1971*b*), possibly because the contribution to synaptic cleft basophilia of neuraminic acid residues is relatively small and/or because most of them are inaccessible to the enzyme. By contrast, dramatic effects are seen after the use of proteolytic enzymes, so that the importance of proteinaceous material in the synaptic cleft is evident. Pepsin and pronase completely remove OsUL-, BIUL-, and E-PTA-stainable (Bloom and Aghajanian, 1968; Pfenninger, 1971*b*) material from both sides of synaptic membranes, and the synaptic contact is opened. If nervous tissue, prefixed in aldehyde, is treated with trypsin and then BIUL stained, the cleft layers are found to persist after the degradation, whereas UL-stainable material (containing anionic groups) has largely disappeared from the synaptic cleft (Pfenninger, 1971*b*). OsUL staining of such trypsinized material is characterized by the loss of synaptic densities except for the resistant osmiophilic middle dense line (Fig. 8). In all these experiments, the synaptic contact is maintained. However, if trypsin digestion precedes aldehyde fixation, the junctions are opened (Fig. 9) and the middle dense line disappears (but the BIUL-positive layers can still be seen). A similar result has not been observed in trypsin-treated E-PTA material (Bloom and Aghajanian, 1968), possibly because of the

←

Figs. 6 and 7. Synaptic contacts dissociated after treatment with high ionic strength (0.8 or 1.2 M $MgSO_4$) BIUL preparation. In Fig. 6, the pre- and postsynaptic elements (pr, po) are not completely disjoined but a wide gap clearly separates the two cleft layers (arrowheads) from each other. In Fig. 7, a putative isolated postsynaptic element (po) with its density contained in a closed, membrane-bounded structure and an isolated presynaptic area (pr) can be seen. Note the intact membrane surface layers (arrowheads). P_2B synaptosome fraction from guinea pig cerebral cortex. Magnifications ×70,000 (Fig. 6) and ×94,000 (Fig. 7), calibration 0.2 μm. Observations by K. H. Pfenninger, P. Marko, and M. Cuenod (1971); illustrations from Pfenninger (1973).

Figs. 8 and 9. Effects of trypsin treatment on synaptic stainability with OsUL (Fig. 8) and BIUL (Fig. 9). Note the disappearance of the postsynaptic density (asterisks) but the persistence of the middle dense line (arrowhead) in the OsUL case. Furthermore, the synaptic junction remained intact, a finding characteristic of the material treated with aldehyde before trypsin digestion. pr, Presynaptic element; d, dendritic element. BIUL staining of presynaptic dense projections (dp), postsynaptic density (po), and external membrane surface material especially on the postsynaptic side (arrowhead) at least partly resists tryptic degradation. Note the dissociation of this synaptic contact in tissue which was subjected to proteolysis first and then aldehyde fixed and BIUL treated. Synapses from the cat subfornical organ. Magnifications ×120,000 (Fig. 8) and ×82,000 (Fig. 9), calibration 0.2 μm. From Pfenninger (1973).

considerable chemical difference between the E-PTA and BIUL reagents (see also Pfenninger, 1973), nor was it obtained in BIUL-treated, isolated synaptic complexes (Cotman and Taylor, 1972). In the latter case, the previous treatment of these subcellular fractions with the detergent Triton X-100 might have caused some of their components to be lost or to have become more susceptible to trypsin. Further studies will be necessary fully to elucidate this problem. As it stands now, the persistence after trypsinization (1) of the BIUL staining in the cleft, (2) of the adhesiveness of the synaptic contact, and (3) of the osmiophilic middle dense line (proteolysis following aldehyde prefixation of tissue in the latter two cases) forms a consistent picture and suggests the presence of at least two major cleft components (Pfenninger, 1971*b*): a trypsin-resistant (although pepsin- and pronase-sensitive) moiety, which bears a large number of BIUL-stainable basic amino groups and a trypsin-sensitive component which appears to bear a large portion of the UL-stainable anionic groups. This second fraction seems to play a role in the mechanism of synaptic adhesion as derived from the trypsin-induced opening of the junctions in unfixed tissue (cross-linkage of cleft components with glutaraldehyde holds the synaptic contact together if prefixed tissue is trypsin treated). Unfortunately, no data are currently available on the effect of proteolytic digestion on carbohydrate staining of the synapse.

2.3. *Synaptic Adhesion Mechanism*

The osmiophilic middle dense line in the synaptic cleft as well as the corresponding stain-free narrow gap between the BIUL-positive membrane surface layers suggests that the most distal region of the synaptic external membrane coat is chemically different from the bulk of the cleft material and/or that it has altered staining properties due to a different chemical environment. In this context, the very surface of the two cleft layers and their interactions are of particular interest. It has been suggested that coordination complexes involving Ca^{2+} and carboxyl groups play an important role in intercellular adhesion. However, it has not been possible to open synaptic junctions even with high concentrations (100 mM) of the chelator ethylene diamine tetraacetate (Pfenninger, 1971*b*; Cotman and Taylor, 1972). By contrast, treatment with solutions of high ionic strength (e.g., 1 M $MgCl_2$) has resulted in the dissociation of 60% of synaptic contacts if fresh slices of brain tissue (subfornical organ) were incubated for 40–60 min (Pfenninger, 1971*b*). It is interesting to note that $NaClO_4$ has proved almost twice as effective in splitting synaptic junctions as other salts at the same ionic strength

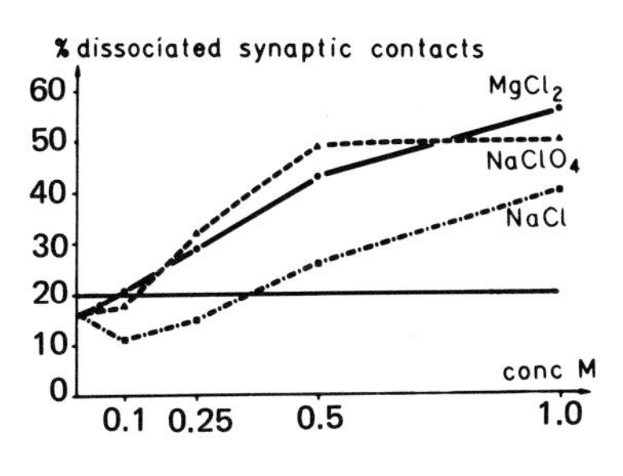

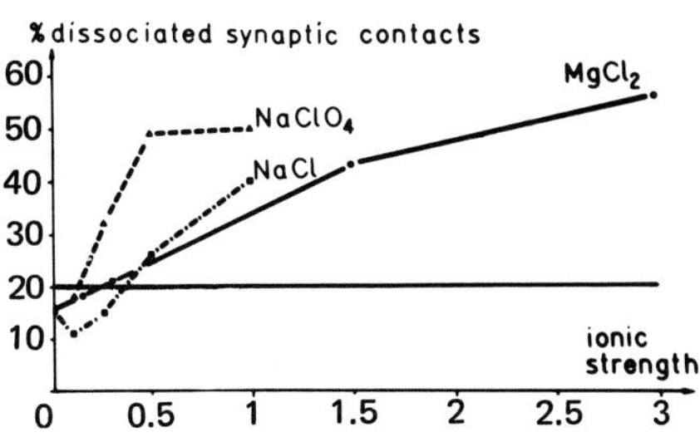

Fig. 10. Graphical presentation of the opening of synaptic contacts (clearly dissociated junctions) as a function of the salt concentration (top) and ionic strength (bottom). Note that $NaClO_4$ is about twice as effective a dissociator as the other salts used. For further description, see text. From Pfenninger (1971*b*).

(Fig. 10). Similar results can be obtained from synaptosomes (Figs. 6 and 7) resuspended in solutions of high ionic strength (K. H. Pfenninger, P. Marko, and M. Cuénod, 1971, reported in Pfenninger, 1973). However, the percentage of opened contacts is somewhat lower (50% for 0.8 or 1.2 M $MgSO_4$) than in the tissue slices. The dissociative action of high ionic strength suggests that the mechanically very strong and chemically very resistant (*cf.* Pfenninger, 1971*b*; Cotman and Taylor, 1972) synaptic adhesiveness is largely if not entirely based on a mechanism of polyionic binding, ionic interactions between charged groups dispersed on the surface of the two protein and carbohydrate layers in the synaptic cleft. The fact that $NaClO_4$, which is assumed to bind particularly strongly to amino groups (Ohlenbusch *et al.*, 1967), is an especially potent junction dissociator stresses the role of these cationic groups in the adhesion mechanism.

So far, we have always referred to the "two cleft layers." This is an oversimplification due to the lack of more precise data; the pre- and postsynaptic external membrane coats must be quite different, as indicated, for example, by the unilateral presence of transmitter release sites and of receptor molecules, respectively. However, as yet the only hints of distinguishing properties come from a somewhat greater resistance of the postsynaptic cleft layer to detergent treatment (Cotman and Taylor, 1972) and tryptic degradation (*cf.* Figs. 45 and 46 in Pfenninger,

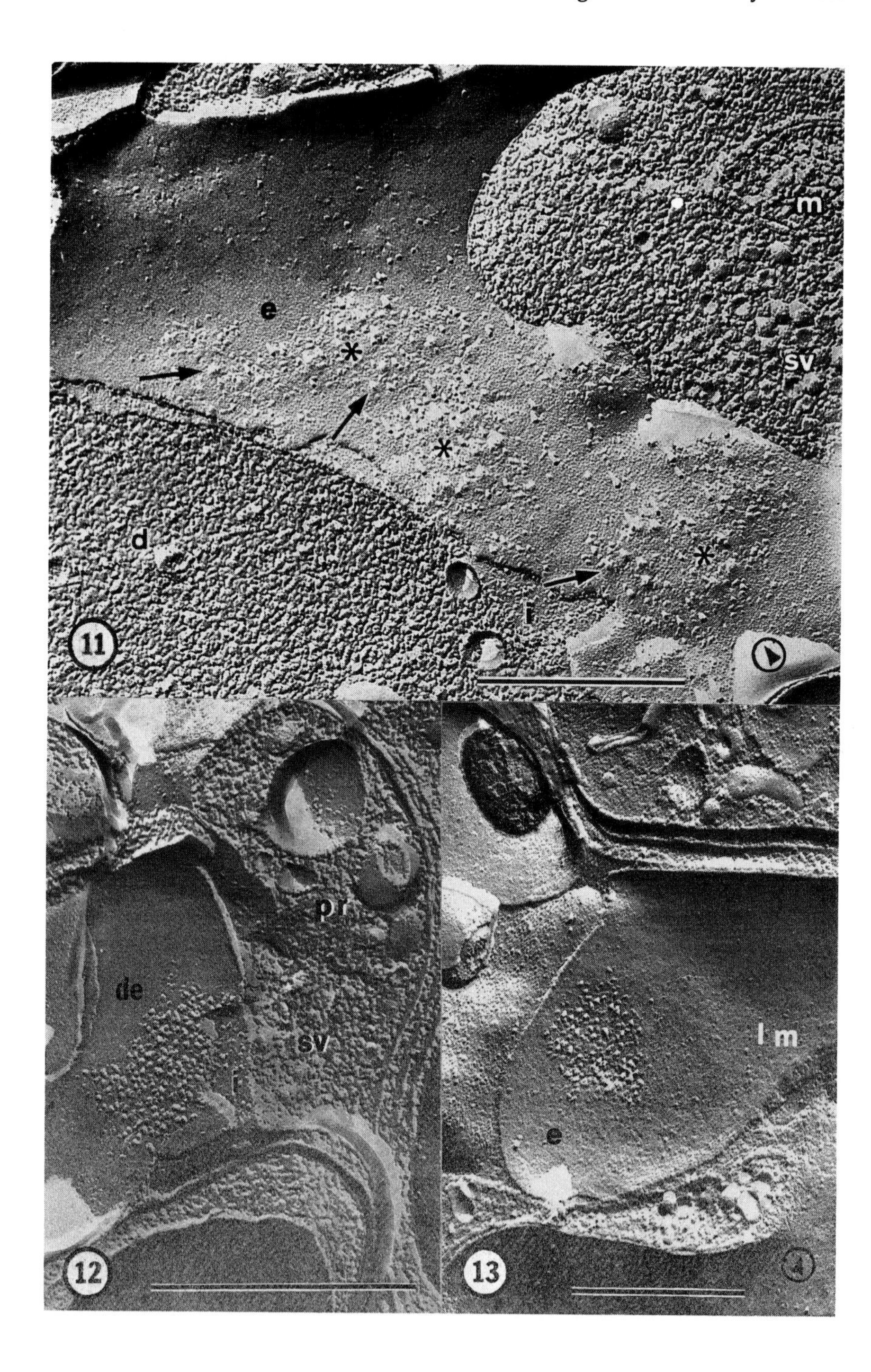
m
e
*
sv
*
d
*
i
11
pr
de
sv
i
12
lm
e
13

1973) and from its somewhat earlier appearance during synapse development (Jones *et al.*, 1974).

2.4. *Inner Structure of Synaptic Membranes*

What are the specializations inside the synaptic membrane? The most striking features revealed by freeze-fracturing of the presynaptic membrane are small protuberances (in the outer membrane leaflet) or indentations toward the cytoplasm (seen as pits in the inner leaflet), of about 200 Å diameter. These so-called vesicle attachment sites represent stimulation- and Ca-dependent points of interaction between synaptic vesicles and the axolemma serving the purpose of transmitter release (Pfenninger *et al.*, 1972; Streit *et al.*, 1972; Pfenninger and Rovainen, 1974) and are illustrated in Fig. 11. Apart from these release sites, comparatively large, 80–110 Å diameter, intramembranous particles seem more frequent in the presynaptic inner and outer membrane leaflets than in the surrounding nonspecialized membrane (Fig. 11). As shown in Figs. 12 and 13, arrays of large (about 100 Å diameter) intramembranous particles are even more striking in the postsynaptic membrane (Sandri *et al.*, 1972). These particle clusters are found in the usually particle-poor outer leaflet and are thus especially conspicuous. However, they may not necessarily represent an obligatory postsynaptic feature; their exclusive occurrence in S-type synapses (putative excitatory junctions with spherical vesicles, wide cleft, thick postsynap-

←

Fig. 11. Freeze-fractured nerve terminal synapsing with a postsynaptic dendrite (d). The external leaflet (e) of the presynaptic membrane is seen from a vantage point inside the nerve terminal whose cytoplasm contains clusters of synaptic vesicles (sv) and mitochondria (m). Note the small protuberances, the vesicle attachment sites (arrows), and clusters of intramembranous particles (asterisks) in the presynaptic membrane. A small area of the inner leaflet (i) of the postsynaptic membrane can also be identified. Circled arrow, shadowing direction. Cat subfornical organ. Magnification ×60,000, calibration 0.5 μm. From Pfenninger *et al.* (1972).

Figs. 12 and 13. Postsynaptic membrane structure as seen in a freeze-fracture replica. In Fig. 12, a patch of large intramembranous particles can be recognized in the otherwise particle-poor outer leaflet of a putative dendritic postsynaptic element (de). The particle aggregation is located just opposite to a cluster of vesicles (sv) in a nerve terminal (pr). i, Inner leaflet of presynaptic membrane. Pigeon optic tectum. From Pfenninger (1973). Figure 13 shows a similar cluster of particles in an outer membrane leaflet (e) in the neuropil of the lamprey spinal cord. However, because further morphological clues such as the exposure of synaptic vesicles in the adjacent profile are missing, the postsynaptic character of the specialized membrane in this picture can only be suggested. From Pfenninger and Rovainen (1972, unpublished observations). Magnifications ×76,000 (Fig. 12) and ×50,000 (Fig. 13), calibration 0.5 μm. Circled arrow, shadowing direction for both figures.

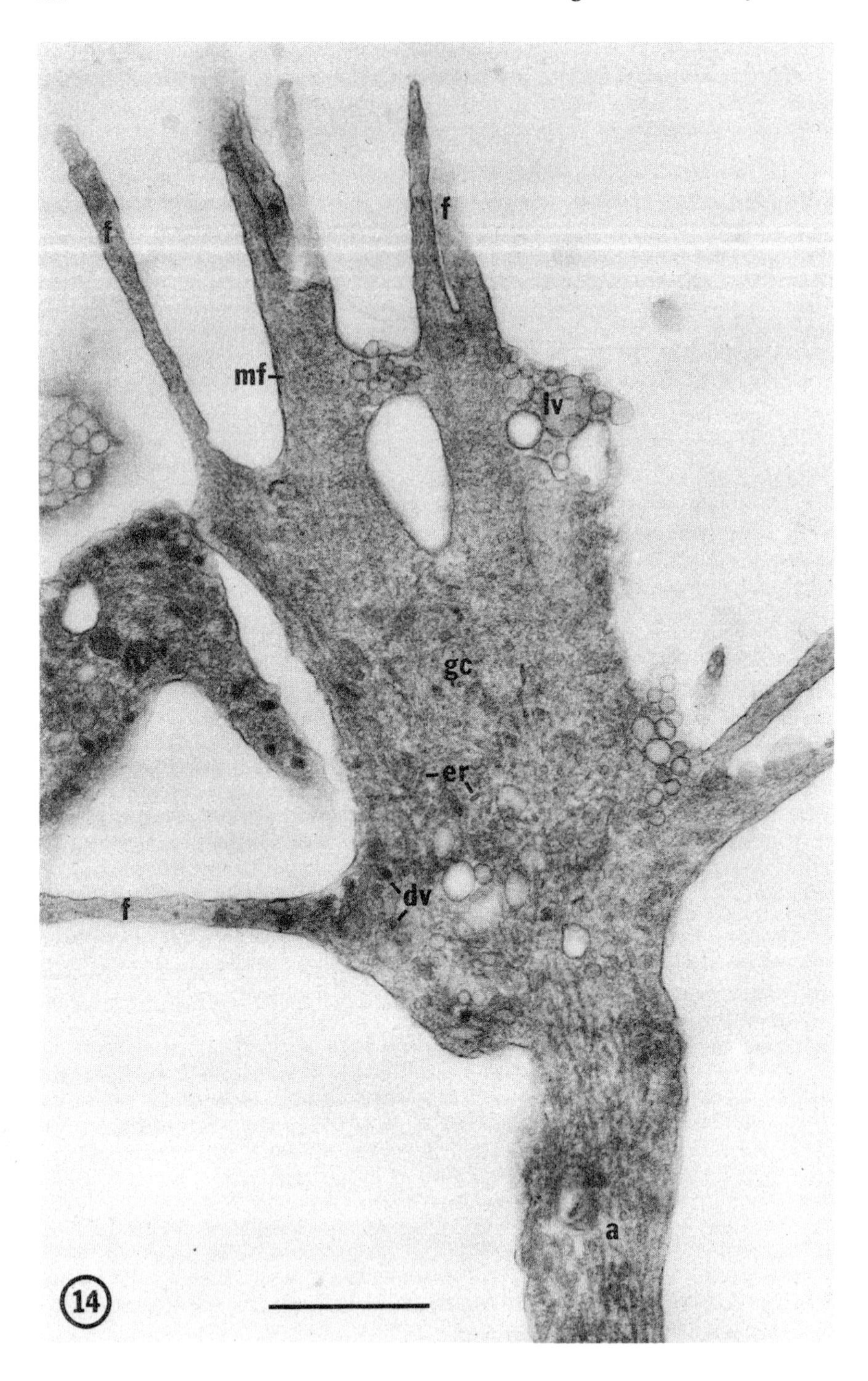
f
f
mf
lv
gc
er
dv
f
a
14

tic density: for review, see Akert *et al.*, 1972; Pfenninger, 1973) has been suggested (Landis and Reese, 1974; Landis *et al.*, 1974). Desmosome-like junctions which bear cytoplasmic densities similar to the postsynaptic density also feature clustered intramembranous particles in the outer leaflets of the junctional membranes (McNutt and Weinstein, 1973; Landis and Reese, 1974; Shienvold and Kelly, 1974). As yet, therefore, it cannot be decided whether the postsynaptic intramembranous particle clusters play a specific role in synaptic function.

3. THE NERVE GROWTH CONE AND SOME OF ITS MEMBRANE PROPERTIES

The nerve growth cone, originally discovered and named in the 1890s by Ramón y Cajal (for review, see Ramón y Cajal, 1929, 1933), is the enlarged irregularly shaped, filopodia-bearing tip of the outgrowing neurite (Fig. 14 and 15). Some of its functional properties were suggested by Ramón y Cajal in his early work before the first successful cultures of nerve tissue were established by Harrison (1907). Harrison observed the outgrowth of fibers bearing growth cones at their tips *in vitro* and, in 1910, in living frog embryonic ectoderm (Harrison, 1910). Under both conditions, he observed how the growth cone emitted and retracted thin projections, the so-called filopodia, and/or gave rise to undulating membranes, the whole structure exhibiting the amoeboid activity which Ramón y Cajal had predicted on the basis of his histological preparations. Observations of the typical shape and the characteristic constant motion of the nerve growth cone were later confirmed in a variety of tissues and species *in vivo* (Speidel, 1933) as well as *in vitro* (Lewis and Lewis, 1912; Hughes, 1953; Nakai, 1956; Pomerat *et al.*, 1967; Nakai and Kawasaki, 1959; Bray, 1970). Please see the note added in proof (p. 357) for further discussion.

3.1. *Growth Cone Morphology*

Following the initial discoveries, about 50 years passed before renewed, significant progress in our understanding of the growth cone

←

Fig. 14. Nerve growth cone in a rat spinal cord culture sectioned in parallel to the supporting collagen substrate. Note the characteristic shape consisting of a slender stem, the axon (a), and the enlarged cone (gc) which bears numerous filopodia (f). Some of the typical organelles can also be seen in this picture: microfilaments (mf), endoplasmic reticulum (er), clusters of large clear vesicles (lv), and large dense-core vesicles (dv). Magnification × 23,000, calibration, 1 μm.

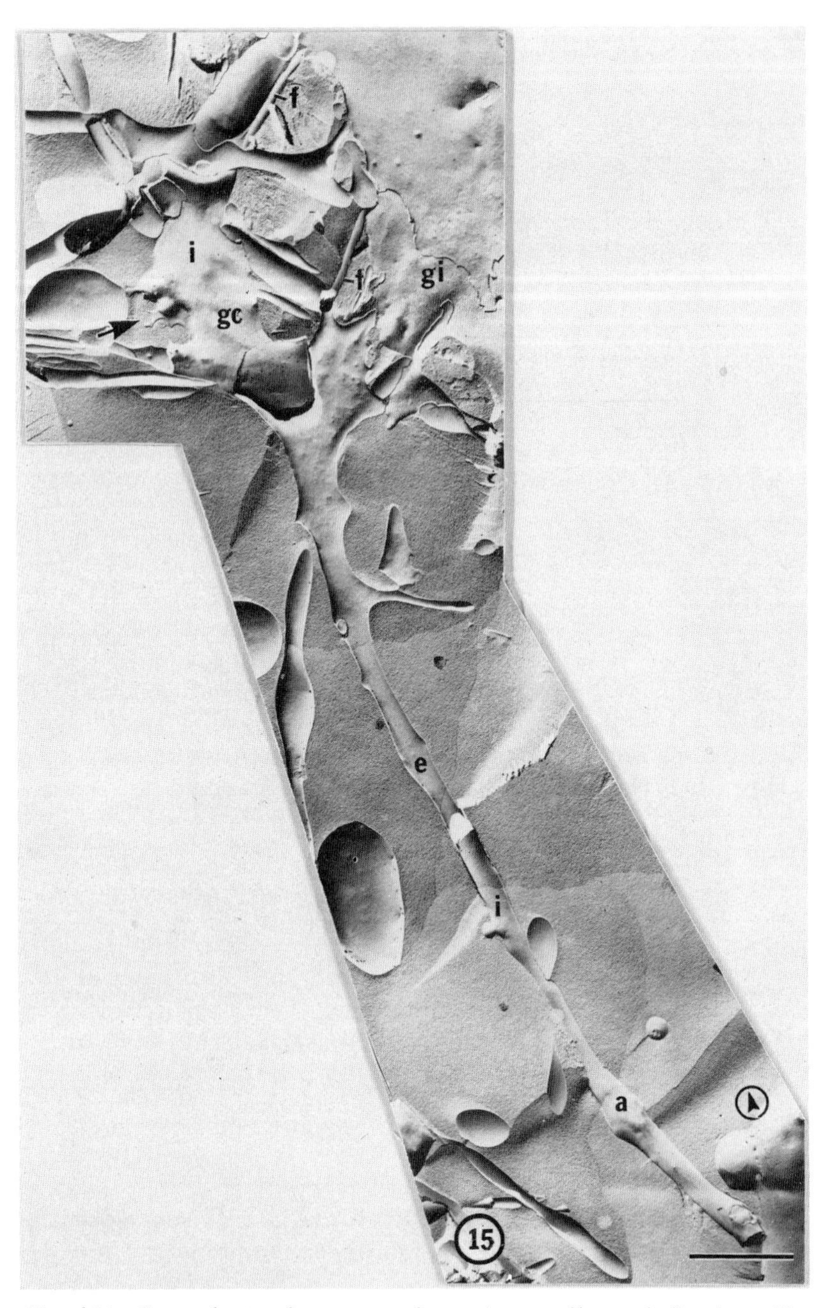

Figs. 15 and 16. Freeze-fractured nerve growth cone in a rat olfactory bulb culture. The survey picture (Fig. 15) shows its typical shape (*cf*. Fig. 14): a segment, here about 17 μm long, of the axon shaft (a) and the growth cone body (gc) with filopodia (f) attached. The arrow points at a ruffling region. e, External, i, inner plasmalemmal leaflets. At higher magnification (Fig. 16), more details of the inner (i) and outer (e) leaflets of the growth

cone plasmalemma can be observed. Note especially the paucity of intramembranous particles (arrowheads) even in the inner leaflet; this is strikingly evident when compared with the inner plasmalemmal leaflet of a neighboring glial process (gi). Magnifications ×7,600 (Fig. 15) and ×22,000 (Fig. 16), calibrations 2 μm (Fig. 15) and 1 μm (Fig. 16). Circled arrows, shadowing direction. From Pfenninger and Bunge (1974).

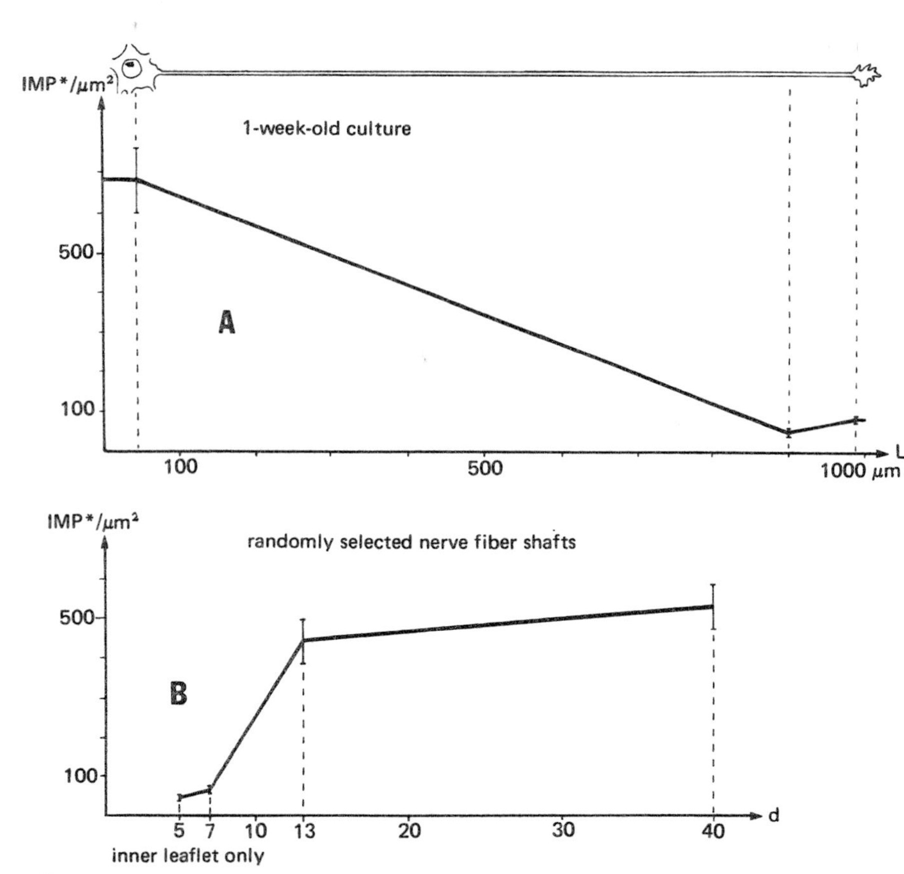

Fig. 17. Density of intramembranous particles in the plasmalemma of the growing neuron as a function of the distance from the perikaryon (A) and, in the axon shaft, as a function of time (B) (schematic representations). For A, an average spinal cord neuron grown in culture for about 1 week has been assumed. Note that the curve is based on three points only and therefore that the shape of the particle density gradient is not known. The figures are given as intramembranous particles per square micrometer plus or minus standard error of the mean. Time scale, d = days. For further description, see text.

was made by the application of electron microscopy (Bodian, 1966; Del Cerro and Snider, 1968; Tennyson, 1970; Kawana *et al.*, 1971; Yamada *et al.*, 1971; Bunge, 1973; Vaughn *et al.*, 1974; Pfenninger and Bunge, 1974; Skoff and Hamburger, 1974). The nerve growth cone (Figs. 14, 15, and 16), a few micrometers in width and length (including the filopodia) but quite variable in size, was found to contain a characteristic set of organelles (Tennyson, 1970; Yamada *et al.*, 1971; Bunge, 1973): immediately subjacent to the plasmalemma, and filling the filopodia, is a dense network of microfilaments which it is thought may consist of actin and be the structural and chemical basis of growth cone motility (Yamada *et al.*, 1970; Bray, 1972; for review, see Pollard and Weihing, 1974). The body of the growth cone is filled with a variety of smooth-surfaced membrane cisternae of various sizes and shapes which may represent smooth endoplasmic reticulum or have some different function; mixed with these cisternae are numerous mitochondria and a striking number of dense-core vesicles with an average diameter of 1200 Å. Lysosomal structures such as multivesicular bodies, dense bodies, and phagolysosomes, as well as coated vesicles, are also frequently encountered. Neuritic microtubules and neurofilaments reach into the more proximal parts of the growth cone body. Clear vesicles identical in size to synaptic vesicles are not found in the growth cone (unless it emerges from an already formed synaptic nerve terminal), but clusters of 1500–2000 Å vesicles frequently occur underneath moundlike protrusions of the plasma membrane, usually in the more proximal region of the cone and near branch points of the nerve fiber (Figs. 14 and 18; Bunge, 1973). At the base of these mounds (Figs. 18 and 19), the plasmalemma forms a number of deep, branched invaginations whose orifices surround the mound usually in a circle (Pfenninger and Bunge, 1973, 1974; Pfenninger *et al.*, 1974). This platelike network of plasmalemmal invaginations compartmentalizes the mound contents including the vesicles from the growth cone cytoplasm. The vesicle-containing mounds are such frequent and characteristic structures of the nerve growth cone that they have even been taken for whole cones (Bodian, 1966; Del Cerro and Snider, 1968; Kawana *et al.*, 1971). However, they are not growth-cone-specific structures and are frequently present in a variety of growing cells such as fibroblasts, glia (Fig. 19), endothelial cells, and blastocysts *in vivo* as well as cultured *in vitro* (Pfenninger, 1973, unpublished observations). During their development, dendrites also bear growth cones (Hinds and Hinds, 1972; Vaughn *et al.*, 1974; Skoff and Hamburger, 1974) which are structurally comparable to their axonal counterparts and contain a similar set of organelles. A remarkable feature of the dendritic growth cone lies in the fact that it can receive synaptic inputs and form postsynaptic membrane specializations at a stage of development when it has not yet acquired dendritic morphology (Vaughn *et al.*, 1974; Skoff and Hamburger, 1974).

3.2. *Intrinsic Structure of the Growth Cone Membrane*

The growth cone plasma membrane, as seen in osmicated, uranyl-lead stained, thin-sectioned specimens, does not exhibit any unusual

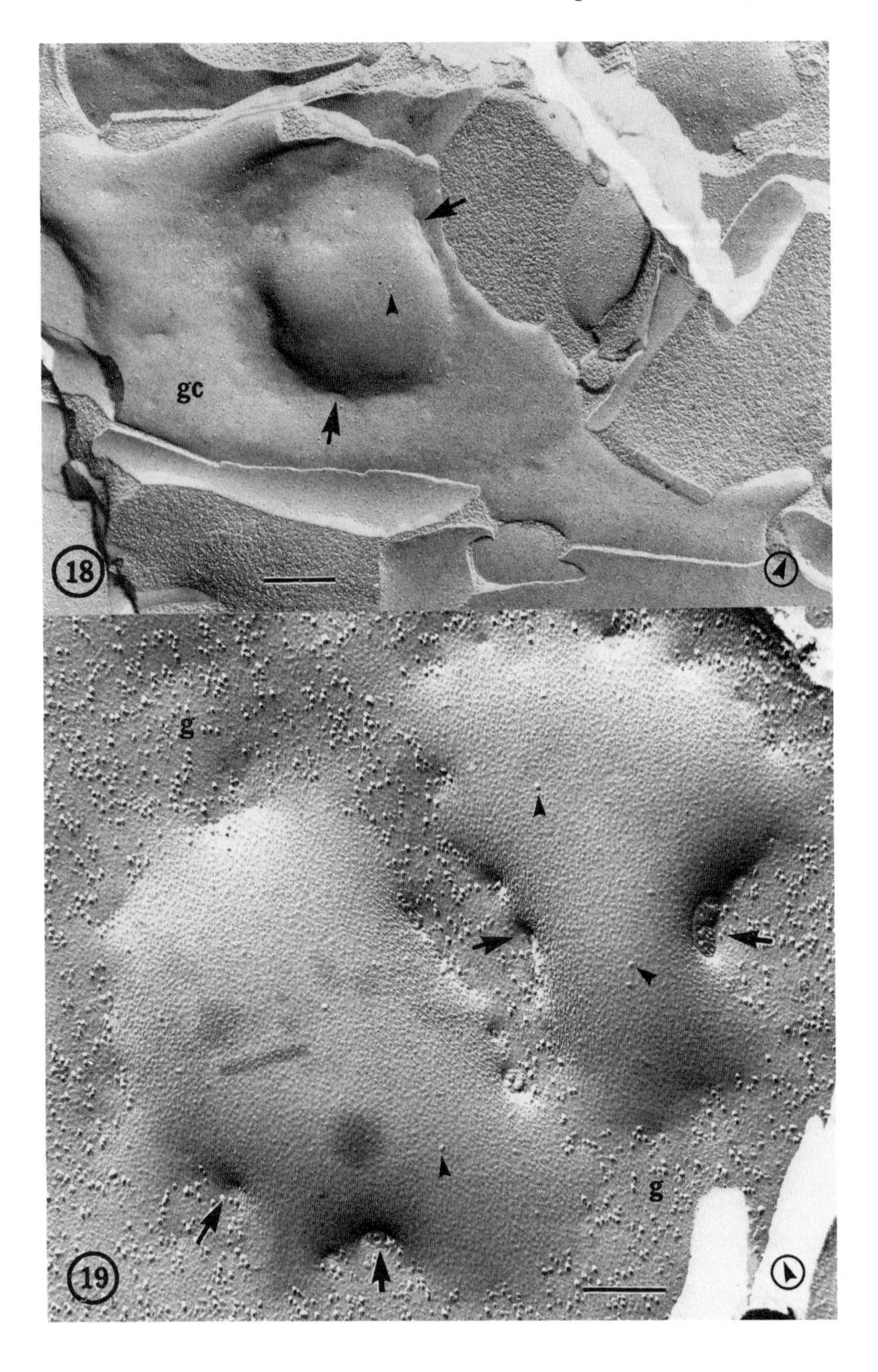
gc
18
g
g
19

features. However, a newly developed technique (Pfenninger, 1974; Pfenninger and Rinderer, 1975) has made it possible to freeze-fracture cultured nerve growth cones (Figs. 15 and 16) and to study the intramembranous properties of their plasmalemma (Pfenninger, 1972; Pfenninger and Bunge, 1974). The main results of this work are concerned with the distribution of intramembranous particles as follows. Regardless of the age of the culture (5–50 days), the particle density in the growth cone plasmalemma (cytoplasmic leaflet) is very low at $78.3 \pm 9.6/\mu m^2$ for rat spinal cord and $93.8 \pm 9.9/\mu m^2$ for rat olfactory bulb fibers (Figs. 16 and 18). In the outer leaflet, the counts amount to about $23/\mu m^2$ in the spinal cord and $32.6 \pm 9.0/\mu m^2$ in olfactory bulb cultures so that the total particle counts in the growth cone plasmalemma average $101/\mu m^2$ and $126/\mu m^2$, respectively. This unusually low particle density contrasts strikingly with the neuronal perikaryal membrane, which contains about 660 intramembranous particles per square micrometer (principal neuron from the superior cervical ganglion, grown for 24 h *in vitro*) in the inner leaflet alone (Fig. 17A). Random samples from the shafts of the neurites show that the particle counts in the inner leaflet increase from $53.3 \pm 3.3/\mu m^2$ (5-day-old cultures) to $527 \pm 60/\mu m^2$ (40-day-old specimens). These data are displayed in graphic form in Fig. 17B. The paucity and the gradual increase of the particle number in nerve fibers can also be observed in native spinal cord (*cf.* Fig. 20) taken from rat embryos at different stages of development. The picture which emerges from these measurements of particle density in the developing nerve fibers is as follows: (1) it is a function of the distance from the perikaryon, with distally decreasing figures, and (2) it is a function of the age of the neurite, the numbers increasing with time. At variance with this notion is the finding that the very young fibers, and in particular their most distal portions, exhibit a lower particle density than the growth cone. Overall, the lowest density values—i.e., just a few single intramembranous particles per square micrometer—are found in the plasmalemma of the mounds (Figs. 18 and 19) and the vesicles underneath (Fig. 20) in nerve growth cones as well as in nonneuronal cells (Pfenninger and Bunge, 1973, 1974; Pfenninger *et al.*, 1974).

←

Figs. 18 and 19. Moundlike, virtually particle-free plasmalemmal protrusions in a nerve growth cone (gc) in a rat spinal cord culture and in a cultured glial cell (g). In both cases, the inner leaflets of the plasma membrane are shown. The mounds are surrounded by dimples or pits (large arrows) which are openings of plasmalemmal invaginations (see text). Arrowheads, intramembranous particles. Magnifications ×51,000 (Fig. 18) and ×58,000 (Fig. 19), calibration 0.2 μm. Circled arrow, shadowing direction.

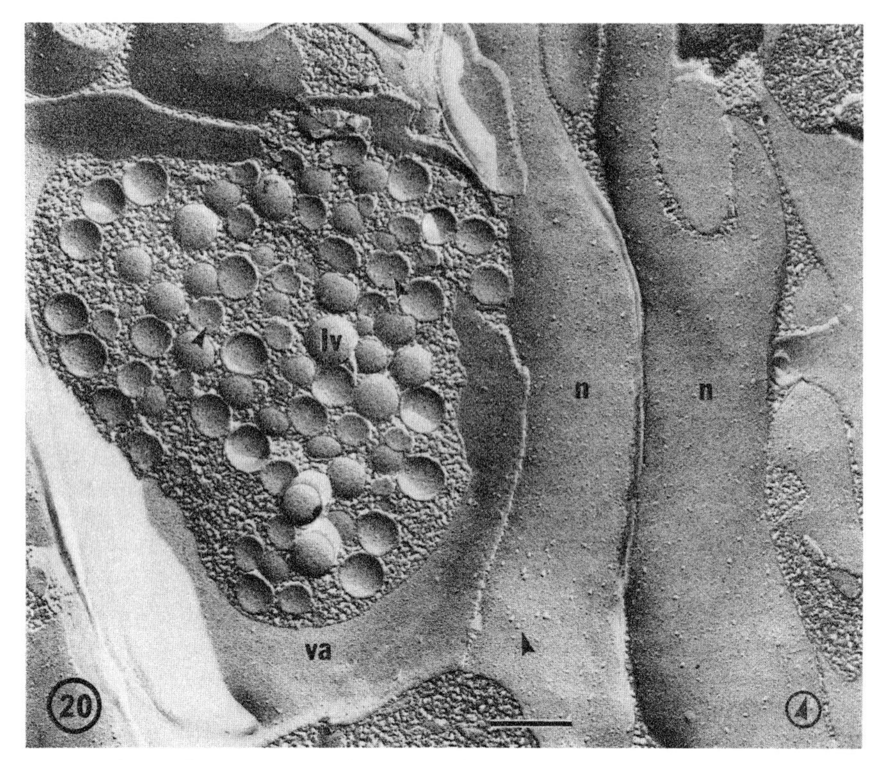

Fig. 20. Freeze-fracture picture of two nerve fibers (n, inner leaflets exposed) and a fiber varicosity (va) in a 14-day-old embryonic rat spinal cord. Note in the varicosity the cluster of large vesicles (lv) characterized by the virtual absence of intramembranous particles (two of which are marked by arrowheads) and the paucity of these particles in the accompanying nerve fibers. Magnification ×58,000, calibration, 0.2 μm. Circled arrow, shadowing direction.

The particle density gradient in the neuron is a very remarkable finding, first, because it dramatically illustrates regional differences in membrane structure in the same cell and, second, because it is maintained over a long period of time despite the well-known fluidity of the plasma membrane (for review, see Singer and Nicolson, 1972). Two explanations are possible: either there is a subplasmalemmal network or an intramembranous component which prevents the lateral diffusion of the intramembranous particles, or, perhaps in combination with the former, particle-free membrane is continuously added at the nerve tip and the equilibration of the particle density in the plasmalemma lags behind. It has been suggested that the particle-free mounds and vesicles are parts of a localized, distal membrane insertion mechanism (Pfenninger and Bunge, 1973; Pfenninger *et al.*, 1974) in agreement with the membrane growth hypothesis proposed earlier by Bray (1970, 1973). The occurrence of these vesicles near, and in connection with, the Golgi apparatus of growing neurons and along the neurite suggests that particle-free membrane "matrix" could be sequestered in the Golgi apparatus and shuttled from there to the peripheral mounds in the form of vesicles; insertion by fusion of these vesicles (at the mounds) would then leave particle-free regions in the plasma membrane. Since the mounds seem to occur mainly distally at the growth cone, such a mechanism would easily explain the observed particle density gradient in the developing neurite. Studies on the uptake of external tracer substances and on the labeling of the growing cell surface (Pfenninger and Bunge, 1973; Pfenninger *et al.*, 1974) have so far yielded results which are compatible with the proposed idea of membrane growth, but conclusive evidence is still lacking.

Whether this hypothesis of membrane expansion is correct or not, the changes in membrane structure clearly indicate that the growing neurites are first bounded by a partially incomplete membrane which matures slowly structurally and biochemically by the gradual insertion, or diffusion into this area, of intramembranous particles to reach adult levels of particle density (Pfenninger and Bunge, 1974). The growth cone itself, however, always exhibits immature plasma membrane.

3.3. *Growth Cone Membrane Cytochemistry*

Evidence for a distinct plasmalemmal structure and composition at the growth cone as revealed by freeze-fracturing makes the further study of membrane properties and the search for a distinct surface biochemistry of great interest. When applied to the growth cone, the BIUL method, bismuth iodide impregnation combined with uranyl-lead staining (which displays synaptic membrane specializations so well), stains only a faint surface coat layer which occasionally bears some longer stainable wisps; no membrane specializations can be detected (Figs. 21, 22 and 23). The cationic complexes of lanthanum nitrate, ruthenium red, and thorium dioxide reveal a heavier coat of negatively charged material on the surface of nerve growth cone and filopodia (James and Tresman, 1972). As these authors state, it is not clear, however, whether this coat has been enhanced artifactually by adsorption to the cell surfaces of substances from the culture medium. Carbohydrate labeling results obtained with the concanavalin A–peroxidase method (Bernhard and Avrameas, 1971) indicate that both the growth cone and the young nerve fiber are more or less uniformly

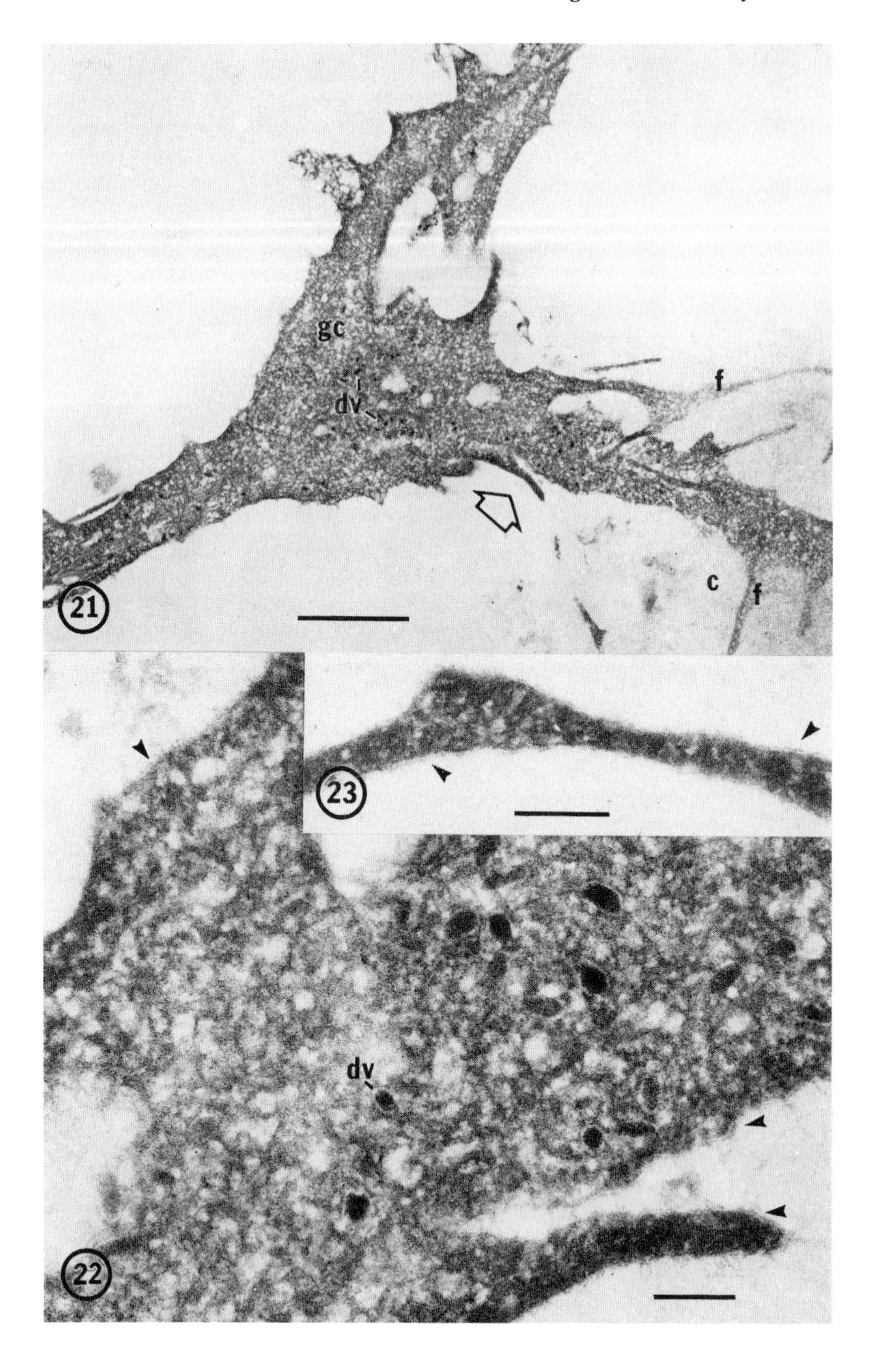
gc
i
dv
f
c
f
21
23
dv
22

Figs. 21, 22, and 23. Bismuth iodide impregnated and uranyl-lead stained (BIUL) growth cone in a rat superior cervical ganglion culture. Note in Fig. 21 the elaborate branching of this cone (gc) and its high content of large dense-core vesicles (dv). The fibrillar material in the lower right-hand corner of this electron micrograph is the collagen culture substrate (c). f, Filopodia. The arrow marks the area enlarged in Fig. 22. At higher magnification (Figs. 22 and 23), it can be seen that BIUL-positive membrane specializations are poorly developed in the growth cone. In particular, only very sparse external surface material is visible (arrowheads). Figure 23 shows an enlarged growth cone filopodium. The tramline pattern of unit membrane structure cannot be seen in these BIUL preparations. Magnifications ×16,000 (Fig. 21), ×59,000 (Fig. 22), and ×68,000 (Fig. 23); calibrations 1 μm (Fig. 21) and 0.2 μm (Figs. 22 and 23).

covered with the reaction product, although this is distributed in a spotty fashion (Fig. 24). The labeling can be prevented with the specific inhibitor α-methylmannoside, so that the presence of significant amounts of mannoside and/or glucoside on the membrane surface can be inferred (Pfenninger and M. B. Bunge, 1973, unpublished observations). The use of peroxidase as a marker, however, does not allow quantitative study of the labeled substrate so that possible subtle differences—e.g., between very young (particle-poorer) and more mature (particle-richer) neuronal plasma membrane—would not become apparent.* Further, indirect information on the growth cone surface chemistry comes from autoradiography studies on cultured nerve tissue (Fig. 25) employing [^{3}H]fucose, a sugar which is a reasonably specific precursor for membrane glycoproteins (fucosyl-glycolipid has not been reported to occur in nerve tissue). At the beginning of such pulse-chase experiments, the radioactivity is concentrated over the usually expanded neuronal Golgi apparatus but no silver grains can be found over peripheral parts of the outgrowing nerve fibers. However, as early as 1–2 h following a precursor pulse of 20 min, radioactivity can be located near the surface, probably over the plasma membrane, of the growth cones and distal fibers, about 1 mm away from the perikarya (Pfenninger, 1974, unpublished). Comparison with supporting cells in cultures indicates that the rate of incorporation in the rapidly expanding neurons is considerably higher than in other cells. The fate of incorporated fucose in these experiments is consistent with the data reported on other systems (Bennett *et al.*, 1974) and, in particular, with axoplasmic flow studies in developing and mature nervous tissue (Droz, 1973; Sjöstrand *et al.*, 1973; Crossland *et al.*, 1974*b*; Gremo *et al.*, 1974; Marchisio *et al.*, 1975). The novel aspect revealed by our data and by the comparable studies by Crossland *et al.* (1974*b*) and by Gremo *et al.* (1974) is that, even at a time when the nerve fiber has not formed or started to form a synaptic

* cf. Note Added in Proof (p. 357).

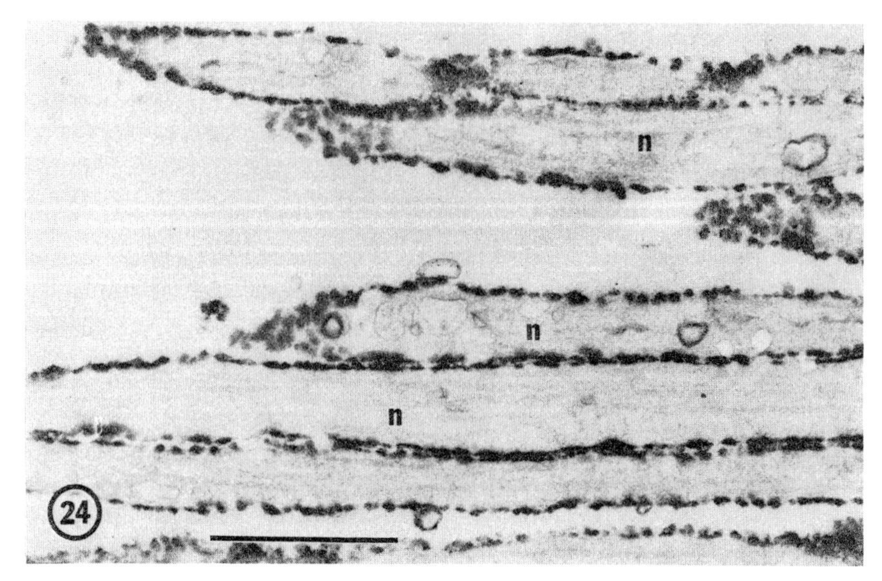

Fig. 24. Concanavalin A–peroxidase labeled nerve fibers (n) in a rat spinal cord culture. The live cultures were first exposed to concanavalin A, then rinsed and left to survive for up to 40 min at 37°C. After aldehyde fixation, they were exposed to peroxidase, which was subsequently cytochemically visualized (*cf*. Bernhard and Avrameas, 1971). Note the spotty, but more or less even, distribution of the label on the nerve fibers. In control experiments with the same protocol except that the incubation with concanavalin A was done in the presence of α-methylmannoside, virtually no reaction product was found on the cell surfaces. Magnification $\times 50{,}000$, calibration 0.5 μm.

contact (in our cultures no target cells are available) and the neuritic plasmalemma is immature (in all three studies the axons are in the process of elongation), significant amounts of fucosyl-glycoprotein are transported at a fast rate to the nerve tip and are probably incorporated into its particle-poor membrane (at present, it is not known whether the insertion into the plasmalemma is confined to the growth cone region).

4. THE FORMATION OF THE SYNAPTIC JUNCTION

A variety of publications on synaptogenesis can be found in the literature (e.g., see Glees and Sheppard, 1964; Aghajanian and Bloom, 1967; Bunge *et al.*, 1967; Bodian, 1968; Larramendi, 1969; Akert *et al.*, 1971; Altman, 1971; Adinolfi, 1972*a*,*b*; Bloom, 1972; Oppenheim and Foelix, 1972; Hayes and Roberts, 1973; Stelzner *et al.*, 1973; Jones *et al.*, 1974). Rather than focusing on the events immediately following the initial contact between axon and target, these studies deal mainly with the later event of the quantitative appearance and the maturation of already present synaptic membrane specializations. New data originate from observations in a culture system utilizing principal neurons isolated from the 18–21 day fetal rat superior cervical ganglion (Rees *et al.*, 1976). The neurons are seeded into a culture dish to which, 48 h later, small explants of 15-day fetal rat thoracic spinal cord are added as a source of presynaptic elements. In the establishment of the culture, both the target principal neurons (Fig. 26) and the presynaptic spinal cord axons have been deprived of previous synaptic contact and display no retained synaptic specializations at the electron microscope level.

4.1. *Initial Encounter of Synaptic Partners*

The initial encounter of the growth cone with the appropriate target cell can be observed under the phase contrast light microscope. As the spinal cord neurites advance across the collagen, the widely exploring and apparently randomly moving filopodia of the growth cones make and break contact with the surface of encountered ele-

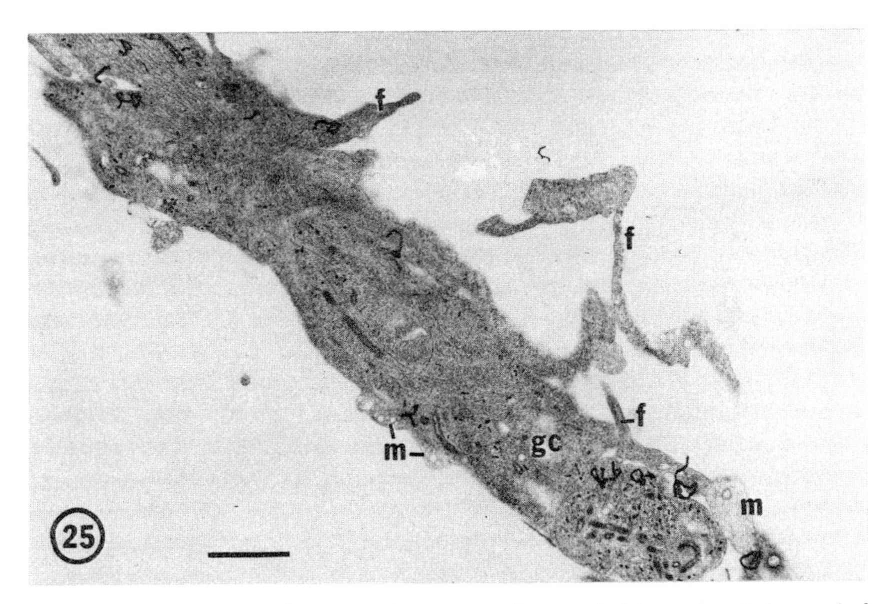

Fig. 25. Autoradiogram of a group of nerve growth cones (gc) in a rat superior cervical ganglion culture which was pulsed with [^{3}H]fucose for 135 min and chased for 110 min. Note the numerous silver grains, almost all of which are found over, or in the immediate neighborhood of, the plasma membrane. f, Filopodia; m, moundlike protrusions containing numerous large clear vesicles. Magnification ×11,000, calibration 1 μm.

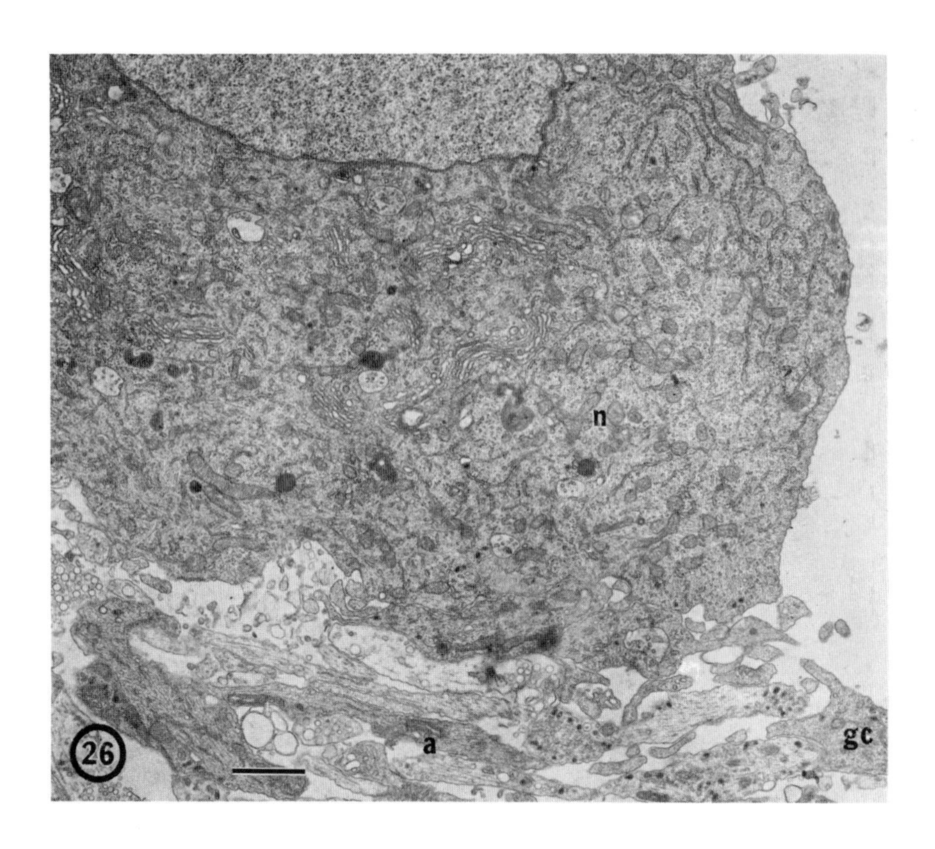
n
a
gc
26

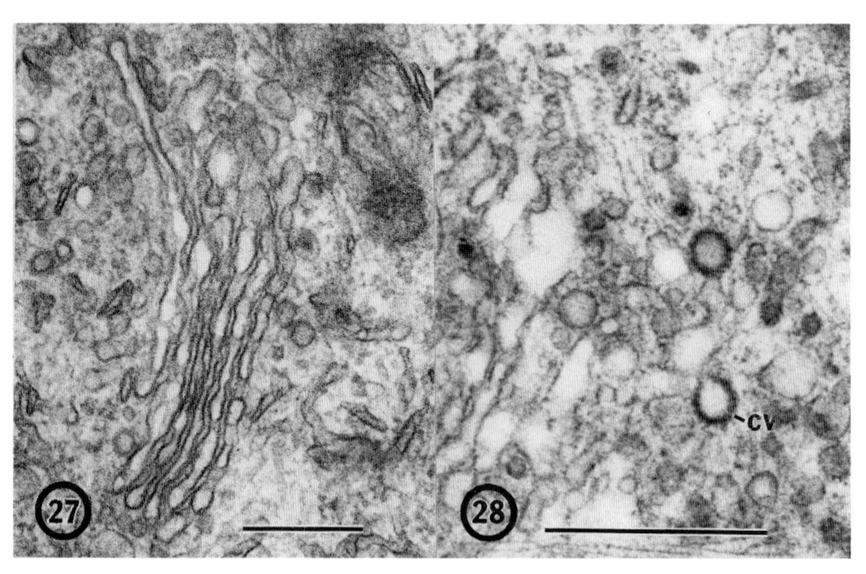
27
28
cv

Fig. 26. Low-power electron micrograph of a cultured isolated superior cervical ganglion principal neuron (n), fixed a few hours after its initial contact by ongrowing spinal cord neurites. Their axons (a) and growth cones (gc) can be seen lying along the lower part of the neuron. Magnification ×9800, calibration 1 μm.

Figs. 27 and 28. Golgi apparatus of an uncontacted (Fig. 27) and another superior cervical ganglion principal neuron which had been contacted by spinal cord neurites a few hours previously (Fig. 28). The Golgi cisternae in the contacted neuron are more extensive and dilated, and an increased number of large coated vesicles (cv) can be seen budding from them. The cell in Fig. 28 has been block-poststained with ethanolic phosphotungstic acid to enhance the contrast of the vesicle coats. Magnifications ×33,000 (Fig. 27) and ×60,000 (Fig. 28), calibration 0.5 μm.

ments. As mentioned in the introduction, such neurites can pass within 5–10 μm of the target neuronal surface, but if no physical contact occurs neuritic growth continues (Fig. 1). Usually a single filopodium makes the initial contact. Within 3–4 min, the undulations of the velamentous growth cone process and the vigorous movement of the other filopodia cease. They appear to withdraw into the growth cone, while the initial contact is maintained and the body of the growth cone moves closer to the target cell. For periods ranging from 10 to 40 min, little activity is observed, but following this quiescent period some nerve growth cones are seen to become active again, to withdraw from the surface of the target cell, and to continue their growth in a slightly altered direction.

Where contact is maintained, electron microscopic observations of serial sections indicate that the first morphological change in the presynaptic element is the close and extensive application to the surface of the neuron of the previously motile growth cone processes in the form of a sheet (see Fig. 29). Numerous small areas of the applied membrane take up a close spatial relationship to the future postsynaptic plasma membrane, leaving an intercellular cleft of only 80 Å. No cytoplasmic densities are present at these points, but some fuzzy material can be seen in the gap.

4.2. *Postsynaptic Response.*

The next structural change is observed in the postsynaptic neuron. Within 6–12 h after initial contact, the Golgi apparatus is increased in size by comparison with uncontacted neighboring neurons. Cisternae are dilated, and there is significant increase in the number of large 100 Å coated vesicles in their vicinity,* which reaches

* It should be noted that in all cultured principal neurons, whether contacted or not, a second, smaller class of coated vesicles 500–700 Å in diameter is seen near the Golgi apparatus; these are also found in relation to lysosomal structures (*cf.* Friend and Farquhar, 1967).

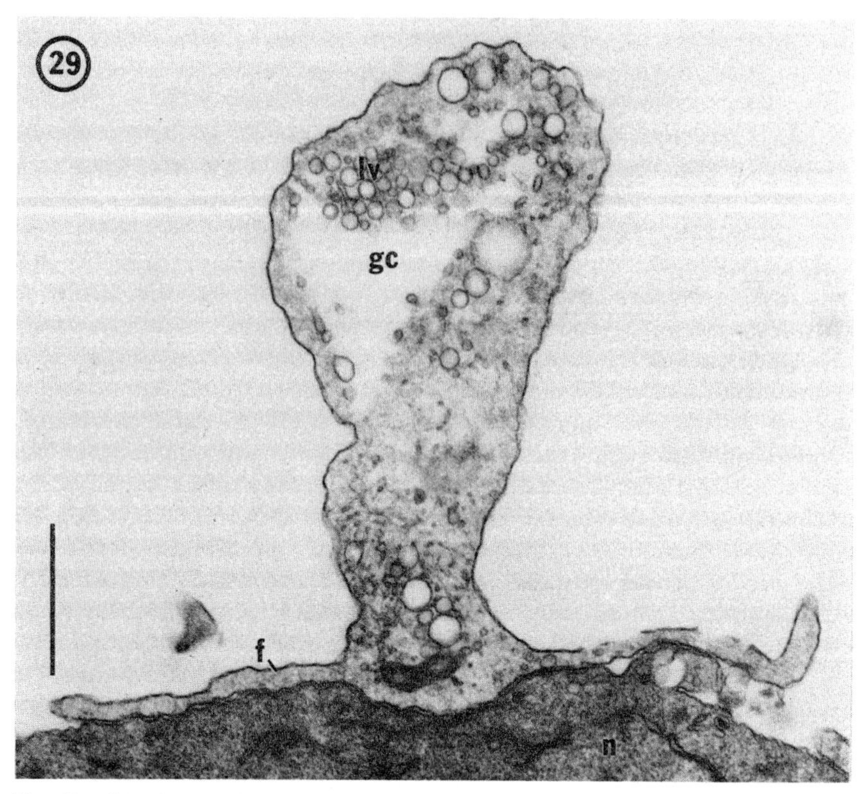

Fig. 29. Distal part of a spinal cord neuritic growth cone (gc) is seen making initial contact with the surface of a neuronal soma (n) in culture. Serial sections of this cone showed the processes (f) to have a flattened, platelike form over some 20 μm^2 of cell surface (*cf.* Rees *et al.*, 1976). The membranes of the opposing elements were separated by a minimum cleft width of approximately 80 Å. Such close apposition occurred at numerous points, between which the membranes curved away from each other. lv, Large clear vesicles. Magnification ×20,000, calibration 1 μm.

a maximum after 72 h (Figs. 27 and 28). These vesicles are also seen in the neuronal cytoplasm, and, within 6–12 h, they appear to be especially frequent at or near the contacted plasmalemma (*cf.* Altman, 1971). Increase of the vesicle count in the Golgi area suggests that, following contact, the 100 Å coated vesicles may be formed in the Golgi apparatus and then shuttled to the contacted surface area, where they would join the plasma membrane by fusion (Rees *et al.*, 1976). This hypothesis, which had already been suggested by Altman (1971) and by Stelzner *et al.* (1973), is supported by the finding that the extracellular tracer horseradish peroxidase is not taken up into such vesicles in the perikaryon and by the presence in the plasmalemma (after 16 h or more of contact) of small membrane patches with a dense cytoplasmic coating of about 110 Å thickness similar to that of the rim of the coated vesicles (Rees *et al.*, 1976).

During these early stages of synaptic development, the postsynaptic density precedes in its appearance all the other synaptic membrane specializations (Figs. 30, 31, and 32) and is increased with time both in length and in width (*cf.* Glees and Sheppard, 1964; Oppenheim and Foelix, 1972; Stelzner *et al.*, 1973). The fusion with the plasmalemma of additional 1000 Å coated vesicles may play an important role in this process, but it is not clear whether the increase in width of the future postsynaptic density represents an expansion of the coat laid down by the coated vesicles. Full extent of the postsynaptic density is found only late in synapse development, when all the other synaptic features are also present. 1000 Å coated vesicles seem to continue to fuse with the postsynaptic membrane until after the completion of synapse formation. Fusion images have been observed during the 3-week maintenance period of these cultures and, occasionally, at principal cell synapses in the intact adult superior cervical ganglion (Rees *et al.*, 1976).

4.3. *Changes in the Intercellular Space*

When the proliferation of the Golgi apparatus takes place, and coated vesicles appear in the contact region, no signs of a synaptic junctional complex can be detected, although the apposition between pre- and postsynaptic membranes is very intimate. As an extension of the study by Rees *et al.* (1976), we measured the cleft width at different stages of synaptic development and found that the initial gap is only 81 ± 2.3 Å wide (average of ten plus or minus standard error of the mean). This contact is considerably closer than in the mature synapse. The width of the intercellular space increases during synapse development to reach 129 ± 1.2 Å, when the first signs of a subjunctional density are clearly evident (but no synaptic vesicles are present), and 182 ± 1.2 Å at the time of the arrival of the first synaptic vesicles. When traces of presynaptic densities are detectable, the postsynaptic density is clearly visible, and synaptic vesicles have appeared in greater number, the cleft measures 222 ± 2.1 Å. The mature synapse (in 3-week-old cultures), by comparison, has a cleft width of 252 ± 3.1 Å. The widening of the intercellular cleft during the early development of the synapse seems to be in disagreement with the data of Jones *et al.* (1974) which indicate a reduction in the average cleft width of synapses in guinea pig cerebral cortex (from about 260 Å at birth to

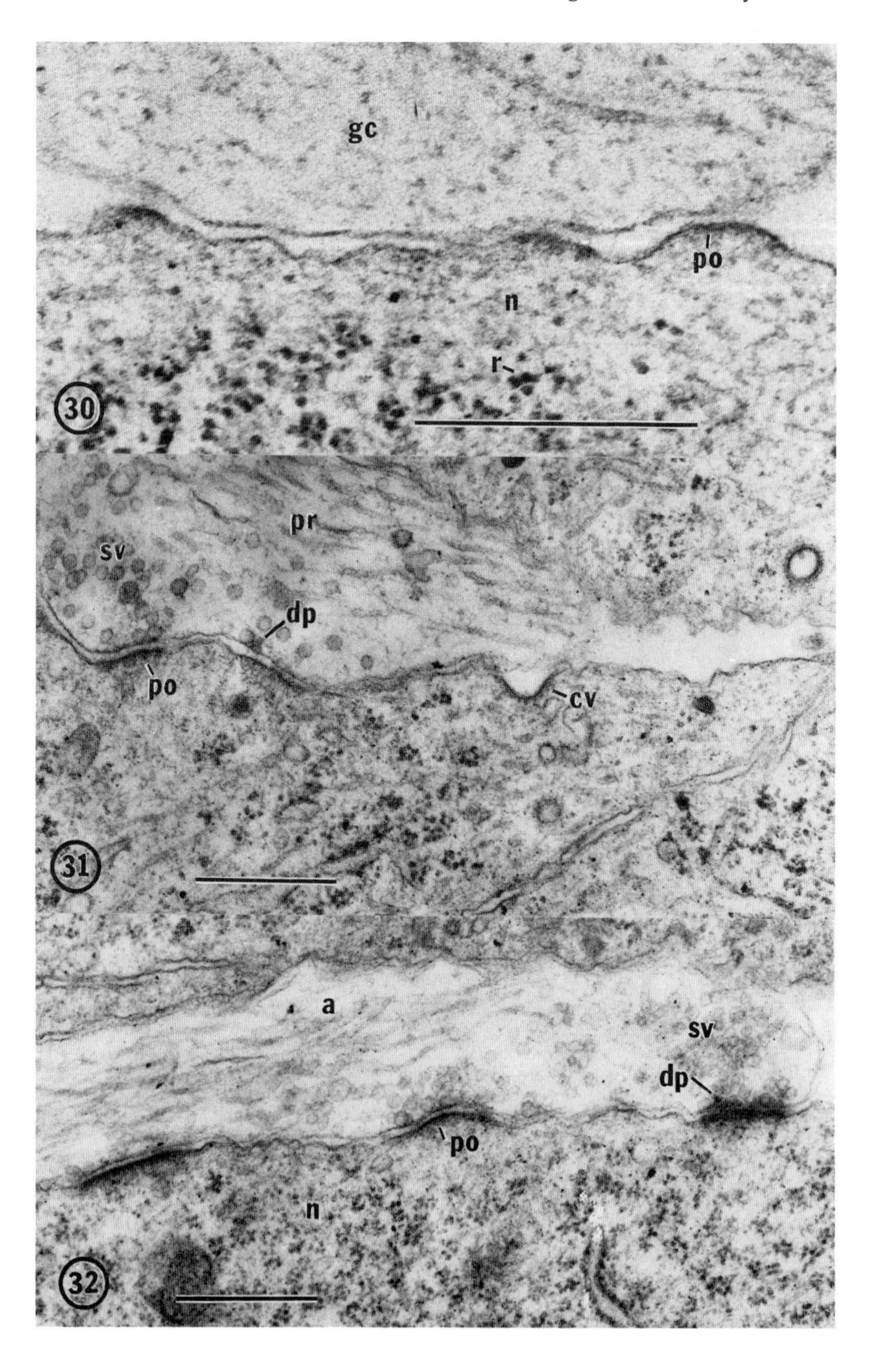
gc
po
n
r
30
pr
sv
dp
po
cv
31
a
sv
dp
po
n
32

225 Å at 11–14 days after birth; measured in E-PTA-treated synapses between the bases of presynaptic dense projections and postsynaptic density, across both synaptic membranes). However, these figures are calculated from data based on a population of different types of young synapses rather than on a homogeneous group and therefore may not be directly comparable to our results. Furthermore, the measurements by Jones *et al.* (1974) originate to a large extent from junctions which exhibit already partly or fully developed presynaptic dense projections and thus are more advanced than the contacts described here. In similarly mature material, Bloom (1972) did not find changes in synaptic cleft width.

Jones *et al.* (1974) reported that the postsynaptic layer of synaptic cleft material appears somewhat earlier than its presynaptic counterpart when visualized with the E-PTA method. As mentioned above, this observation stresses the integrity of, if not a difference between, the two cleft layers and is consistent with the general picture of early postsynaptic and delayed presynaptic differentiation.

4.4. *Presynaptic Developments*

Structural changes are not apparent in the presynaptic growth cone prior to the appearance of the subjunctional entity. However, once this is established, gradual replacement of growth cone organelles by synapse-typical elements occurs (Figs. 31, 32, 33, and 34). Small groups of synaptic vesicles appear in the interior of the growth cone near the membranous reticulum and gather in the area apposed

←

Figs. 30, 31, and 32. Stages of synapse formation between spinal cord neurites and isolated superior cervical ganglion neurons in culture. In Fig. 30, part of a growth cone (gc) can be seen in close association with a neuronal perikaryon (n). A first indication of postsynaptic densities (po) is evident. r, Ribosomes. Magnification ×80,000. Figure 31 shows a more advanced stage with, at the left, a well-developed postsynaptic density (po), a widened intercellular space, and a cluster of vesicles (sv) presynaptically (pr). A less mature contact can be recognized to the right, where the postsynaptic density is only partly formed but a small presynaptic density (dp) is already evident. Still farther to the right, a coated vesicle (cv) is in continuity with the postsynaptic plasma membrane. Note that, in the neurite, growth cone organelles have been replaced by typical nerve terminal elements. Magnification ×40,000. Figure 32 shows three synaptic contacts between the same axon (a) and neuronal soma (n) at different stages of maturation. Note especially the increased presynaptic differentiation (presynaptic densities and synaptic vesicles) from the left to the right. This picture indicates that the formation of multiple synaptic junctions between the same pre- and postsynaptic elements need not be synchronized. Magnification ×40,000, calibration in all three figures 0.5 μm. Figure 31 from Rees *et al.* (1976).

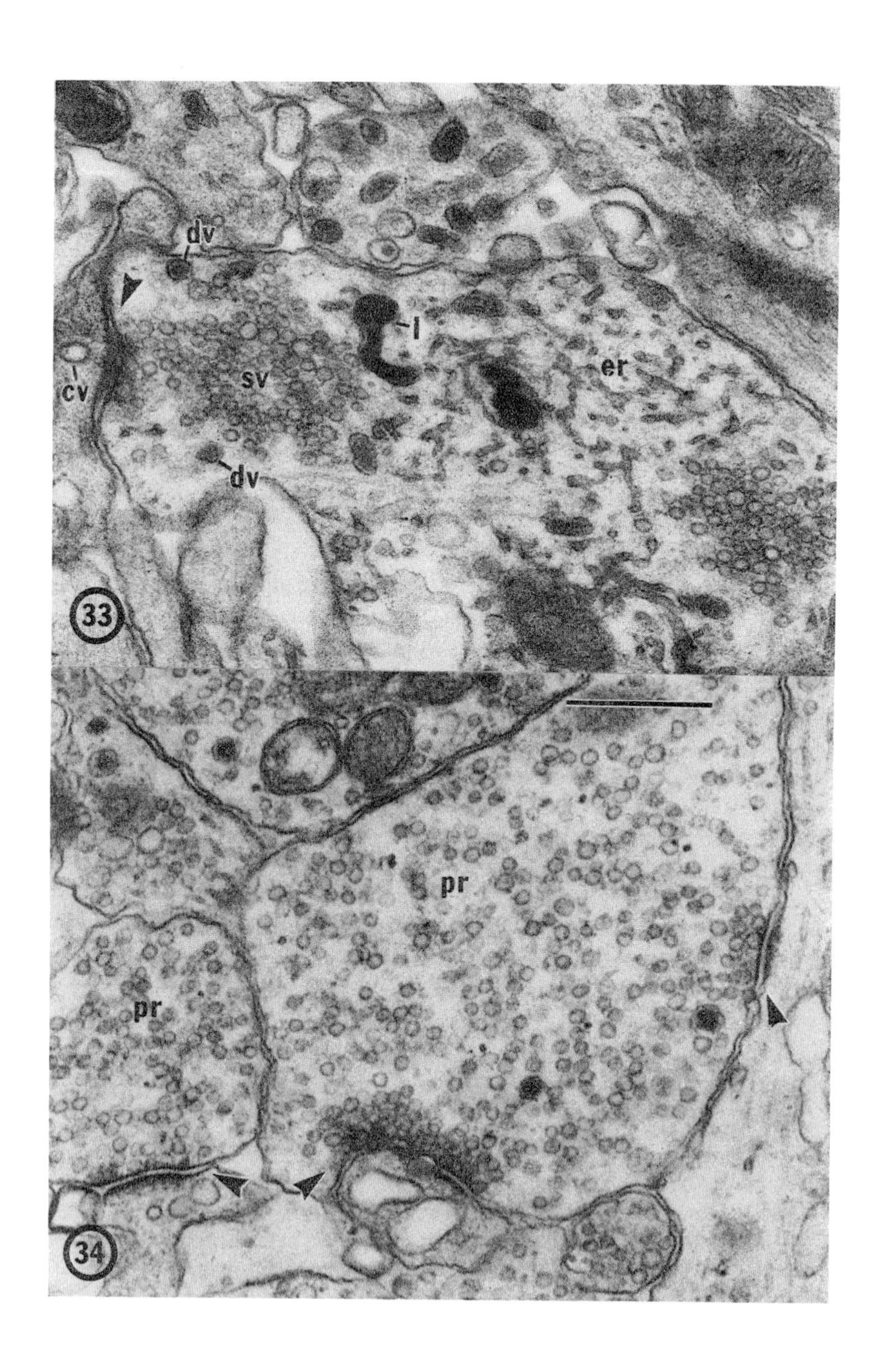
dv
l
er
sv
cv
dv
33
pr
pr
34

to the early postsynaptic membrane specialization (Figs. 32, 33, and 34). Their origin remains uncertain, although images suggestive of their budding off from branched and smooth endoplasmic reticulum in the interior of the growth cones are seen. Portions of this endoplasmic reticulum and most of the gathering synaptic vesicles are ZIO positive (Kawana *et al.*, 1971). Some 24–36 h following the initial contact, when synaptic vesicles, mitochondria, and the postsynaptic density are already present, presynaptic dense projections also become visible (Figs. 33 and 34). Delay in the time of the appearance of the presynaptic dense projections could be due to a difficulty of visualization since their stainability in osmicated, uranyl-lead-stained preparations is known to be inferior to that of the subjunctional density (*cf.* Pfenninger, 1973). E-PTA staining of these cultures, however, has shown a distinct postsynaptic specialization to be present before presynaptic dense projections appear (Rees *et al.*, 1976), a finding largely consonant with that of Bloom, Akert, and Jones and their collaborators in a variety of developing nerve tissues (Aghajanian and Bloom, 1967; Akert *et al.*, 1971; Adinolfi 1972*b*; Bloom, 1972; Jones *et al.*, 1974). Whether the presence of the synaptic vesicles is a prerequisite for the formation of the dense projections is not known; they seem to appear almost simultaneously (Bunge *et al.*, 1967). During synapse maturation, the size and the number of dense projections per synapse increase (Adinolfi, 1972*b*; Bloom, 1972; Jones *et al.*, 1974), suggesting gradual enlargement of the total synaptic zone.

The role of the presynaptic large dense-core vesicles remains uncertain. In the advancing spinal cord neurites, they are invariably present in the more proximal parts of the growth cones, scattered among the branched membranous reticulum (Figs. 14, 21, 22), and some appear to relate to lysosomes. They are absent from the motile filopodia, and they are not seen in the filopodia or other growth cone portions which are in contact with the postsynaptic neuron until *after* the subjacent density is established (Rees *et al.*, 1976). However, as the clear synaptic vesicles gather at the forming synaptic site, the counts of the large dense-core vesicles increase from 0 to 1.45 ± 0.26 per section within

←

Figs. 33 and 34. Transformation of the growth cone into a fully matured presynaptic ending in culture. Whereas the formation of the junction (arrow) in Fig. 33 is almost completed, some of the elaborate endoplasmic reticulum (er) typical of growth cones, lysosomes (l), and many dense-core vesicles (dv) are still present in this terminal. cv, Coated vesicles. The nerve endings (pr) in Fig. 34, by contrast, are filled with an abundance of synaptic vesicles, and only a few dense-core vesicles can be seen. Note the three almost mature synaptic contacts (arrows). Magnification ×40,000, calibration 0.5 μm. Figure 34 from Rees *et al.* (1976).

5000 Å of the synaptic cleft (average counts plus or minus standard error of the mean). When the presynaptic dense projections make their appearance, this figure reaches a maximum of 1.9 ± 0.18 (*cf.* Fig. 33). In the mature synapses present in older cultures, the same average number is greatly reduced again to 0.35 ± 0.11, a fraction of the maximal value (these quantitative data were obtained from the experimental material accumulated by Rees *et al.*, 1976). The precisely timed increase in the number of large dense-core vesicles in the immediate neighborhood of the developing junction (*cf.* Oppenheim and Foelix, 1972) suggests that they may play a specific role in the process of synapse formation (*cf.* also Pfenninger *et al.*, 1969).

4.5. *Five Steps of Synapse Formation*

In summary, it has been determined that the complex mechanism of transformation of the nerve growth cone into a synaptic nerve terminal is completed within 24–48 h *in vitro* and that the following, morphologically distinct stages can be recognized in the culture system:

1. Initial contact with scattered sites of close, 80 Å apposition of growth cone and lamellopodia. Proliferation of postsynaptic Golgi apparatus.
2. Postsynaptic density present, but no synaptic vesicles presynaptically. Synaptic clefts enlarged to 130 Å.
3. Few synaptic vesicles and postsynaptic density present. Synaptic cleft enlarged to 180 Å. Large numbers of large dense-core vesicles.
4. Presynaptic dense projections appearing; many synaptic vesicles and postsynaptic density present. Maximal number of large dense-core vesicles within 5000 Å of synaptic cleft. Cleft width 220 Å.
5. Mature synapse with fully developed pre- and postsynaptic densities and wide, 250 Å synaptic cleft. Number of large dense-core vesicles decreased.

Comparative studies by Bloom and collaborators (Bloom, 1972) on the first occurrence of synaptic junctions and on electrophysiological and pharmacological developments indicate that the appearance of synaptic paramembranous densities parallels the awakening of chemical transmission in cerebral and cerebellar cortex and, probably, in all nervous tissues.

5. DISCUSSION: THE SPECIFIC NATURE OF THE RECOGNIZING MEMBRANE

Let us consider the known and hypothetical events leading to the formation of a synaptic contact. First of all, the growth cone must find its way to its target cell using a mechanism of *guidance* which, from observations of nerve tips in culture, is probably based on membrane surface properties as much as the recognition mechanism proper. Whether the mode of guidance is represented by differential adhesion between growth cone and cellular environment (possibly in the form of a gradient: *cf.* Sperry, 1963), perhaps similar to the haptotactic mechanism as proposed by Carter (1967), is not known. Evidence for alternate mechanisms is not available, either. Following probing, interaction between membranes of the growth cone and an appropriate target may cause conformational changes of interacting surface molecules and/or the release of signal substances which result in the subsequent formation of pre- and postsynaptic membrane specializations. This mechanism of *recognition** proper which probably takes place during the early close association between the growth cone and the target cell evidently is not a unilateral event. It has been noted (Pfenninger, 1973) that the formation of any type of symmetrical or asymmetrical intercellular junction where specializations occur at exactly juxtaposed membrane sites requires some sort of mutual recognition process (probably less specific than that required for the synapse), involving the transfer of signals from cell to cell. The data of Rees *et al.* (1976) on synapse formation indicate that prior to changes in the growth cone the postsynaptic element reacts to the apposition of prospective pre- and postsynaptic membranes by a proliferation of the Golgi apparatus, fusion of coated vesicles with or near the future postsynaptic membrane, and formation of the first traces of a postsynaptic density. On the presynaptic side, this process appears more complex: Following interaction of the growth cone with a target cell, it may be necessary for a signal to travel retrogradely to the perikaryon, where the bulk of protein synthesis occurs. Newly synthesized material would then have to be transported down the axon to the nerve terminal for the formation of presynaptic membrane specializations. (Is the enhanced occurrence of large dense-core vesicles a sign of this process?) Time occupied by such retrograde and anterograde transport could explain

* One might define *cellular recognition* as selective interaction between cells which causes specific morphological and/or functional changes (such as formation of a junction) in one or both of the interacting partners.

why developments on the presynaptic side lag behind the postsynaptic event.

Gradual local insertion of membrane components forms the synaptic membrane specializations, enhances the thickness of synaptic cleft material, and increases the particle density in pre- and postsynaptic membranes. It is remarkable that the locally inserted and concentrated junctional components do not move away from the prospective synaptic site by lateral diffusion; undoubtedly, with the incorporation of new constituents at the synaptic site, membrane structure is further altered in such a way that the loss of these new elements is prevented. During synapse maturation, the function of both outer membrane "coats"—i.e., of the cleft material—is certainly changed. Recognition is no longer required, but *maintenance* of the synaptic contact (via the exchange of "trophic" factors?) and its adherence become essential, and the architecture of the cleft substance may possibly be altered to facilitate diffusion of the neurotransmitter across the intercellular space. As we have seen above, adhesiveness of the synaptic junction relies on the external membrane surface materials filling the synaptic cleft. The experiments using treatment with high ionic strength and with proteolytic enzymes have shown the involvement of polar groups of both types (polyionic binding) and of at least one trypsin-sensitive cleft component.

The membrane surface of the advancing growth cone has proved to be devoid of any such specializations. Overall, it becomes evident that there is a total difference between the growth cone during its approach to the postsynaptic target and the final product, the apposed presynaptic nerve terminal (*cf.* Figs. 2 and 14). Not only do gross morphology and the sets of organelles of the two elements differ, but there are also clear distinctions in their plasma membrane properties which are of great importance in synaptogenesis. It should be noted that there is also a marked difference between the growth cone membrane and the nonspecialized plasma membrane, the only clearly distinguishing parameter known so far being the density of intramembranous particles. The various membrane properties are summarized in Table I.

It is surprising that the plasmalemma of the growth cone, participating as it does in the formidable task of guidance and recognition, appears in freeze-fracture as a morphologically simple, immature membrane, in structure similar only to the myelin sheath (Branton, 1967). As we have pointed out before (Pfenninger and Bunge, 1974), complex membranes with high protein content have generally been found to be rich in intramembranous particles, and, in certain cells, a correlation has been demonstrated between the regional distribution of such particles and glycoproteins (Tillack *et al.*, 1972). Thus the suggestion that

Table I. Properties of Nonspecialized Neuronal Plasma Membrane, Growth Cone Plasmalemma, and Presynaptic Membrane

	Nonspecialized plasmalemma	Growth cone plasmalemma	Presynaptic membrane
Intramembranous particles per μm^2 [a]	660	86	735
External membrane "coat"			
Thickness[b]	~60 Å	40 Å	125 Å
BIUL stainability	++	+	++++
Cytoplasmic membrane specializations	Thin uniform layer	No distinct specializations	Presynaptic dense projections

[a] Density in the inner leaflet.
[b] One-half of width of intercellular space.

glycoproteins function as key molecules in the recognition process (e.g., see Roth *et al.*, 1971, and this volume; Barondes, this volume) as well as in other mechanisms where surface interactions lead to changes in cellular behavior presents an apparent contradiction to our finding of low intramembranous particle content in the growth cone. Its resolution may lie in one or more of the following possibilities:

1. Glycoproteins are virtually absent from growth cone membranes and thus are not important for recognition.
2. Glycoproteins are present but are not related to, or represented as, intramembranous particles because (a) their peptides do not permeate the plasmalemma and/or (b) their peptides (which *per se* are probably too small to form 80 Å particles) do not aggregate with other intramembranous proteins to form particles.
3. Gangliosides whose lipid tails are probably not seen as special structures in freeze-fracture replicas represent a large fraction of the carbohydrate-containing membrane components and play an important role in the recognition mechanism.

It seems possible to exclude the first proposition for two reasons. The presence of carbohydrate on the growth cone surface is suggested by its acidophilic staining properties (James and Tresman, 1972) and by the specific binding to it of concanavalin A (Pfenninger and M. B. Bunge, 1973, unpublished observations). In addition, the rapid incorporation of [^{3}H]fucose into the growth cone membrane (Pfenninger, 1974, unpublished) indicates the presence there of fucosyl-glycoprotein (the brain is not known to contain significant amounts of fucosyl-glycolipid). A judgment of the third possibility cannot yet be made, but our data could be satisfactorily explained by the second possibility, the presence in the growth cone membrane of glycoproteins not related to intramembranous particles.

It is not clear whether our present concepts of the paucity of intramembranous particles and lack of other, cytochemically demonstrable specializations in growth cone plasmalemma represent prerequisites designed for the neural recognition mechanism or whether they are just accompanying features consequent to the local addition of new particle-free membrane according to our growth hypothesis (Pfenninger *et al.*, 1974; *cf.* Bray, 1970, 1973). It is likely that the crude methods used so far for the investigation of growth cone membrane properties are inadequate to reveal the expected specific recognition features. However, we can say today that the growth cone is bounded by immature membrane, which may bear further distinctive, as yet unrecognized properties, especially on its surface.

It was our endeavor in this chapter to describe some of the phenomena related to neural guidance and recognition and some of the properties of the membranes directly involved in this process. A thorough understanding of the crucial mechanisms will have to be based on our knowledge of the nature of, and the interactions between, the surface molecules of the apposed membranes of two meeting cellular processes. Obviously, we are still far from this goal and further extensive and greatly refined studies on cellular behavior, growth cone biochemistry, and cytochemistry (perhaps utilizing immunological techniques) will be required for future progress in this field.

Acknowledgments

The authors wish to express their gratitude to Drs. Richard P. and Mary B. Bunge of the Department of Anatomy, Washington University School of Medicine, with whose collaboration most of the work on nerve growth cones and synapse development was carried out. Some of the related concepts presented here have evolved during discussions with the Drs. Bunge. Dr. Richard P. Bunge was Rosemary P. Rees's thesis advisor on her work on synapse formation in culture. Helpful criticism and suggestions concerned with the improvement of the manuscript were offered by Dr. W. M. Cowan of the Department of Anatomy, Washington University School of Medicine, and by Drs. George E. Palade and Sally Zigmond of the Section of Cell Biology, Yale University School of Medicine, and are herewith gratefully acknowledged. It is Karl H. Pfenninger's pleasure to thank the staff of his present host institution, the Section of Cell Biology, Yale University

School of Medicine, for assistance without which the completion of this chapter would not have been possible.

The following publishers gave their permission to reproduce figures which had previously appeared in their journal or books: Academic Press, New York; Chapman and Hall, London; Gustav Fischer Verlag, Stuttgart; Rockefeller University Press, New York.

This work was supported by a fellowship of the Swiss Academy for Medical Science awarded to Karl H. Pfenninger and United States Public Health Service Grant No. NS-09923 awarded to Dr. Richard P. Bunge.

6. REFERENCES

Adinolfi, A. M., 1972*a*, Morphogenesis of synaptic junctions in layers I and II of somatic sensory cortex, *Exp. Neurol.* **34**:372.

Adinolfi, A. M., 1972*b*, The organization of paramembranous densities during postnatal maturation of synaptic junctions in the cerebral cortex, *Exp. Neurol.* **34**:383.

Aghajanian, G. K., and Bloom, F. E., 1967, The formation of synaptic junctions in developing rat brain: A quantative electron microscopic study, *Brain Res.* **6**:716.

Akert, K., and Pfenninger, K., 1969, Synaptic fine structure and neural dynamics, in: *Cellular Dynamics of the Neuron* (S. H. Barondes, ed.). pp. 245–260, I.S.C.B. Symposium, Paris, Academic Press, New York.

Akert, K., Moor, H., Pfenninger K., and Sandri C., 1969, Contributions of new impregnation methods and freeze-etching to the problem of synaptic fine structure, in: Mechanisms of Synaptic Transmission (K. Akert and P. Waser, eds.), *Prog. Brain Res.* **31**:223.

Akert, K., Moor, H., and Pfenninger, K., 1971, Synaptic fine structure, in: *First International Symposium on Cell Biology and Cytopharmacology* (F. Clementi and B. Ceccarelli, eds.), pp. 273–290, Advances in Cytopharmacology, Vol. 1, Raven Press, New York.

Akert, K., Pfenninger, K., Sandri, C., and Moor, H., 1972, Freeze-etching and cytochemistry of vesicles and membrane complexes in synapses of the central nervous system, in: *Structure and Function of Synapses* (G. D. Pappas and D. P. Purpura, eds.), pp. 67–86, Raven Press, New York.

Altman, J., 1971, Coated vesicles and synaptogenesis: A developmental study in the cerebellar cortex of the rat, *Brain Res.* **30**:311.

Attardi, D. G., and Sperry, R. W., 1963, Preferential selection of central pathways by regenerating optic fibers, *Exp. Neurol.* **7**:46.

Barondes, S. H., 1974, Synaptic macromolecules: Identification and metabolism, *Ann. Rev. Biochem.* **43**:147.

Bennett, G., Leblond, C. P., and Haddad, A., 1974, Migration of glycoprotein from the Golgi apparatus to the surface of various cell types as shown by radioautography after labeled fucose injection to rats, *J. Cell Biol.* **60**:258.

Bernhard, W., and Avrameas, S., 1971, Ultrastructural visualization of cellular carbohydrate components by means of concanavalin A, *Exp. Cell Res.* **64**:232.

Bittiger, H., and Schnebli, H. P., 1974, Binding of concanavalin A and ricin to synaptic junctions of rat brain, *Nature (London)* **249**:370.

Bloom, F. E., 1972, The formation of synaptic junctions in developing rat brain, in: *Structure and Function of Synapses* (G. D. Pappas and D. P. Purpura eds.), pp. 101–120, Raven Press, New York.

Bloom, F. E., and Aghajanian, G. K., 1968, Fine structural and cytochemical analysis of the staining of synaptic junctions with phosphotungstic acid, *J. Ultrastruct. Res.* **22**:361.

Bodian, D., 1966, Development of fine structure of spinal cord in monkey fetuses. I. Motoneuron neuropil at time of onset of reflex activity, *Bull. Johns Hopkins Hosp.* **119**:129.

Bodian, D., 1968, Development of the fine structure of spinal cord in monkey fetuses. II. Pre-reflex period to period of long intersegmental reflexes, *J. Comp. Neurol.* **133**:113.

Bondareff, W., 1967, An intercellular substance in rat cerebral cortex, submicroscopic distribution of ruthenium red, *Anat. Rec.* **157**:527.

Bondareff, W., and Sjöstrand, J., 1969, Cytochemistry of synaptosomes, *Exp. Neurol.* **24**:450.

Branton, D., 1967, Fracture faces of frozen myelin, *Exp. Cell Res.* **45**:703.

Bray, D., 1970, Surface movements during the growth of single explanted neurons, *Proc. Natl. Acad. Sci. USA* **65**:905.

Bray, D., 1972, Cytoplasmic actin: A comparative study, *Cold Spring Harbor Symp. Quant. Biol.* **37**:567.

Bray, D., 1973, Branching patterns of individual sympathetic neurons in culture, *J. Cell Biol.* **56**:702.

Bretscher, M. S., 1973, Membrane structure: Some general principles, *Science* **181**:622.

Brunngraber, E. G., Dekirmenjian, H., and Brown, B. D., 1967, The distribution of protein-bound *N*-acetylneuraminic acid in subcellular fractions of rat brain, *Biochem. J.*, **103**:73.

Bunge, M. B., 1973, Fine structure of nerve fibers and growth cones of isolated sympathetic neurons in culture, *J. Cell Biol.* **56**:713.

Bunge, M. B., Bunge, R. P., and Peterson, E. R., 1967, The onset of synapse formation in spinal cord cultures as studied by electron microscopy, *Brain Res.* **6**:728.

Burger, M. M., 1971, Cell surface in neoplastic transformation, in: *Current Topics in Cellular Regulation*, (Vol. 3,B. L. Horecker and E. R. Stadtman, eds.), pp. 135–193, Academic Press, New York.

Carter, S. B., 1967, Haptotaxis and the mechanism of cell motility, *Nature (London)* **213**:256.

Chamley, J. H., Goller, I., and Burnstock, G., 1973, Selective growth of sympathetic nerve fibers to explants of normally densely innervated autonomic effector organs in tissue culture, *Dev. Biol.* **31**:362.

Cotman, C. W., and Taylor, D., 1972, Isolation and structural studies on synaptic complexes from bat brain, *J. Cell Biol.* **55**:696.

Cotman, C. W., and Taylor, D., 1974, Localization and characterization of concanavalin A receptors in the synaptic cleft, *J. Cell Biol.* **62**:236.

Cotman, C. W., Mahler, H. R., and Hugli, T. E., 1968, Isolation and characterization of insoluble proteins of the synaptic plasma membrane, *Arch. Biochem. Biophys.* **126**:821.

Cotman, C. W., Banker, G., Churchill, L., and Taylor, D., 1974, Isolation of postsynaptic densities from rat brain, *J. Cell Biol.* **63**:441.

Crossland, W. J., Cowan, W. M., Rogers, L. A., and Kelly, J. P., 1974*a*, The specification of the retino-tectal projection in the chick, *J. Comp. Neurol.* **155**:127.

Crossland, W. J., Currie, J. R., Rogers, L. A., and Cowan, W. M., 1974*b*, Evidence for a rapid phase of axoplasmic transport at early stages in the development of the visual system of the chick and frog, *Brain Res.* **78**:483.

Del Cerro, M. P., and Snider, D. S., 1968, Studies on the developing cerebellum: Ultrastructure of the growth cones, *J. Comp. Neurol.* **133**:341,

De Robertis, E., 1964, *Histophysiology of Synapses and Neurosecretion,* Pergamon Press, New York

Droz, B., 1973, Renewal of synaptic proteins, *Brain Res.* **62**:383.

Dunn, G. A., 1971, Mutual contact inhibition of extension of chick sensory nerve fibres *in vitro, J. Comp. Neurol.* **143**:491.

Edelman, G. M., Spear, P. G., Rutishauser, U., and Yahara, I., 1974, Receptor specificity and mitogenesis in lymphocyte populations, in: *The Cell Surface in Development* (A. A. Moscona, ed.), pp. 141–164, Wiley, New York.

Forssman, J., 1900, Zur Kenntnis des Neurotropismus, *(Ziegler's) Beitr. pathol. Anat. allgem. Pathol.* **27**:407.

Friend, D. S., and Farquhar, M. G., 1967, Functions of coated vesicles during protein absorption in the rat vas deferens, *J. Cell Biol.* **35**:357.

Gaze, R. M., 1970, *The Formation of Nerve Connections,* Academic Press, London.

Glees, P., and Sheppard, B. L., 1964, Electron microscopical studies of the synapse in the developing chick spinal cord, *Z. Zellforsch.* **62**:356.

Grainger, E., and James, D. W., 1970, Association of glial cells with the terminal parts of neurite bundles extending from chick spinal cord *in vitro, Z. Zellforsch.* **108**:93.

Gray, E. G., 1959, Axosomatic and axodendritic synapses of cerebral cortex: An electron microscopic study, *J. Anat.* **93**:420.

Gray, E. G., 1966, Problems of interpreting the fine structure of vertebrate and invertebrate synapses, *Int. Rev. Gen. Exp. Zool.* **2**:139.

Gray, E. G., 1969, Electron microscopy of excitatory and inhibitory synapses: A brief review, in: Mechanisms of Synaptic Transmission (K. Akert and P. G. Waser, eds.), *Prog. Brain. Res.* **31**:141.

Gremo, F., Sjöstrand, J., and Marchisio, P. C., 1974, Radioautographic analysis of ^{3}H-fucose labelled glycoproteins transported along the optic pathway of chick embryos, *Cell Tissue Res.* **153**:465.

Harrison, R. G., 1907, Observations on the living developing nerve fiber, *Anat. Rec.* **1**:116.

Harrison, R. G., 1910, The outgrowth of the nerve fiber as a mode of protoplasmic movement, *J. Exp. Zool.* **9**:787.

Hayes, B. P., and Roberts, A., 1973, Synaptic junction development in the spinal cord of an amphibian embryo: An electron microscope study, *Z. Zellforsch.* **137**:251.

Hibbard, E., 1967, Visual recovery following regeneration of the optic nerve through oculomotor nerve root in *Xenopus, Exp. Neurol.* **19**:350.

Hinds, J. W., and Hinds, P. L., 1972, Reconstruction of dendritic growth cones in neonatal mouse olfactory bulbs, *J. Neurocytol.* **1**:169.

Hughes, A., 1953, The growth of embryonic neurites: A study on cultures of chick neural tissues, *J. Anat.* **87**:150.

Jacobson, M., 1970, *Developmental Neurobiology,* Holt, Rinehart and Winston, New York.

James, D. W., and Tresman, R. L., 1972, The surface coats of chick dorsal root ganglion cells *in vitro, J. Neurocytol.* **1**:383.

Jones, D. G., Dittmer, M. M., and Reading, L. C., 1974, Synaptogenesis in guinea pig cerebral cortex: A glutaraldehyde-PTA study, *Brain Res.* **70**:245.

Kawana, E., Sandri, C., and Akert, K., 1971, Ultrastructure of growth cones in the cerebral cortex of the neonatal rat and cat, *Z. Zellforsch.* **115**:284.

Landis, D. M. D., and Reese, T. S., 1974, Differences in membrane structure between excitatory and inhibitory synapses in the cerebellar cortex, *J. Comp. Neurol.* **155:**93.

Landis, D. M. D., Reese, T. S., and Raviola, E., 1974, Differences in membrane structure between excitatory and inhibitory components of the reciprocal synapses in the olfactory bulb, *J. Comp. Neurol.* **155:**67.

Larramendi, L. M. H., 1969, Analysis of synaptogenesis in the mouse, in: *Neurobiology of Cerebellar Evolution and Development* (R. Llinas, ed.), pp. 803–845, American Medical Association Educational and Research Foundation, Chicago.

Lehninger, A. L., 1968, The neuronal membrane, *Proc. Natl. Acad. Sci. USA* **60:**1069.

Lewis, W. H., and Lewis, M. R., 1912, The cultivation of sympathetic nerves from the intestine of chick embryos in saline solution, *Anat. Rec.* **6:**7.

Marchisio, P. C., Gremo, F., and Sjöstrand, J., 1975, Axonal transport in embryonic neurons: The possibility of a proximo-distal axolemmal transfer of glycoproteins, *Brain Res.* **85:**281.

McNutt, S. N., and Weinstein, R. S., 1973, Membrane ultrastructure at mammalian intercellular junctions, *Prog. Biophys. Mol. Biol.* **26:**45.

Meyer, W. J., 1969, The localisation of extracellular substances in the brain with phosphotungstic acid, *Anat. Rec.* **163:**229 (abst.).

Meyer, W. J., 1970, Distribution of phosphotungstic acid-stained carbohydrate moieties in the brain, dissertation, University of California at Los Angeles.

Moscona, A. A., 1974, Surface specification of embryonic cells: Lectin receptors, cell recognition, and specific cell ligands, in: *The Cell Surface in Development* (A. A. Moscona, ed.), pp. 67–99, Wiley, New York.

Nakai, J., 1956, Dissociated dorsal root ganglia in tissue culture, *Am. J. Anat.* **99:**81.

Nakai, J., 1960, Studies on the mechanism determining the course of nerve fibers in tissue culture. II. The mechanism of fasciculation, *Z. Zellforsch.* **52:**427.

Nakai, J., and Kawasaki, Y., 1959, Studies on the mechanism determining the course of nerve fibers in tissue culture. I. The reaction of the growth cone to various obstructions, *Z. Zellforsch.* **51:**108.

Nakajima, S., 1965, Selectivity in fasciculation of nerve fibres *in vitro*, *J. Comp. Neurol.* **125:**193.

Ohlenbusch, H. H., Olivera, B. M., Tuan, D., and Davidson, N., 1967, Selective dissociation of histones from calf thymus nucleoprotein, *J. Mol. Biol.* **25:**299.

Olson, M. I., and Bunge, R. P., 1973, Anatomical observations on the specificity of synapse formation in tissue culture, *Brain Res.* **59:**19.

Oppenheim, R. W., and Foelix, R. F., 1972, Synaptogenesis in the chick embryo spinal cord, *Nature (London)* **235:**126.

Pappas, G. D., and Waxman, S. G., 1972, Synaptic fine structure—Morphological correlates of chemical and electrotonic transmission, in: *Structure and Function of Synapses* (G. D. Pappas and D. P. Purpura, eds.), pp. 1–43, Raven Press, New York.

Pease, D. C., 1966, Polysaccharides associated with the exterior surface of epithelial cells: Kidney, intestine, brain, *J. Ultrastruct. Res.* **15:**555.

Pease, D. C., 1970, Phosphotungstic acid as a specific electron stain for complex carbohydrates, *J. Histochem. Cytochem* **18:**455.

Pfenninger, K. H., 1971*a*, The cytochemistry of synaptic densities. I. An analysis of the bismuth iodide impregnation method, *J. Ultrastruct. Res,* **34:**103.

Pfenninger, K. H., 1971*b*, The cytochemistry of synaptic densities. II. Proteinaceous components and mechanism of synaptic connectivity, *J. Ultrastruct. Res.* **35:**451.

Pfenninger, K. H., 1972, Freeze-cleaving of outgrowing nerve fibers in tissue culture, *J. Cell Biol.* **55:**203a.

Pfenninger, K. H., 1973, Synaptic morphology and cytochemistry, *Prog. Histochem. Cytochem.* **5**:1.

Pfenninger, K. H., 1974, Freeze-cleavage of cell monolayers grown *in vitro,* in: *Proceedings of the 8th International Congress on Electron Microscopy,* Canberra, Vol. II, pp. 38–39, Australian Academy of Science, Canberra.

Pfenninger, K. H., and Bunge, M. B., 1973, Observations on plasmalemmal growth zones in developing neural tissue, *J. Cell Biol.* **59**:264a.

Pfenninger, K. H., and Bunge, R. P., 1974, Freeze-fracturing of nerve growth cones and young fibers: A study of developing plasma membrane, *J. Cell Biol.* **63**:180.

Pfenninger, K. H., and Rinderer, E. R., 1975, Methods for the freeze-fracturing of nerve tissue cultures and cell monolayers, *J. Cell Biol.* **65**:15.

Pfenninger, K. H., and Rovainen, C. M., 1974, Stimulation- and calcium-dependence of vesicle attachment sites in the presynaptic membrane; a freeze-cleave study on the lamprey spinal cord, *Brain Res.* **72**:1.

Pfenninger, K., Sandri, C., Akert, K., and Eugster, C. H., 1969, Contribution to the problem of structural organization of the presynaptic area, *Brain Res.* **12**:10.

Pfenninger, K., Akert, K., Moor, H., and Sandri, C., 1972, The fine structure of freeze-fractured presynaptic membranes, *J. Neurocytol.* **1**:129.

Pfenninger, K. H., Bunge, M. B., and Bunge, R. P., 1974, Nerve growth cone plasmalemma—Its structure and development, in: *Proceedings of the 8th International Congress on Electron Microscopy,* Canberra, Vol. II, pp. 234–235, Australian Academy of Science, Canberra.

Pollard, T. D., and Weihing, R. R., 1974, Actin and myosin and cell movement, *CRC Crit. Rev. Biochem.* **2**:1.

Pomerat, C. M., Hendelman, W. J., Raiborn, C. W., Jr., and Massey, J. F., 1967, Dynamic activities of nervous tissue *in vitro,* in: *The Neuron* (H., Hyden, ed.), pp. 119–178, American Elsevier, New York.

Raisman, G., and Field, P. M., 1973, A quantitative investigation of the development of collateral reinnervation after partial deafferentation of the septal nuclei, *Brain Res.* **50**:241.

Rambourg, A., and Leblond, C. P., 1967, Electron microscopic observation on the carbohydrate-rich cell coat present at the surface of cells in the rat, *J. Cell Biol.* **32**:27.

Ramón y Cajal, S., 1913, *Estudios Sobre la Degeneracion y Regeneracion del Systema Nervioso,* 2 vols. (trans. and ed. by R. M. May, 1928: *Degeneration and Regeneration of the Nervous System),* Hafner, New York, reprinted 1959.

Ramón y Cajal, S., 1929, *Etudes sur la Neurogénèse de Quelques Vertébrés* (trans. by L. Guth: *Studies on Vertebrate Neurogenesis),* Thomas, Springfield, Ill., 1960.

Ramón y Cajal, S., 1933, *Neuronismo o Reticularismo?* (trans. and ed. by M. U. Purkiss and C. A. Fox: *Neuron Theory or Reticular Theory?),* Consejo Superior de Investigaciones Cientificas, Instituto "Ramón y Cajal," Madrid, 1954.

Rees, R. P., and Bunge, R. P., 1974, Morphological and cytochemical studies of synapses formed in culture between isolated rat superior cervical ganglion neurons, *J. Comp. Neurol.* **157**:1.

Rees, R. P., Bunge, M. B., and Bunge, R. P., 1976, Morphological changes in the neuritic growth cone and target neuron during synaptic junction development in culture, *J. Cell Bio.* **68**:240.

Revel, J. P., and Ito, S., 1967, The surface components of cells, in: *The Specificity of Cell Surfaces* (Davis and Warren, eds.), pp. 211–234, Prentice-Hall, Englewood Cliffs, N.J.

Roth, S., McGuire, E. J., and Roseman, S., 1971, Evidence for cell surface glycosyltransferases: Their potential role in cellular recognition, *J. Cell Biol.* **51:**536.
Sandri, C., Akert, K., Livingston, R. B., and Moor, H., 1972, Particle aggregations at specialized sites in freeze-etched postsynaptic membranes, *Brain Res.* **41:**1.
Shienvold, F. L., and Kelly, D. E., 1974, Desmosome structure revealed by freeze-fracture and tannic acid staining, *J. Cell Biol.* **63:**313a.
Sidman, R. L., 1974, Contact interaction among developing mammalian brain cells, in: *The Cell Surface in Development* (A. A. Moscona, ed.), pp. 221–253, Wiley, New York.
Sidman, R. L., and Rakic, P., 1973, Neuronal migration, with special reference to developing human brain: A review, *Brain Res.* **62:**1.
Singer, S. J., and Nicolson, G. L., 1972, The fluid mosaic model of the structure of cell membranes, *Science* **175:**720.
Sjöstrand, J., Karlsson, J.-O., and Marchisio, P. C., 1973, Axonal transport in growing and mature retinal ganglion cells, *Brain Res.* **62:**395.
Skoff, R. P., and Hamburger, B., 1974, Fine structure of dendritic and axonal growth cones in embryonic chick spinal cord, *J. Comp. Neurol.* **153:**107.
Speidel, C. C., 1933, Studies of living nerves. II. Activities of ameboid growth cones, sheath cells, and myelin segments, as revealed by prolonged observation of individual nerve fibers in frog tadpoles, *Am. J. Anat.* **52:**1.
Sperry, R. W., 1944, Optic nerve regeneration with return of vision in anurans, *J. Neurophysiol.* **7:**57.
Sperry, R. W., 1963, Chemoaffinity in the orderly growth of nerve fiber patterns and connections, *Proc. Natl. Acad. Sci. USA* **50:**703.
Steck, T. L., 1974, The organization of proteins in the human red blood cell membrane, *J. Cell Biol.* **62:**1.
Stelzner, D. J., Martin, A. H., and Scott, G. L., 1973, Early stages of synaptogenesis in the cervical spinal cord of the chick embryo, *Z. Zellforsch.* **138:**475.
Streit, P., Akert, K., Sandri, C., Livingston, R. B., and Moor, H., 1972, Dynamic ultrastructure of presynaptic membranes at nerve terminals in the spinal cord of rats: Anesthetized and unanesthetized preparations compared, *Brain Res.* **48:**11.
Tani, E., and Ametani, T., 1971, Extracellular distribution of ruthenium red-positive substance in the cerebral cortex, *J. Ultrastruct. Res.* **34:**1.
Tennyson, V. M., 1970, The fine structure of the axon and growth cone of the dorsal root neuroblast of the rabbit embryo, *J. Cell Biol.* **44:**62.
Tillack, T. W., Scott, R. E., and Marchesi, V. T., 1972, The structure of erythrocyte membranes studied by freeze-etching. II. Localization of receptors for phytohemagglutinin and influenza virus to the intramembranous particles, *J. Exp. Med.* **135:**1209.
Weiss, P., and Taylor, A. C., 1944, Further experimental evidence against "neurotropism" in nerve regeneration, *J. Exp. Zool.* **95:**233.
Wolfe, L. S., 1961, The distribution of gangliosides in subcellular fractions of guinea pig cerebral cortex, *Biochem. J.* **79:**348.
Van der Loos, H., 1963, Fine structure of synapses in the cerebral cortex, *Z. Zellforsch.* **60:**815.
Vaughn, J. E., Henrikson, C. K., and Grieshaber, J. A., 1974, A quantitative study of synapses on motor neuron dendritic growth cones in developing mouse spinal cord, *J. Cell Biol.* **60:**664.
Yamada, K. M., Spooner, B. S., and Wessells, N. K., 1970, Axon growth: Roles of microfilaments and microtubules, *Proc. Natl. Acad. Sci. USA* **66:**1206.
Yamada, K. M., Spooner, B. S., and Wessells, N. K., 1971, Ultrastructure and function of growth cones and axons of cultured nerve cells, *J. Cell Biol.* **49:**614.

6

Biochemical Studies of Synaptic Macromolecules

Are There Specific Synaptic Components?

I. G. MORGAN and G. GOMBOS

1. INTRODUCTION

It is highly likely that the relevant specificities of the nerve ending region are localized in the synaptic plasma membrane since it is presumably contact between the external surfaces of membranes which is important in neuronal recognition. But even with this initial precision several levels of specificity have to be considered. The specificities of the synaptosomal plasma membrane relative to the plasma membranes of nonneural cells, to those of nonneuronal neural cells, and to other parts of the neuronal plasma membrane such as the dendritic, perikaryal, and axonal plasma membranes must be analyzed. And within the synaptic region the synaptosomal plasma membrane consists of nonjunctional and junctional regions, the latter consisting of pre- and postsynaptic membranes.

The former, relatively gross, specificities probably involve those basic contituents which distinguish one type of structure from another.

I. G. MORGAN and G. GOMBOS · Chargés de Recherche au Centre National de la Recherche Scientifique. Centre de Neurochimie du CNRS, Strasbourg, France.

I. G. Morgan's present address: Department of Behavioural Biology, Research School of Biological Sciences, Australian National University, Canberra, A.C.T. 2601, Australia.

But there is another level of specificity which is more difficult to study: that which defines a neuron of a given transmitter type or that which distinguishes between functional and nonfunctional synapses. An even finer specificity, correspondingly even more difficult to study, is that of the properties of neurons and their plasma membranes which enable them to enter into possibly transient contact with other neurons or glial cells during development to regulate their final position, morphology, and contacts. Most important would be the extremely fine differences between neurons which would bring a given neuron into contact with other neurons in a specific manner.

Other authors will discuss the extent to which neurons are capable of independent development and the extent to which their development is regulated by the complex interplay of spatial and temporal factors rather than by precisely defined biochemical differences in membranes. We intend to review our knowledge of the structure of neural plasma membranes and to analyze to what extent present data throw some light on the biochemical basis of the development of the nervous system.

2. ISOLATION OF NEURAL PLASMA MEMBRANES

At the present time, the only plasma membrane of acceptable purity which can be isolated from the central nervous system is that derived from synaptosomes, i.e., from pinched-off nerve endings (for review, see Whittaker, 1969; De Robertis and Rodriguez de Lores Arnaiz, 1969). This membrane will thus be taken as the reference membrane to which other membranes will be compared. Several methods are now available for the preparation of these *synaptosomal* or *synaptic* plasma membranes at around 70–80% purity as plasma membrane (Morgan *et al.*, 1971, 1973*a*; Cotman and Matthews, 1971; Levitan *et al.*, 1972; Gurd *et al.*, 1974; Jones and Matus, 1974). The preparations obtained are highly enriched in putative markers of the neuronal plasma membrane such as gangliosides and (Na+K)ATPase (Morgan *et al.*, 1971, 1973*a*; Levitan *et al.*, 1972; Gurd *et al.*, 1974; Cotman, personal communication). There is limited (5–10%) contamination with external mitochondrial membrane (Morgan *et al.*, 1971; Gurd *et al.*, 1974) and some (5–20%) contamination with elements of the endoplasmic reticulum and the Golgi apparatus (Cotman *et al.*, 1971; Levitan *et al.*, 1972; Morgan *et al.*, 1973*a**; Gurd *et al.*, 1974). Finally, there may be slight

* These values are higher than those reported in our first paper on these preparations (Morgan *et al.*, 1971). A more detailed account of these results will be given elsewhere.

contamination with lysosomes and soluble proteins (Cotman and Matthews, 1971; Gurd *et al.*, 1974; Morgan *et al.*, in preparation*). Preparations can be obtained which are essentially devoid of myelin and inner mitochondrial membrane (Morgan *et al.*, 1971). While it is difficult to rule out by enzymatic and chemical markers the presence of large amounts of glial, axonal, and perikaryal plasma membrane, the nature of the original synaptosomal fractions and some of the morphological and biochemical characteristics of the final plasma membranes (Morgan *et al.*, 1972, 1973*a*; Jones and Matus, 1974) suggest that the preparations are primarily derived from the nerve terminal regions. The contamination with glial material has been estimated to be somewhere in the region of 5–20%, but such estimations are subject to numerous errors, not least of which are the uncertainties about what constitute markers of glial membranes (Morgan *et al.*, 1973*a*; Barondes, 1974). It should be stressed that the new methods are significant improvements over the classical methods, which were developed for analytical rather than preparative purposes (Whittaker *et al.*, 1964; Rodriguez de Lores Arnaiz *et al.*, 1967), and as such are to be preferred for meaningful compositional studies.

Accepting reasonably pessimistic assessments of contamination, the synaptosomal plasma membrane preparations could be only around 50% pure, whereas, accepting reasonably optimistic assessments, they may be up to 80% pure. The real figure lies somewhere between these limits. But whatever the precise degree of purity it must always be borne in mind that the preparations are contaminated, and as such can be used only with caution for studies of neuronal specificity. In particular, little significance can be placed on minor components, among which, unfortunately, the interesting specificities are likely to be found, unless the minor components are purified and localized by immunohistochemical techniques. It also needs to be stressed that the published methods should be followed as precisely as possible, and that even then the preparations should be rigorously checked for contamination in any experiments in which contamination is likely to be crucial. This necessity is reinforced when slight changes are made in the methods used. A final caveat concerning these preparations is that they are intrinsically heterogeneous since they are probably derived from a series of different types of neurons. This can only further complicate the interpretation of minor components.

Synaptosomal plasma membranes can be further fractionated to give synaptic junctional fractions (Cotman and Taylor, 1972; Davis and

* See footnote p. 180.

Bloom, 1973; Morgan *et al.*, in preparation), whose purity remains to be investigated. However, preliminary morphological and biochemical studies suggest that the synaptic junctional fractions are at least 60–70% pure (Churchill *et al.*, in preparation). Synaptic junctional fractions are isolated from the synaptosomal plasma membranes by extraction with Triton X-100. This must always be borne in mind in interpreting analyses of these fractions since the detergent could preferentially extract certain of the components of the synaptic junctions, even if these structures remain morphologically intact. Thus we have evidence that acetylcholinesterase, which seems to be preferentially associated with postsynaptic material in synaptosomes (McBride and Cohen, 1972), is almost completely extracted from isolated synaptic junctions (Morgan *et al.*, in preparation).

Another method for preparing synaptic junctions has been developed based on extraction with sodium deoxycholate (Walters and Matus, 1975). These "junctions" appear to consist almost exclusively of the postsynaptic density. A method for preparation of the postsynaptic density has been reported which involves extraction of either synaptosomal plasma membranes or synaptic junctions with sodium lauroyl *N*-sarcosinate (Cotman *et al.*, 1974). The postsynaptic densities have been reported to be around 80–90% pure on morphological criteria.

Axonal plasma membranes have been prepared from various nonmyelinated nerves, particularly those of the squid (Camejo *et al.*, 1969; Fischer *et al.*, 1970; Marcus *et al.*, 1972), the lobster (Denburg, 1972; Barnola *et al.*, 1973; Fosset *et al.*, in preparation), and the garfish (Grefrath and Reynolds, 1973; Chacko *et al.*, 1974). While these preparations are less well characterized than are the synaptosomal plasma membranes, some seem to be reasonably pure as plasma membrane. However, no estimate of the contamination with glial or Schwann cell plasma membranes has been made.

Unfortunately, it is not possible at the present time to prepare axonal plasma membrane from the central nervous system of the mammals most used for studying the synaptosomal plasma membrane. Methods have been reported for the isolation of axonal fragments from rat brain (Lemkey-Johnston and Dekirmenjian, 1970) and bovine brain (de Vries *et al.*, 1972). However, the latter preparation has given rather unexpected results upon biochemical analysis (in particular, the presence of large amounts of cerebrosides and sulfatides, which are normally associated with myelin (de Vries and Norton, 1974)), and more clarity as to their status is necessary before those axonal fragments can be used as a source of axonal plasma membrane. Several fractions enriched in parallel fiber fragments have been isolated from the cerebellum (Lemkey-Johnston and Larramendi, 1968; Dekirmenjian *et al.*,

1969), but these preparations, although enriched, are far from pure, and they require further purification before axonal plasma membranes could conceivably be isolated from them.

Isolation of cell body plasma membrane is, at the present, an equally formidable problem. Several methods have been reported for isolating nerve cell bodies reasonably free from contamination with intact glial cells (Freysz *et al.*, 1968; Rose, 1967; Norton and Poduslo, 1970; Sellinger *et al.*, 1971; Blomstrand and Hamberger, 1969). However, the characterization of these fractions of neuronal perikarya is still rudimentary and much more work needs to be done. A method for preparing neuronal cell body plasma membrane from such preparations has been published (Henn *et al.*, 1972), but, on the evidence so far presented, the enrichment of neuronal plasma membrane in the final fractions is rather limited. A method for preparing Purkinje cell bodies with well-preserved plasma membranes has also been reported (Cohen *et al.*, 1974), but plasma membrane has not yet been isolated from such preparations. Methods for preparing plasma membranes from dissociated embryonic neuroblasts have been reported (Hemminki, 1973), but the purity of these fractions is still to be thoroughly investigated. Some information on the characteristics of the neuronal perikaryal plasma membrane can be obtained by studying neuroblastoma cell lines in tissue culture (Truding *et al.*, 1974), but no well-characterized preparations of neuroblastoma cell plasma membrane have as yet been reported.

In contrast to the neuronal fractions, there is considerable evidence that the glial cell fractions prepared by the methods of Rose (1967), Blomstrand and Hamberger (1969), and Sellinger *et al.* (1971) are significantly contaminated with neuronal fragments, probably with synaptosomes. Thus high activities of acetylcholinesterase, choline acetyltransferase, and glutamate decarboxylase, which would reasonably be expected to be localized in neurons, have been detected in glial cell preparations (Hemminki *et al.*, 1973; Arbogast and Arsenis, 1974; Nagata *et al.*, 1974; Freysz *et al.*, in preparation). This could throw some doubt upon the significance of the high levels of (Na+K)ATPase (Medzihradsky *et al.*, 1971) and gangliosides (Norton and Poduslo, 1971; Hamberger and Svennerholm, 1971) reported in isolated glial cells. This contamination is presumably a problem with most of the preparations of glial cells reported in the literature, and it is thus evident that any attempt to purify glial plasma membranes from these preparations will run into problems of contamination with neuronal material.

A further problem is that of the possible differences between astrocyte and oligodendrocyte plasma membranes, even though methods for preparing astrocytes and oligodendrocytes separately have been reported (Poduslo and Norton, 1972). Some approach to the nature of glial cell plasma membranes may be possible by using tissue culture systems of tumoral glial cells, immature astroblasts, or astrocytes in primary culture, but as yet plasma membranes have not been prepared from these sources. The myelin sheath, which is a specialization of the oligodendrocyte plasma membrane, may also throw some light upon this problem.

3. CHEMICAL COMPOSITION OF NEURAL PLASMA MEMBRANES

Although a number of studies have been devoted to the lipids of synaptosomal and axonal plasma membranes (Cotman *et al.*, 1969; Breckenridge *et al.*, 1972; Camejo *et al.*, 1969; Zambrano *et al.*, 1971; Chacko *et al.*, 1972; de Vries and Norton, 1974), which appear to be characterized by highly unsaturated, and therefore probably highly fluid lipid phases, it seems *a priori* unlikely that these compounds are responsible for neuronal recognition. The necessary specificity for recognition is more likely to be found in the proteins of the membranes, and perhaps even more in the glycoproteins. While they may not have a precise specificity, it is probable that the gangliosides, which are highly concentrated in neuronal plasma membranes, and membrane-associated mucopolysaccharides may also have important roles in regulating neural cell contacts.

3.1. *Proteins*

The protein composition of the synaptosomal plasma membrane has been studied by several groups, with results in reasonable agreement (Waehneldt *et al.*, 1971; Banker *et al.*, 1972; McBride and Van Tassel, 1972; Morgan *et al.*, 1973*a*,*b*). The protein profiles (Fig. 1) of synaptosomal plasma membranes in SDS-polyacrylamide gel electrophoresis are relatively simple, being dominated by three bands which correspond in molecular weight to those of the subunits of bovine brain (Na+K)ATPase (Uesugi *et al.*, 1971). The protein profiles of brain microsomal fractions are similar to these of synaptosomal plasma membranes (Gurd *et al.*, 1974), but it must not be forgotten that "microsomal" fractions are not pure endoplasmic reticulum. In fact, the levels of (Na+K)ATPase and gangliosides in brain microsomal fractions suggest that 20–30% of the protein of these fractions could be derived from neuronal plasma membrane. In our hands, while the protein and glyco-

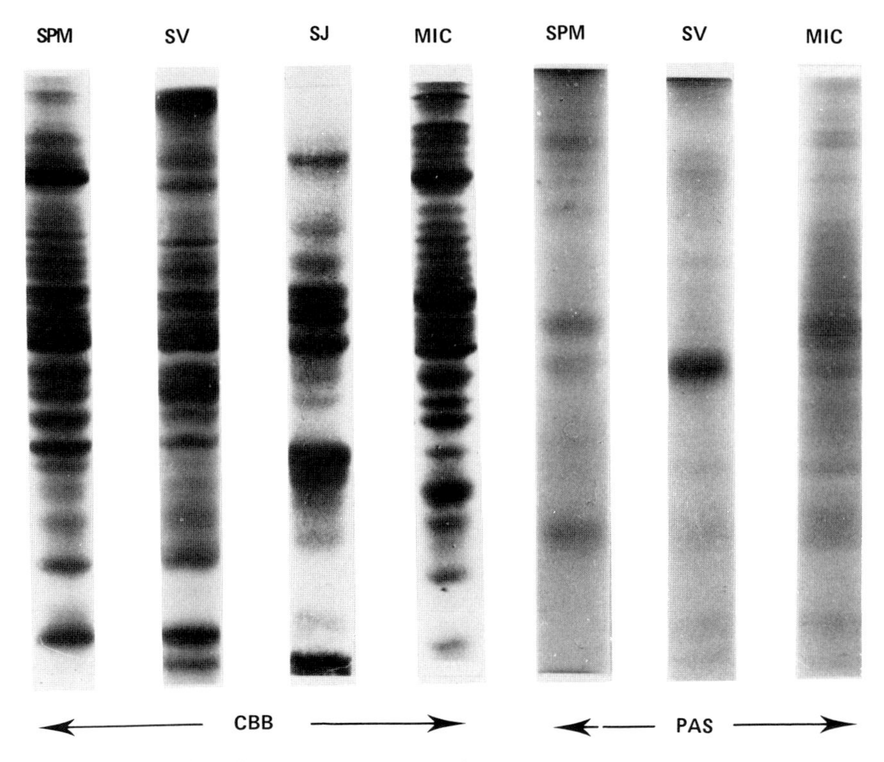

Fig. 1. Polyacrylamide gel electrophoresis of synaptosomal plasma membranes (SPM), synaptic vesicles (SV), synaptic junctions (SJ), and the microsomal fraction (MIC) in 12% polyacrylamide gels in a sodium dodecylsulfate system. Subcellular fractions were extracted with chloroform–methanol prior to treatment. Gels were stained with Coomassie brilliant blue (CBB) or by a periodic acid–Schiff procedure (PAS). The synaptic junction fraction did not show any bands after PAS staining.

protein profiles of the synaptosomal plasma membrane and microsomal fractions are similar, the microsomal fraction is more complex, and the putative (Na+K)ATPase bands are less dominant (Fig. 1), in agreement with the enzymatic characteristics of the fractions.

Axonal plasma membranes isolated from garfish olfactory nerve (Grefrath and Reynolds, 1973) or crab nerves (Fosset *et al.*, in preparation) seem to have some similarities in protein profile to the synaptosomal plasma membrane, despite the marked species differences. In particular, the dominance of the putative (Na+K)ATPase bands is evident. Little is known about the neuronal cell body plasma membrane, but it has been reported to have a protein profile similar to that of the synaptosomal plasma membrane (Karlsson *et al.*, 1973). Studies on the synaptosomal plasma membranes of different regions have not shown up any marked regional differences.

Few preparations of glial cell plasma membrane are available, but it has been reported that the glial cell plasma membrane has a protein profile remarkably similar to that of the synaptosomal and perikaryal plasma membrane (Karlsson *et al.*, 1973). However, we must stress that the problem of synaptosomal contamination of the initial glial fraction could be of crucial importance in this case.

The protein profiles of synaptic junction fractions (Fig. 1) are significantly simpler than those of the initial synaptosomal plasma membrane (Morgan *et al.*, 1973*b*; Banker *et al.*, 1974; Davis and Bloom, 1970). The (Na+K)ATPase bands seem to be extracted, and depending on the Triton X-100: protein ratios, and the exposure of the membranes

Table I. Amino Acid Compositions of Subcellular Fractions[a]

	SPM[b]	SPM[c]	SJ[b]	PSD[c] (0.4% NLS[d])	PSD[c] (3% NLS[d])
Ala	8.87	7.5–8.3	7.61	7.8	8.7
Arg	3.96	4.0–4.8	4.67	6.0	5.7
Asp	10.05	9.6–10.5	10.18	9.0	10.0
Cys	0.90	—	1.06	—	—
Glu	11.59	11.0–12.2	18.75	12.5	13.0
Gly	7.71	7.0–7.8	6.33	7.8	8.6
His	0.81	1.8–2.1	1.40	2.9	2.9
Ile	3.68	4.3–5.3	4.40	4.4	3.7
Leu	10.59	8.3–9.3	9.08	8.8	7.9
Lys	4.67	5.9–6.6	4.62	5.8	5.7
Met	0.90	1.9–3.0	0.85	2.1	1.3
Phe	5.20	3.6–4.6	4.44	3.7	3.1
Pro	4.31	4.4–5.1	3.07	6.3	6.7
Ser	8.45	7.1–9.8	6.33	8.6	8.3
Thr	7.00	5.4–6.1	5.69	5.6	6.0
Tyr	5.29	2.6–3.2	6.13	3.1	2.8
Val	4.66	5.7–7.3	5.46	5.6	4.9
Hy-Lys[e]	1.08	—	n.d.[f]	—	—
Hy-Pro[e]	0.27	—	n.d.[f]	—	—

[a] Determined as previously described (Zanetta *et al.*, 1975) after lipid extraction with chloroform–methanol.
[b] Morgan *et al.* (in preparation).
[c] Banker *et al.* (1974).
[d] N-Lauryl sarcosine.
[e] Tentative identification.
[f] Not detectable.

to Ca^{2+}, very simple profiles can be obtained (Morgan *et al.*, in preparation). The postsynaptic densities prepared by deoxycholate extraction contain a major band which corresponds in molecular weight and in comigration experiments to tubulin (Walters and Matus, 1975). After sodium *N*-lauroyl sarcosinate extraction, a major protein band was also observed, which could correspond in molecular weight to tubulin (Banker *et al.*, 1974). Thus there is good agreement as to the simplicity of the synaptic junction or postsynaptic density in experiments performed by several methods. This tubulin could correspond to the particulate colchicine-binding (Feit and Barondes, 1970) and tubulinlike proteins (Feit *et al.*, 1971) reported in nerve endings.

Isolated synaptic junctions and postsynaptic densities react well with the ethanolic phosphotungstic acid (Bloom and Aghajanian, 1966) and BIUL (Pfenninger, 1971) reagents, which specifically stain them *in vivo*. However, the amino acid compositions (Table I) of the fractions (Morgan *et al.*, in preparation; Banker *et al.*, 1974) give no indication that they are enriched in the basic amino acids which were believed to be the basis of the selective staining (Bloom and Aghajanian, 1968; Pfenninger, 1971). Thus the most obvious morphological synaptic specificity has yet to be correlated with defined proteins, or with the gross properties of some protein fractions.

3.2. *Glycoproteins*

The glycoprotein composition of the synaptosomal plasma membrane has been extensively studied by SDS-polyacrylamide gel electrophoresis (Waehneldt *et al.*, 1971; Banker *et al.*, 1972; Morgan *et al.*, 1973*a*,*b*). There are five or six major PAS-positive bands which seem to be common to synaptosomal plasma membranes isolated from many brain regions, but little is known about the chemical composition of these glycoproteins. Both synaptic junctional complexes and postsynaptic densities are less rich in glycoproteins than are the synaptosomal plasma membranes from which they are derived (Morgan *et al.*, in preparation; Churchill *et al.*, in preparation), but the possibility of preferential extraction of glycoproteins by the detergents used during the preparation cannot be excluded. There appear to be some differences in the carbohydrate composition of these fractions (Table II). The relatively low glycoprotein content of synaptic junctional complexes contrasts with the histochemical evidence for abundant glycoproteins in synaptic junctions *in vivo* (Rambourg and Leblond, 1967; Bondareff and Sjöstrand, 1969; McBride *et al.*, 1970) and *in vitro* (Matus *et al.*, 1973; Bittiger and Schnebli, 1974; Cotman and Taylor, 1974). In addition to

Table II. Carbohydrate Composition of Subcellular Fractions[a]

	μg total sugar/mg protein	Molar ratio						
		Fucose	Galactose	Mannose	Glucose	*N*-Acetyl-glucosamine	*N*-Acetyl-galactosamine	*N*-Acetyl-neuraminic acid
SPM	66	0.220	0.34	0.91	~0.700	1.00	0.190	0.480
SJ	2030	0.772	9.15	1.32	1425.000	1.00	0.188	0.191
MIC	45	0.260	0.59	1.16	0.215	1.00	0.136	0.440
SV	26	0.410	0.94	1.07	0.770	1.00	0.150	0.350

[a] Determined as described previously (Zanetta *et al.*, 1975) after lipid extraction with chloroform–methanol. Large amounts of glucose were detected in the synaptic junction fraction. This could be derived from the sucrose and Ficoll used during preparation of the fractions. When this sugar is not considered, then synaptic junctions contain less sugar than other fractions.

the possibility of extraction, minor differences in preparative techniques appear to have marked effects on the properties of isolated synaptic junctions, and these may be sufficient to explain the differences.

Little is known about the glycoproteins of axonal, cell body, or glial cell plasma membranes. However, it is highly probable that these plasma membranes contain glycoproteins, from histochemical and very incomplete biochemical studies (Rambourg and Leblond, 1967; Henn *et al.*, 1972). Myelin contains a very limited range of high molecular weight glycoproteins, including a major glycoprotein of molecular weight 110,000 (Brady and Quarles, 1973), but these must still be fully characterized.

Synaptosomal plasma membrane glycoproteins tend to be oriented toward the exterior since sialidase eliminates most of the glycoprotein sialic acid from synaptosomes (Sellinger *et al.*, 1969; Bosmann and Carlson, 1972). Moreover, synaptosomes can be agglutinated by a range of lectins (Bosmann, 1972; Gombos and Zanetta, unpublished results), and electron microscopic studies of the binding sites suggest that they are concentrated on the outer surface of the synaptosomal plasma membrane (Matus *et al.*, 1973; Bittiger and Schnebli, 1974; Cotman and Taylor, 1974).

The interaction of the synaptosomal plasma membrane glycoproteins with lectins has been used in several attempts to purify some of the glycoproteins (Gurd and Mahler, 1974; Gombos *et al.*, 1974*a*,*b*; Zanetta *et al.*, 1975). While no synaptosomal plasma membrane glycoprotein has yet been obtained pure, fractions highly enriched in glycoproteins have been obtained. The fractions, in addition to the major synaptosomal plasma membrane glycoproteins, have a large number of minor PAS-positive bands. Unfortunately, given the low, but not negligible, levels of contamination detected in synaptosomal plasma membrane preparations, it is difficult, without immunohistochemical localization, to attribute these bands to the synaptosomal plasma membrane itself.

In addition to the presence of glycoproteins in the synaptosomal plasma membrane which are capable of reacting with lectins of different specificities, Churchill *et al*. (in preparation) have found that synaptic junction fractions have some binding affinity for sugars, which may suggest that there are lectinlike molecules in these areas.

3.3. *Gangliosides*

Gangliosides are highly concentrated in synaptosomal plasma membranes (Breckenridge *et al.*, 1972), as concentrated as is

(Na+K)ATPase activity (Morgan *et al.*, 1971). There do not appear to be any specific synaptosomal gangliosides, but there is a slightly decreased level of G_{M1} (Avrova *et al.*, 1973), which may be in part associated with the myelin sheath (Suzuki *et al.*, 1967). It is not known if axonal and cell body plasma membranes contain gangliosides, but almost certainly gangliosides are not specifically localized in synapses. Based on microdissection studies (Derry and Wolfe, 1967), it was for a long time believed that gangliosides were specific to neurons, and it was therefore assumed that the axonal and cell body plasma membranes would also contain these lipids. Studies based on isolated neuronal and glial cells have claimed that glial cells are as rich in gangliosides as are the neurons (Norton and Poduslo, 1971; Hamberger and Svennerholm, 1971) and moreover that glial cells contain the same complex polysialogangliosides as do neurons (Hamberger and Svennerholm, 1971; Abe and Norton, 1974). The validity of these results is obviously critically dependent on the purity of the glial cell fractions obtained, particularly on the possibility of synaptosomal contamination previously evoked. In fact, glial cell lines in tissue culture, under a variety of growth conditions, lack the polysialogangliosides associated with brain and the synaptosomal plasma membrane (Robert and Rebel, 1975).

3.4. *Mucopolysaccharides*

Relatively little is known about brain mucopolysaccharides. The early literature has been reviewed by Margolis (1969). Histochemical studies indicate that the extracellular ground substance might be mucopolysaccharide, and there are limited data which suggest that these mucopolysaccharides might be concentrated in the synaptic cleft region. Margolis and Margolis (1974) have reported that astrocytes, oligodendrocytes, and neurons contain mucopolysaccharides in approximately equal amounts, and it has been reported that mucopolysaccharides are associated with synaptosomes, and more specifically with synaptic vesicles (Vos *et al.*, 1969).

4. ENZYME LOCALIZATION IN NEURAL PLASMA MEMBRANES

One of the most widely discussed theories for the control of cellular interactions is the transglycosylation theory of Roseman (1970). This theory implies that on plasma membranes there are glycosyltransferases capable of interacting with carbohydrate chains on other plasma membranes, thus modifying their properties. This possibility has been much discussed for the synaptic region in the context of studies on the possibility of synaptic glycoprotein synthesis (see Morgan *et al.*, 1974, 1976, for a detailed discussion).

Various studies have suggested that there may be synthesis in nerve endings on the basis of *in vitro* incorporation experiments (Festoff *et al.*, 1971) or on the basis of a detailed analysis of the *in vivo* kinetics of incorporation of sugars into nerve ending fractions (Barondes, 1968; Zatz and Barondes, 1970; de Vries and Barondes, 1971), but other experiments tend to contradict both the former data (Ramirez, 1974) and the latter data (Zatz and Barondes, 1971; Marinari *et al.*, 1972; Morgan *et al.*, 1974). The most convincing evidence for synaptic glycoprotein synthesis is that of Dutton *et al.* (1973), who have shown that nerve endings label an apparently uniquely limited range of glycopeptides. In view of the limited resolution of current techniques for the separation of glycopeptides, these findings need to be extended, since if appropriately extended they could provide extremely good evidence of such synaptic synthesis. Unfortunately, in the context of the Roseman (1970) hypothesis the results of Dutton *et al.* (1973) are of limited relevance since the glycoprotein synthesis examined appears to be primarily, albeit not exclusively, mitochondrial.

More relevant are the studies concerned with the synaptosomal glycosyltransferases. Both glycoprotein and glycolipid glycosyltransferases have been reported to be localized in nerve ending fractions (Den and Kaufman, 1968; Den *et al.*, 1970), or more specifically in synaptic vesicles and synaptosomal plasma membranes (Broquet and Louisot, 1971). More recent studies using improved methods of isolation of synaptosomal fractions have tended to diminish the glycosyltransferase activities attributable to the synaptic fractions (Reith *et al.*, 1972; Ko and Raghupathy, 1971, 1972; Raghupathy *et al.*, 1972). However, with the exception of the studies of Raghupathy *et al.* (1972), it has not been possible to conclude that synaptosomal glycosyltransferases do not exist. At the present time, it is possible to attribute these glycosyltransferases to contaminating Golgi apparatus (the interpretation we favor) or to intrinsic plasma membrane activities. While biochemical parsimony favors the former interpretation, the latter cannot be excluded by the present data in the literature.

Although the localization of the synthetic enzymes for carbohydrate chains is controversial, the localization of at least one of the degradative enzymes is well established. Sialidase, like the gangliosides, has been shown to be overwhelmingly concentrated in synaptosomal plasma membranes (Schengrund and Rosenberg, 1970), a finding which has been corroborated and extended to show that synaptic vesicles and synaptic mitochondria are virtually devoid of this activity (Tettamanti *et al.*, 1972). The other glycosidases are less associated with synaptosomal plasma membranes, although their precise localization has yet to be established. A desialylation–sialylation cycle is possible in nerve ending plasma membranes, although the presence of sialyltransferases has not been investigated. However, several lines of indirect evidence suggest that this attractive possibility is probably not true (for detailed discussion, see Morgan *et al.*, 1976).

5. IMMUNOLOGICAL STUDIES OF NEURAL PLASMA MEMBRANES

Several studies have been devoted to the effect of antisynaptosome or antisynaptosomal plasma membrane sera on the physiological or morphological properties of neurons (De Robertis *et al.*, 1966, 1968; Wald *et al.*, 1968; Jarosch and Precht, 1972; Raiteri *et al.*, 1972; Costin *et al.*, 1972), but few conclusions can be drawn from these papers.

A more productive approach has been to prepare antibodies to synaptosomes or synaptosomal plasma membranes and to localize the corresponding antigens immunochemically. Kornguth *et al.* (1969) used antibodies to intact synaptosomes to show that the antigens were primarily localized in synapse-rich regions of the cerebral cortex and cerebellum. No reaction was observed with cervical spinal cord. Livett *et al.* (1974*a*) performed similar experiments using isolated synaptosomal plasma membranes as antigens. Reaction was again observed in synapse-rich areas of the cerebral cortex and cerebellum, but in addition the white matter and the molecular layer reacted in a manner which suggested that the axonal plasma membrane contained the appropriate antigen. However, the perikaryal plasma membrane, at least of Purkinje cells, did not appear to react. Similar results have been obtained by Matus (personal communication), except that the antisynaptosomal plasma membrane sera did not appear to react with the axonal plasma membrane. From these results, it seems clear that immunologically the synaptosomal plasma membrane is quite distinct from the cell body plasma membrane, but it is not yet clear whether it is also distinct from the axonal plasma membrane. This immunological specificity contrasts with the failure of biochemical techniques to detect marked differences in the major components.

Livett *et al.* (1974*b*) have prepared antibodies to synaptic vesicles and have shown that there is extensive cross-reaction between synaptic vesicles and synaptosomal plasma membranes and antibodies raised against them. More recently, Bock and Jorgensen (1975) have reported that the major antigenic constituent of the synaptic vesicles can be detected in synaptosomal plasma membrane fractions. These results agree with the limited biochemical data that there may be common constituents in synaptic vesicles and synaptosomal plasma membranes (Breckenridge and Morgan, 1972; Morgan *et al.*, 1973*a,b*), but cross-contamination is difficult to exclude in these experiments. However, these results contrast with the observation of Herschman *et al.* (1972) that antisera prepared by injection of intact synaptosomes reacted only with synaptosomal plasma membranes and not with myelin, nuclei, soluble protein, mitochondria, and synaptic vesicles. At first sight, this is surprising since synaptosomes contain soluble protein, synaptic vesicles, and mitochondria, but the authors suggest that since the synaptosomes were injected without Freund's adjuvant only the outer surface of the synaptosomal plasma membrane was antigenic. Other workers, by contrast, have reported common antigens in myelin, synaptosomal plasma membranes, mitochondria, and microsomal fractions (Lim and Hsu, 1971; De Robertis *et al.*, 1966; Mickey *et al.*, 1971; McMillan *et al.*, 1971).

Jorgensen and Bock (1974) have prepared antibodies against relatively purified synaptosomal plasma membranes and have found six brain-specific antigens to be present. Unfortunately, they did not control the purity of the fractions used, and in view of their demonstration of the danger of trace contamination with extremely antigenic components (Bock *et al.*, 1974) these results must still be viewed with some caution. In general, antisera of this type ought to be routinely adsorbed against soluble proteins, and highly purified myelin and mitochondrial fractions, in order to reduce the problem.

This approach has, to some extent, been followed by Orosz *et al.* (1973, 1974), who sequentially adsorbed antisera to phospholipid-free cerebral cortex extracts against serum, liver, kidney, and finally brain soluble proteins. The antiserum then reacted specifically with the post-synaptic web of isolated synaptosomes, and the antigen was localized in the Triton-soluble fraction of crude synaptosomal plasma membrane preparations. No antigen was detected in myelin, nuclei, synaptic vesicles, microsomes, or mitochondria. This antigen thus joins the very limited group of compounds which have been detected in synaptosomal plasma membrane but not in synaptic vesicle and microsomal fractions such as the GP-350 (Van Nieuw Amerongen and Roukema, 1974) and a protein phosphorylated by intrinsic synaptosomal plasma membrane protein kinase (Ueda *et al.*, 1973). It would be tempting to relate these compounds to the morphological specialization of the junctional complex, but this will be difficult to test experimentally in view of the possible extraction of some proteins and glycoproteins from isolated junctions.

Some interesting immunological studies have been performed on cells in tissue culture. Akeson and Herschman (1974) have defined a set of antigens specific to differentiated neuroblastoma cells, which seem to be concentrated in particulate matter from brain. It would be most interesting to know whether these antigens are localized in synaptosomal plasma membranes. Is the antigen of Orosz *et al.* (1973, 1974) related to these? Schachner (1974) has reported glial cell-specific membrane antigens which should also prove to be extremely useful, but they are as yet relatively uncharacterized.

6. CONCLUSIONS

The data in the literature, although far from complete, go some way toward dealing with the lower levels of synaptic specificity in that major common glycoproteins and proteins of the synaptosomal plasma membrane have been specifically defined. There are some indications that the synaptosomal plasma membrane may have much in common with the axonal and perhaps the cell body plasma membrane from biochemical experiments, whereas the immunological approach has tended to suggest that the synaptosomal plasma membrane may be distinct from the other parts of the neuronal plasma membrane. These results are not necessarily contradictory, since biochemically major components can be immunologically minor components and *vice versa*.

It certainly seems logical to relate these specific antigens to the morphological specialization of the junctional complex and to the other synaptosomal plasma membrane specific components reported in the literature. However, as mentioned above, these correlations have not yet been tested experimentally (except for the antigen of Orosz *et al.*, 1973, 1974), and this may prove to be difficult.

The studies so far performed are hardly even relevant to higher levels of specificity. Moreover, very little is known about the properties of neural plasma membranes, which, without being directly responsible for specificity, may be involved in developmental mechanisms. Thus the existence of plasma membrane glycosyltransferases in the central nervous system is, to say the least, unproven. Although there is some suggestive evidence from tissue culture (Pyke and Dent, 1971; Treska-Ciesielski *et al.*, 1971; Gombos *et al.*, 1972), our knowledge of lectinlike and lectin-binding macromolecules in neural plasma membranes is limited.

Despite this rather pessimistic survey, there are some grounds for optimism. While it cannot be said that the problem of contamination of subcellular and cellular fractions has been eliminated, considerable progress in this area has been made. In particular, synaptosomal subfractions of acceptable and, perhaps more importantly, defined purity are available. By use of these techniques and the application of similar characterization procedures to other fractions, it seems feasible to define the lower levels of synaptosomal specificity. It might be a useful approach to attempt to identify the proteins reportedly synthetized by synaptosomes (Bosmann, 1971) or synaptosomal plasma membranes (Ramirez *et al.*, 1972), as well as those rapidly transported to the nerve ending (for review, see Jeffrey and Austin, 1973).

But it must be borne in mind that synaptosomal specificity, even if investigated at the final degree of analysis, may have little to do with neuronal recognition, since the synapse is the result of recognition phenomena rather than an active agent of these processes. Furthermore, the immense complexity of recognition processes in the central nervous system suggests that more progress is likely to be made in defining the basic principles involved in these phenomena by study of sponges and slime molds, or in a more neural context by study of grosser specificities such as those which come into play, at least at one level, in the innervation of muscle or the development of the visual system.

7. REFERENCES

Abe, T., and Norton, W. T., 1974, The characterization of sphingolipids from neurons and astroglia of immature rat brain, *J. Neurochem.* **23:**1025.

Akeson, R., and Herschman, H., 1974, Neural antigens of morphologically differentiated neuroblastoma cells, *Nature (London)* **249:**620.

Arbogast, B. W., and Arsenis, C., 1974, The enzymatic ontogeny of neurons and glial cells isolated from postnatal rat cerebral gray matter, *Neurobiology* **4:**21.

Avrova, N. F., Chenykaeva, E. Y., and Obukhova, E. L., 1973, Ganglioside composition and content of rat-brain subcellular fractions, *J. Neurochem.* **20:**997.

Banker, G., Crain, B., and Cotman, C. W., 1972, Molecular weights of the polypeptide chains of synaptic plasma membranes, *Brain Res.* **42:**508.

Banker, G., Churchill, L., and Cotman, C. W., 1974, Proteins of the postsynaptic density *J. Cell Biol.* **63:**456.

Barnola, F. V., Villegas, R., and Camejo, G., 1973, Tetrododoxin receptors in plasma membranes isolated from lobster nerve fibers, *Biochim. Biophys. Acta* **298:**84.

Barondes, S. H., 1968, Incorporation of radioactive glucosamine into macromolecules at nerve endings, *J. Neurochem.* **15**:699.

Barondes, S. H., 1974, Synaptic macromolecules: Identification and metabolism, *Annu. Rev. Biochem.* **43**:147.

Bittiger, H., and Schnebli, H. P., 1974, Binding of concanavalin A and ricin to synaptic junctions of rat brain, *Nature (London)* **249**:370.

Blomstrand, C., and Hamberger, A., 1969, Protein turnover in cell-enriched fractions from rabbit brain, *J. Neurochem.* **16**:1401.

Bloom, F. E., and Aghajanian, G. K., 1966, Cytochemistry of synapses: Selective staining for electron microscopy, *Science* **154**:1575.

Bloom, F. E., and Aghajanian, G. K., 1968, Fine structural and cytochemical analysis of the staining of synaptic junctions with phosphotungstic acid, *J. Ultrastruct. Res.* **22**:361.

Bock, E., and Jørgensen, O. S., 1975, Rat brain synaptic vesicles and synaptic plasma membranes compared by crossed immunoelectrophoresis, *FEBS Lett.* **52**:37.

Bock, E., Jørgensen, O. S., and Morris, S. J., 1974, Antigen–antibody crossed electrophoresis of rat brain synaptosomes and synaptic vesicles: Correlation to water-soluble antigens from rat brain, *J. Neurochem.* **22**:1013.

Bondareff, W., and Sjöstrand, J., 1969, Cytochemistry of synaptosomes, *Exp. Neurol.* **24**:450.

Bosmann, H. B., 1971, Identification of proteins and glycoproteins synthesized by isolated rat cortex synaptosomes *in vitro*, *Neurobiology* **1**:144.

Bosmann, H. B., 1972, Sialic acid on the synaptosome surface and the effect of concanavalin A and trypsin on synaptosome electrophoretic mobility, *FEBS Lett.* **22**:97.

Bosmann, H. B., and Carlson, W., 1972, Identification of sialic acid at the nerve-ending periphery and electrophoretic mobility of isolated synaptosomes, *Exp. Cell Res.* **72**:436.

Brady, R. O., and Quarles, R. H., 1973, The enzymology of myelination, *Mol. Cell Biochem.* **2**:23.

Breckenridge, W. C., and Morgan, I. G., 1972, Common glycoproteins of synaptic vesicles and the synaptosomal plasma membrane, *FEBS Lett.* **22**:253.

Breckenridge, W. C., Gombos, G., and Morgan, I. G., 1972, The lipid composition of adult rat brain synaptosomal plasma membranes, *Biochim. Biophys. Acta* **266**:695.

Broquet, P., and Louisot, P., 1971, Biosynthèse des glycoprotéines cérébrales. II. Localisation subcellulaire des transglycosylases cérébrales, *Biochimie* **53**:921.

Camejo, G., Villegas, G. M., Barnola, F. V., and Villegas, R., 1969, Characterization of two different membrane fractions isolated from the first stellar nerve of the squid *Dosidicus gigas*, *Biochim. Biophys. Acta* **193**:247.

Chacko, G. K., Goldman, D. E., and Pennock, B. E., 1972, Composition and characterization of the lipids of garfish *(Lepisosteus osseus)* olfactory nerve: A tissue rich in axonal membranes, *Biochim. Biophys. Acta* **280**:1.

Chacko, G. K., Goldman, D. E., Malhotra-H., C., and Dewey, M. M., 1974, Isolation and characterization of plasma membrane fractions from garfish *(Lepisosteus osseus)* olfactory nerve, *J. Cell Biol.* **62**:831.

Cohen, J., Dutton, G. R., Wilkin, G. P., Wilson, J. E., and Balazs, R., 1974, A preparation of viable cell perikarya from developing rat cerebellum with preservation of a high degree of morphological integrity, *J. Neurochem.* **23**:899.

Costin, A., Cotman, C., Hafemann, D. R., and Herschman, H. R., 1972, Effect of anti-brain synaptosomal fraction serum and complement on evoked potentials and impedance, *Experientia* **28**:411.

Cotman, C. W., and Matthews, D. A., 1971, Synaptic plasma membranes from rat brain synaptosomes: Isolation and partial characterization, *Biochim. Biophys. Acta* **249**:380.

Cotman, C. W., and Taylor, D., 1972, Isolation and structural studies on synaptic complexes from rat brain, *J. Cell Biol.* **55**:696.

Cotman, C. W., and Taylor, D., 1974, Localization and characterization of concanavalin A receptors in the synaptic cleft, *J. Cell Biol.* **62**:236.

Cotman, C. W., Blank, M. L., Moehl, A., and Snyder, F., 1969, Lipid composition of synaptic plasma membranes isolated from rat brain by zonal centrifugation, *Biochemistry* **8**:4606.

Cotman, C. W., McCaman, R. E., and Dewhurst, S. A., 1971, Subsynaptosomal distribution of enzymes involved in the metabolism of lipids, *Biochim. Biophys. Acta* **249**:395.

Cotman, C. W., Banker, G., Churchill, L., and Taylor, D., 1974, Isolation of post-synaptic densities from rat brain, *J. Cell Biol.* **63**:441.

Davis, G. A., and Bloom, F. E., 1970, Proteins of synaptic junctional complexes, *J. Cell Biol.* **47**:46A.

Davis, G. A., and Bloom, F. E., 1973, Isolation of junctional complexes from rat brain, *Brain Res.* **62**:135.

Dekirmenjian, H., Brunngraber, E. G., Lemkey-Johnston, N., and Larramendi, L. M. H., 1969, Distribution of gangliosides, glycoprotein-NANA and acetylcholinesterase in axonal and synaptosomal fractions of cat cerebellum, *Exp. Brain Res.* **8**:97.

Den, H., and Kaufman, B., 1968, Ganglioside and glycoprotein glycosyltransferases in synaptosomes, *Fed. Proc.* **27**:346.

Den, H., Kaufman, B., and Roseman, S., 1970, Properties of some glycosyltransferases in embryonic chicken brain, *J. Biol. Chem.* **245**:6607.

Denburg, J. L., 1972, An axon plasma membrane preparation from the walking legs of the lobster *Homarus americanus*, *Biochim. Biophys. Acta* **282**:453.

De Robertis, E., and Rodriguez de Lores Arnaiz, G., 1969, Structural components of the synaptic region, in: *Handbook of Neurochemistry*, Vol. II (A. Lajtha, ed.), pp. 365–392, Plenum Press, New York.

De Robertis, E., Lapetina, E., Pecci Saavedra, J., and Soto, E. F., 1966, *In vivo* and *in vitro* action of antisera against isolated nerve-endings of the brain cortex, *Life Sci.* **5**:1979.

De Robertis, E., Lapetina, E. G., and Wald, F., 1968, The effect of antiserum against nerve-ending membranes from cat cerebral cortex on the ultrastructure of isolated nerve-endings and mollusc neurons, *Exp. Neurol.* **21**:322.

Derry, D. M., and Wolfe, L. S., 1967, Gangliosides in isolated neurons and glial cells, *Science* **158**:1450.

de Vries, G. H., and Barondes, S. H., 1971, Incorporation of ^{14}C-*N*-acetylneuraminic acid into brain glycoproteins and gangliosides *in vivo*, *J. Neurochem.* **18**:101.

de Vries, G. H., and Norton, W. T., 1974, The lipid composition of axons from bovine brain, *J. Neurochem.* **22**:259.

de Vries, G. H., Norton, W. T., and Raine, C. S., 1972, Axons: Isolation from mammalian central nervous system, *Science* **175**:1370.

Dutton, G. R., Haywood, P., and Barondes, S. H., 1973, Carbon-14 labeled glucosamine incorporation into specific products in the nerve-ending fraction *in vivo* and *in vitro*, *Brain Res.* **57**:397.

Feit, H., and Barondes, S. H., 1970, Colchicine-binding activity in particulate fractions of mouse brain, *J. Neurochem.* **17**:1355.

Feit, H., Dutton, G. R., Barondes, S. H., and Shelanski, M. L., 1971, Microtubule protein: Identification and transport to nerve-endings, *J. Cell Biol.* **51**:138.

Festoff, B. W., Appel, S. H., and Day, E., 1971, Incorporation of ^{14}C-glucosamine into synaptosomes *in vitro*, *J. Neurochem.* **18:**1871.

Fischer, S., Cellino, M., Zambrano, F., Zampighi, G., Tellez Nagel, M., Marcus, D., and Canessa-Fischer, M., 1970, The molecular organization of nerve membranes. I. Isolation and characterization of plasma membranes from the retinal axons of the squid: an axolemma rich preparation, *Arch. Biochem. Biophys.* **138:**1.

Freysz, J., Bieth, R., Judes, C., Sensenbrenner, M., Jacob, M., and Mandel, P., 1968, Quantitative distribution of phospholipids in neurons and glial cells isolated from rat cerebral cortex, *J. Neurochem.* **15:**307.

Gombos, G., Hermetet, J. C., Reeber, A., Zanetta, J. P., and Treska-Ciesielski, J., 1972, The composition of glycopeptides, derived from neural membranes which affect neurite growth *in vitro*, *FEBS Lett.* **24:**247.

Gombos, G., Reeber, A., Zanetta, J. P., and Vincendon, G., 1974*a*, Fractionation of nervous tissue membrane glycoproteins, in: *Actes du Colloque International No. 221 du CNRS sur les Glycoconjugués*, pp. 829–844, Villeneuve d'Ascq, 20–27 June 1973, Editions du CNRS, Paris.

Gombos, G., Zanetta, J. P., Reeber, A., Morgan, I. G., and Vincendon, G., 1974*b*, Affinity chromatography of brain membrane glycoproteins, *Biochem. Soc. Trans.* **2:**627.

Grefrath, S. P., and Reynolds, J. A., 1973, Polypeptide components of an excitable plasma membrane, *J. Biol. Chem.* **248:**6091.

Gurd, J. W., and Mahler, H. R., 1974, Fractionation of synaptic plasma membrane glycoproteins by lectin affinity chromatography, *Biochemistry* **13:**5193.

Gurd, J. W., Jones, L. R., Mahler, H. R., and Moore, W. J., 1974, Isolation and partial characterization of rat brain synaptic plasma membranes, *J. Neurochem.* **22:**281.

Hamberger, A., and Svennerholm, L., 1971, Composition of gangliosides and phospholipids of neuronal and glial cell enriched fractions, *J. Neurochem.* **18:**1821.

Hemminki, K., 1973, Plasma membranes isolated from immature brain cells, Commentat. Biol. Soc. Fenn. **70:**1.

Hemminki, K., Hemminki, E., and Giacobini, E., 1973, Activity of enzymes related to neurotransmission in neuronal and glial fractions, *Int. J. Neurosci.* **5:**87.

Henn, F. A., Hansson, H. A., and Hamberger, A., 1972, Preparation of plasma membrane from isolated neurons, *J. Cell Biol.* **533:**654.

Herschman, H. R., Cotman, C. W., and Matthews, D. H. 1972, Serologic specificities of brain subcellular organelles. I. Antisera to synaptosomal fractions, *J. Immunol.* **108:**1362.

Jarosch, E., and Precht, W., 1972, Effects of antibodies directed toward membrane fragments of synaptosomes on cerebellar field potentials, *Brain Res.* **42:**225.

Jeffrey, P. L., and Austin, L., 1973, Axoplasmic transport, *Prog. Neurobiol.* **2:**205.

Jones, D. H., and Matus, A. I., 1974, Isolation of synaptic plasma membrane from brain by combined flotation–sedimentation density gradient centrifugation, *Biochim. Biophys. Acta* **356:**276.

Jørgensen, O. S., and Bock, E., 1974, Brain specific synaptosomal membrane proteins demonstrated by crossed immunoelectrophoresis, *J. Neurochem.* **23:**879.

Karlsson, J. O., Hamberger, A., and Henn, F. A., 1973, Polypeptide composition of membranes derived from neuronal and glial cells, *Biochim. Biophys. Acta* **298:**219.

Ko, G. K. W., and Raghupathy, E., 1971, Glycoprotein biosynthesis in the developing rat brain. I. Microsomal galactosyltransferase utilizing endogenous and exogenous protein acceptors, *Biochim. Biophys. Acta* **244:**396.

Ko, G. K. W., and Raghupathy, E., 1972, Glycoprotein biosynthesis in the developing rat brain. II. Microsomal galactosaminyltransferase utilizing endogenous and exogenous protein acceptors, *Biochim. Biophys. Acta* **264:**129.

Korneguth, S. E., Anderson, J. W., and Scott, G., 1969, Isolation of synaptic complexes in a caesium chloride density gradient: Electron microscopic and immunohistochemical studies, *J. Neurochem.* **16:**1017.

Lemkey-Johnston, N., and Dekirmenjian, H., 1970, The identification of fractions enriched in nonmyelinated axons from rat whole brain, *Exp. Brain Res.* **11:**392.

Lemkey-Johnston, N., and Larramendi, L. M. H., 1968, The separation and identification of fractions of non-myelinated axons from the cerebellum of the cat, *Exp. Brain Res.* **5:**326.

Levitan, I. B., Mushynski, W. E., and Ramirez, G., 1972, Highly purified synaptosomal membranes from rat brain: Preparation and characterization, *J. Biol. Chem.* **247:**5376.

Lim, R., and Hsu, L. W., 1971, Studies on brain-specific membrane proteins, *Biochim. Biophys. Acta* **249:**569.

Livett, B. G., Rostas, J. A. P., Jeffrey, P. L., and Austin, L., 1974*a*, Antigenicity of isolated synaptosomal membranes, *Exp. Neurol.* **43:**330.

Livett, B. G., Fenwick, E. M., and Eadie, J., 1974*b*, Ca^{++}-dependent redistribution of antigenic components during membrane fusion, *Proc. Aust. Biochem. Soc.* **7:**57.

Marcus, D., Canessa-Fischer, M., Zampighi, G., and Fischer, S., 1972, The molecular organization of nerve membranes. VI. The separation of axolemma from Schwann cell membranes of giant and retinal squid axons by density gradient centrifugation, *J. Membr. Biol.* **9:**209.

Margolis, R. U., 1969, Mucopolysaccharides, in: *Handbook of Neurochemistry*, Vol. I (A. Lajtha, ed.), pp. 245–260, Plenum Press, New York.

Margolis, R. U., and Margolis, R. K., 1974, Distribution and metabolism of mucopolysaccharides and glycoproteins in neuronal perikarya, astrocytes and oligodendroglia, *Biochemistry* **13:**2849.

Marinari, U. M., Morgan, I. G., Mack, G., and Gombos, G., 1972, Synthesis of synaptic glycoproteins. II. Delayed labeling of the glycoproteins of synaptic vesicles and synaptosomal plasma membranes, *Neurobiology* **2:**176.

Matus, A., De Petris, S., and Raff, M. C., 1973, Mobility of concanavalin A receptors in myelin and synaptic membranes, *Nature (London) New Biol.* **244:**278.

McBride, W. J., and Cohen, H., 1972, Cytochemical localization of acetylcholinesterase on isolated synaptosomes, *Brain Res.* **41:**489.

McBride, W. J., and Van Tassel, J., 1972, Resolution of proteins from subfractions of nerve-endings, *Brain Res.* **44:**177.

McBride, W. J., Mahler, H. R., Moore, W. J., and White, F. P., 1970, Isolation and characterization of membranes from rat cerebral cortex, *J. Neurobiol.* **2:**73.

McMillan, P. N., Mickey, D. D., Kaufman, B., and Day, E. D., 1971, Specificity and cross-reactivity of antimyelin antibodies as determined by sequential adsorption analysis, *J. Immunol.* **107:**1611.

Medzihradsky, F., Nandhasri, P. S., Idoyaga-Vargas, V., and Sellinger, O. Z., 1971, A comparison of the ATPase activity of the glial cell fraction and the neuronal perikaryal fraction isolated in bulk from rat cerebral cortex, *J. Neurochem.* **18:**1599.

Mickey, D. D., McMillan, P. N., Appel, S. H., and Day, E. D., 1971, Specificity and cross-reactivity of antisynaptosome antibodies as determined by sequential adsorption analysis, *J. Immunol.* **107:**1599.

Morgan, I. G., Wolfe, L. S., Mandel, P., and Gombos, G., 1971, Isolation of plasma membranes from rat brain, *Biochim. Biophys. Acta* **241:**737.

Morgan, I. G., Reith, M., Marinari, U., Breckenridge, W. C., and Gombos, B., 1972, The isolation and characterization of synaptosomal plasma membranes, in: *Advan. Exp. Med. Biol.* **25**:209.

Morgan, I. G., Breckenridge, W. C., Vincendon, G., and Gombos, G., 1973*a*, The proteins of nerve-ending membranes, in: *Proteins of the Nervous System* (Schneider, D. Johnson, ed.), pp. 171–192, Raven Press, New York.

Morgan, I. G., Zanetta, J. P., Breckenridge, W. C., Vincendon, G., and Gombos, G., 1973*b*, The chemical structure of synaptic membranes, *Brain Res.* **62**:405.

Morgan, I. G., Marinari, U. M., Dutton, G. R., Vincendon, G., and Gombos, G., 1974, The biosynthesis of synaptic glycoproteins, in: *Actes du Colloque International No. 221 du CNRS sur les Glycoconjugués,* pp. 1043–1052, Villeneuve d'Ascq, 20–27 June 1973, Editions du CNRS, Paris.

Morgan, I. G., Gombos, G., and Tettamanti, G., 1976, Glycoproteins and glycolipids of the nervous system, in: *Mammalian Glycoproteins and Glycolipids* (W. Pigman and M. I. Horowitz, eds.), in press.

Nagata, Y., Mikoshiba, K., and Tsukada, Y., 1974, Neuronal cell body-enriched and glial cell-enriched fractions from young and adult rat brains: Preparations, and morphological and biochemical properties, *J. Neurochem.* **22**:493.

Norton, W. T., and Poduslo, S. E., 1970, Neuronal soma and whole neuroglia of rat brain: A new isolation technique, *Science* **167**:1144.

Norton, W. T., and Poduslo, S. E., 1971, Neuronal perikarya and astroglia of rat brain: Chemical composition during myelination, *J. Lipid Res.* **12**:84.

Orosz, A., Hamori, J., Falus, A., Madarasz, E., Lakos, I., and Adam, G., 1973, Specific antibody fragments against the postsynaptic web, *Nature (London) New Biol.* **245**:18.

Orosz, A., Madarasz, E., Falus, A., and Adam, G., 1974, Demonstration of detergent-soluble antigen specific for the synaptosomal membrane fraction isolated from the cat cerebral cortex, *Brain Res.* **76**:119.

Pfenninger, K. H., 1971, The cytochemistry of synaptic densities. I. An analysis of the bismuth iodide impregnation method, *J. Ultrastruct. Res.* **34**:103.

Poduslo, S. E., and Norton, W. T., 1972, Isolation and some chemical properties of oligodendroglia from calf brain, *J. Neurochem.* **19**:727.

Pyke, K. W., and Dent, P. B., 1971, The *in vitro* effects of phytohaemagglutinin (PHA) on human neuroblastoma and chick embryo ganglion cells, *Fed. Proc.* **30**:398A.

Raghupathy, E., Ko, G. K. W., and Peterson, N. A., 1972, Glycoprotein biosynthesis in the developing rat brain. III. Are glycoprotein glycosyl transferases present in synaptosomes? *Biochim. Biophys. Acta* **286**:339.

Raiteri, M., Bertollini, A., and la Bella, R., 1972, Synaptosome antisera effect permeability of synaptosomal membranes *in vitro, Nature (London) New Biol.* **238**:242.

Rambourg, A., and Leblond, C. P., 1967, Electron microscope observations on the carbohydrate-rich cell coat present at the surface of cells in the rat, *J. Cell Biol.* **32**:27.

Ramirez, G., 1974, Nonspecific incorporation of glucosamine into rat brain synaptosomes and endoplasmic reticulum, *Biochim. Biophys. Acta* **338**:337.

Ramirez, G., Levitan, I. B., and Mushynski, W. E., 1972, Highly purified synaptosomal membranes from rat brain. Incorporation of amino-acids into membrane proteins *in vitro, J. Biol. Chem.* **247**:5382.

Reith, M., Morgan, I. G., Gombos, G., Breckenridge, W. C., and Vincendon, G., 1972, Synthesis of synaptic glycoproteins. I. The distribution of UDP-galactose: *N*-

acetylglucosamine galactosyl transferase and thiamine diphosphatase in adult rat brain subcellular fractions, *Neurobiology* **2:**169.

Robert, J., and Rebel, G., 1975, Gangliosides of cultured glial cells, *Abstracts Vth Int. Meeting ISN Barcelona*, p. 219.

Rodriguez de Lores Arnaiz, G., Alberici, M., and De Robertis, E., 1967, Ultrastructural and enzymic studies of cholinergic and non cholinergic synaptic membranes isolated from brain cortex, *J. Neurochem.* **14:**215.

Rose, S. P. R., 1967, Preparation of enriched fractions from cerebral cortex containing isolated, metabolically active neuronal and glial cells, *Biochem. J.* **102:**33.

Roseman, S., 1970, The synthesis of complex carbohydrates by multiglycosyl transferase systems and their potential function in intercellular adhesion, *Chem. Phys. Lipids* **5:**270.

Schachner, M., 1974, NS-1 (nervous system antigen-1), a glial-cell-specific antigenic component of the surface membrane, *Proc. Natl. Acad. Sci. USA* **71:**1795.

Schengrund, C. L., and Rosenberg, A., 1970, Intracellular location and properties of bovine brain sialidase, *J. Biol. Chem.* **245:**6196.

Sellinger, O. Z., Borens, R. N., and Nordrum, L. M., 1969, The action of trypsin and neuraminidase on the synaptic membranes of brain cortex, *Biochim. Biophys. Acta* **173:**185.

Sellinger, O. Z., Azcurra, J. M., Johnson, D. E., Ohlsson, W. G., and Lodin, Z., 1971, Independence of protein synthesis and drug uptake in nerve cell bodies and glial cells isolated by a new technique, *Nature (London) New Biol.* **230:**253.

Suzuki, K., Poduslo, S. E., and Norton, W. T., 1967, Gangliosides in the myelin fraction of developing rats, *Biochim. Biophys. Acta* **144:**375.

Tettamanti, G., Morgan, I. G., Gombos, G., Vincendon, G., and Mandel, P., 1972, Subsynaptosomal localization of brain particulate neuraminidase, *Brain Res.* **47:**516.

Treska-Ciesielski, J., Gombos, G., and Morgan, I. G., 1971, Effet de la concanavaline A sur les neurones de ganglions spinaux d'embryons de poulet en culture, *C. R. Acad. Sc. Paris* **273D:**1041.

Truding, R., Shelanski, M. L., Daniels, M. P., and Morell, P., 1974, Comparison of surface membranes isolated from cultured murine neuroblastoma cells in the differentiated or undifferentiated state, *J. Biol. Chem.* **249:**3973.

Ueda, T., Maeno, H., and Greengard, P., 1973, Regulation of endogenous phosphorylation of specific proteins in synaptic membrane fractions from rat brain by adenosine 3′-5′ monophosphate, *J. Biol. Chem.* **248:**8295.

Uesugi, S., Dulac, N. C., Dixon, J. F., Hexum, T. D., Dahl, J. L., Perue, J. F., and Hokin, L. E., 1971, Studies on the characterization of the sodium-potassium transport adenosine triphosphatase. VI. Large scale partial purification and properties of a lubrol-solubilized bovine brain enzyme, *J. Biol. Chem.* **246:**531.

van Nieuw Amerongen, A., and Roukema, P. A., 1974, GP-350, a sialoglycoprotein from calf brain: Its subcellular localization and occurrence in various brain areas, *J. Neurochem.* **23:**85.

Vos, J., Kuriyama, K., and Roberts, E., 1969, Distribution of acid mucopolysaccharides in subcellular fractions of mouse brain, *Brain Res.* **12:**172.

Waehneldt, T. V., Morgan, I. G., and Gombos, G., 1971, The synaptosomal plasma membrane: Protein and glycoprotein composition, *Brain Res.* **34:**403.

Wald, F., Mazzuchelli, A. N., Lapetina, E. G., and De Robertis, E., 1968, The effect of antiserum against nerve-ending membranes from cat cerebral cortex on the bioelectrical activity of mollusc neurons, *Exp. Neurol.* **21:**336.

Walters, B. B., and Matus, A. I., 1975, Proteins of the synaptic junction, *Biochem. Soc. Trans.* **3:**109.

Whittaker, V. P., 1969, The synaptosome, in: *Handbook of Neurochemistry*, Vol. II (A. Lajtha, ed.), pp. 327–364, Plenum Press, New York.

Whittaker, V. P., Michaelson, I. A., and Kirkland, R. J. A., 1964, The separation of synaptic vesicles from nerve ending particles (synaptosomes), *Biochem. J.* **90:**293.

Zambrano, F., Cellino, M., and Canessa-Fischer, M., 1971, The molecular organization of nerve membranes. IV. The lipid composition of plasma membranes from squid retinal axons, *J. Membr. Biol.* **6:**289.

Zanetta, J. P., Morgan, I. G., and Gombos, G., 1975, Synaptosomal plasma membrane glycoproteins: Fractionation by affinity chromatography on concanavalin A, *Brain Res.* **83:**337.

Zatz, M., and Barondes, S. H., 1970, Fucose incorporation into glycoproteins of mouse brain, *J. Neurochem.* **17:**157.

Zatz, M., and Barondes, S. H., 1971, Particulate and solubilized fucosyl transferases from mouse brain, *J. Neurochem.* **18:**1625.

III

Toward a Molecular Basis of Neuronal Recognition

7

Cell Recognition in Embryonic Morphogenesis and the Problem of Neuronal Specificities

A. A. MOSCONA

1. INTRODUCTION

Embryonic morphogenesis depends on the aggregation and organization of individual cells and cell groups into characteristic multicellular patterns which give rise to the definitive tissues and organs. During the early development of the nervous system, many of its precursor cells emigrate from their sites of origin toward their final locations; there, they associate with similar or other cell types into characteristic patterns and eventually become functionally linked. The experimental facts strongly suggest that the emergence in the embryo of the complex organization of the nervous system depends on cell–cell recognition or cellular affinities, i.e., on the ability of the cells to distinguish, through contact, one kind from another, and to display selectivities in forming morphogenetic associations. This capacity arises and evolves in the cells with their progressive differentiation. We assume that, as the various cell lines and sublines arise in the embryo, cell surfaces

A. A. MOSCONA·Departments of Biology and Pathology, and the Committee on Developmental Biology, University of Chicago, Chicago, Illinois.

become specified and encoded with molecular "labels" of increasing subtlety and that these contribute to the mechanisms which mediate cell recognition and selective cell adhesion. There are excellent indications from histological and experimental studies on neurological mutants of mice that failure in the mechanisms of cell recognition prevents cells from becoming correctly organized and may lead to defective morphogenesis in the affected parts of the nervous system (Sidman, 1974).

It should be noted that cell recognition involves affinities not only between identical cells (*homotypic* cell recognition) but also between different cells that cooperate developmentally (*allotypic* cell recognition). Examples of the latter are the contacts between glia and neurons, between neurons and muscle cells, and associations between different kinds of neurons (e.g., retina ganglion cells and optic tectum cells). Allotypic recognition can be as highly specific as homotypic affinity: in some cases, a single neuron seems to be recognized out of many (Baylor and Nicholls, 1971).

The mechanisms which directly mediate cell recognition and cell association into histogenetic groupings reside in the cell surface. Their expressions or characteristics change as cells undergo differentiation, and this alters the contact properties and "social interactions" of cells. These changes are undoubtedly regulated by genomic and phenotypic controls, and they may also be modulated by conditions in the cell's microenvironment, such as contact (or absence of contact) with other cells and with various extracellular constituents.

The elucidation of these changes will require a better understanding of the composition, structure, and biogenesis of embryonic cell membranes—matters about which there is, as yet, insufficient information. However, recently there has been significant progress in this area. It is now evident that, contrary to older views, the cell surface is a dynamic-mosaic structure, with a *variegated* architecture and composition, and diverse in different kinds of cells. Furthermore, there is increasing evidence that the surface plays an important role also in the regulatory circuitry of the cell, in that it contributes to the control of cell growth, cell metabolism, cell movement, and cell differentiation. It is reasonable to suppose that this nature of the cell surface has much to do with the mechanisms and processes of cell recognition and selective cell association in embryonic morphogenesis. Thus the stage seems to be set for venturing speculations, formulating new experimental questions in this area, and reviewing past information in the context of present concepts.

That sorting out of cells during the embryonic development of the nervous system reflects the existence of neuronal specificities was first suggested by the work and views of Sperry (1943); this led to the concept that unique chemical labels are present on each neuron and that these enable cell recognition and the establishment of highly selective interneuronal connections (Sperry, 1951, 1965). This concept has been extensively investigated on the visual system (reviews by Gaze, 1974; Hunt and Jacobson, 1974). As Gaze pointed out, virtually all experimental work in this system supports the existence of intercellular recognition based on neuronal chemospecificity, and the arguments are mainly about the mechanism.

Our own work (Moscona, 1952, 1962, 1963, 1968) led to the suggestion that embryonic cell recognition, in general, involves the function of specific molecular components of the cell surface. It indicated that the initial formation of specific neuronal connections may be governed by the same fundamental principles which apply to morphogenetic cell affinities and selective cell associations also in other developing systems. Consent with this view was expressed more recently by Barondes, by Gerisch, and by Roth, along with their associates (Rosen *et al.*, 1973, 1974; Gerisch *et al.*, 1972; Barbera *et al.*, 1973), on the basis of their studies on morphogenetic cell aggregations. The origins of these views go back to Lillie (1913), Weiss (1947), and Tyler (1946), and they draw on analogies from concepts derived from studies on immune systems (Jerne, 1967) and immunoglobulins (Edelman, 1967).

For obvious reasons, it has not been practical or feasible to analyze the mechanisms of cell recognition in intact embryos; therefore, we have developed experimental approaches utilizing reaggregation *in vitro* of dissociated embryonic cells. These experiments made it possible to construct tissues from single-cell suspensions under controlled conditions and thus to examine in considerable detail various aspects of cell recognition and morphogenesis. This approach was first applied to embryonic avian and mammalian nonneural cells (Moscona, 1952, 1957; Moscona and Moscona, 1952), then extended to neural retina cells (Moscona, 1960, 1961) and brain cells (Garber and Moscona, 1969, 1972*a*,*b*). The work has led to the isolation from embryonic and other cells of specific "cell aggregating" factors; and it has suggested, as a working hypothesis, that the effect of these factors is due to components derived from the cell surface which function as "cell-ligands" and mediate cell recognition and histotypic cell associations (Moscona, 1962, 1968, 1974; Hausman and Moscona, 1975).

In this chapter I will describe, in turn, certain of the key experiments on morphogenetic reaggregation of embryonic neural retina cells, the isolation and biochemical characterization of a retina-specific cell-aggregating factor, and finally the cell-ligand hypothesis and its possible implications for the problem of specific cell contacts in neurogenesis. Considering how little is known about these problems, speculations are indeed hazardous; nevertheless, much of this essay is devoted to raising questions, experimental possibilities, and hypothetical considerations, rather than to reviewing past information in detail.

2. CELL AGGREGATION

The embryonic neural retina of a 7–10 day chick embryo consists of several morphologically distinguishable kinds of retinocytes stratified in three major layers in a characteristic pattern. This tissue can be readily and cleanly isolated and maintained *in vitro*. When an isolated embryonic retina is briefly treated with pure trypsin, the enzyme cleaves intercellular bonds, degrades proteins at the cell surface and between cells, increases the fluidity of the cell membrane, and thus makes it possible for the cells to be released live from the tissue into suspension. Most of the dispersed single cells assume spherical shapes, but some retain their original shapes and cytoplasmic processes. Electronmicroscopic analysis has shown that the cell types distinguishable in the intact embryonic retina are all represented in the cell suspension (Sheffield and Moscona, 1970). Other embryonic tissues can be similarly dissociated into discrete, live cells; from the time of its introduction (Moscona, 1952), this "trypsinization" procedure has been widely used for the preparation of cell suspensions from a variety of sources.

The cell-separating effect of trypsin depends on its enzymatic activity, therefore on the presence of susceptible substrates at the cell surface and between cells. Since some of these materials are involved in linking the cells into a tissue, therefore when they are degraded, altered, or removed the cells disassociate from each other. As will be pointed out later on, reassociation of these cells into a tissue depends on regeneration of the removed components and their functional organization on the cell surface.

Under suitable culture conditions, the dissociated embryonic cells tend to reaggregate into multicellular clusters within which they progressively reconstruct tissue. Such histotypic cell reaggregation is most readily accomplished by means of the procedure of "aggregation by rotation" (Moscona, 1961). The cell suspension (in a suitable culture medium which need not contain serum or other proteins) is gently swirled in an Erlenmeyer flask on a gyratory shaker at 37°C under standardized conditions. If the cells are mutually adhesive, they reaggregate. Their aggregation kinetics (rate of cell inclusion in clusters and of the increase in the size of aggregates) depends on various cellular and experimental factors, most of which can be controlled and examined (Moscona, 1961). In general, greater mutual adhesiveness of the cells results in faster formation of larger aggregates.

Studies on the reaggregation kinetics of cells from different embryonic tissues and ages and on the shape and structure of aggregates led to several fundamental findings. Cells from different tissues and embryonic ages yield reproducible and characteristic "aggregation patterns" which can be used for examining the mechanisms of cell reassociation and the effects of various experimental variables. These aggregation patterns change with the embryonic age of the cells (i.e., with differentiation); in general, the ability of dissociated cells to reaggregate and reconstruct their typical tissue declines with age, more rapidly and markedly in some types of cells than in others. Finally, morphogenetic cell reaggregation does not require simultaneous cell replication; basically, it represents reassembly of cellular units into multicellular complexes, sorting out of the cells executed through cell movements, and organization of the cells.

In the early aggregates of neural retina cells (0–2 h of aggregation), the various types of retinocytes are interspersed mostly at random (Sheffield and Moscona, 1969); this nonselective stickiness of trypinized cells indicates "despecification" of the cell surface due to the effects of the enzyme. This primary phase of aggregation transits gradually into the second phase, during which the cells sort out, assume their characteristic positions, and achieve a retinotypic organization: the photoreceptor cells, bipolar cells, and ganglion cells segregate and align in their respective layers. As this process reaches completion (within 24 h after the beginning of aggregation), the reconstructed retina tissue resumes its characteristic biochemical and morphological differentiation, eventually including the formation of synapses (Sheffield and Moscona, 1970).

Additional evidence that in such reconstructed retina tissue the cells reestablish normal relationships comes from studies on induction by hydrocortisone of glutamine synthetase (GS). This enzyme is a marker of differentiation in the embryonic retina and an indicator of the normality of its intercellular contacts. GS is readily inducible in intact retina tissue (in vivo and in vitro); however, it is not inducible in separated retina cells (e.g., in primary monolayer cultures), or in cells that were caused to reassociate atypically, or were prevented from developing their normal retinotypic associations (Morris and Moscona, 1971; Moscona, 1974). On the other hand, in regular aggregates of retina cells the enzyme becomes inducible as soon as tissue reconstruction is achieved and retinotypic cell relationships are reestablished.

The above experiments raise an additional point of general relevance to the problem of specific cell contacts. Differentiation of neural retina cells, of which this enzyme induction is an indicator, evidently depends on the maintenance of histotypic cell associations, i.e., on the cell surfaces being in correct contact relationships. This suggests that conditions at the cell surface arising from the presence or absence of histotypic cell contacts may affect the ability of the cells to pursue phenotypic developments. Thus, in considering the problem of specific cell contacts in neuronal development, it might be worthwhile to keep in mind their possible relationship to sustaining or promoting neuronal differentiation. The phenomenon of degeneration of unconnected or misconnected neurons during brain development may be an aspect of this problem.

Similarly to the aggregation of retina cells, suspensions of cells from various regions of the brain (cerebrum, diencephalon, optic tectum, cerebellum, and medulla) from mouse or chick embryos of appropriate ages reaggregate and reconstruct brainlike tissues (Garber and Moscona, 1969, 1972*a*; DeLong, 1970). By whatever mechanisms the cells had originally acquired their positions in the embryo, following their dissociation and reaggregation they are again able to construct their typical tissue patterns. It is a reasonable assumption that both in normal development and in cell aggregates the sorting out of cells is mediated by similar or identical mechanisms and that the information for cell organization resides in the nature of these mechanisms.

In this connection, of particular interest is the work on aggregation of cerebrum cells from "reeler" mutant mouse embryos (DeLong, 1970; DeLong and Sidman, 1970; Sidman, 1974). The reeler's disorder is fundamentally a problem of failure in cell recognition (Sidman, 1968), and it results in malpositioning of cells in certain specific areas of the brain. This lesion is reflected also in cell reaggregation *in vitro:* while cells dissociated from normal cerebral cortex sort out correctly in reaggregates and reestablish a pattern which closely resembles that of the normal embryonic cortex, reeler cells form in aggregates atypical patterns similar to those of the embryonic reeler cortex; i.e., the reeler cells do not sort out normally and do not become organized into obvious laminae (DeLong and Sidman, 1970). These differences between normal and mutant patterns of cell sorting out are consistent with the view that the mechanisms of cell recognition and selective adhesion reflect gene action and thus involve products of gene expression (Moscona and Moscona, 1962; Moscona, 1974).

Other tissues of the reeler mouse show no obvious histological abnormalities; therefore, it appears that the activity of these mutant gene(s) relates narrowly to cell recognition in the brain. There is considerable evidence that in other tissues of the mouse embryo, cell associations involve the activities of other genes (Bennett, 1964; Bennett *et al.*, 1972; Artzt and Bennett, 1976). Thus *tissue-specific* cell recognition may reflect the function of tissue-specific components, or "labels" in cell surface, coded for by different genes. This possibility is consistent with the existence of tissue-specific cell surface antigens on embryonic cells (Goldschneider and Moscona, 1972).

It has long been known that histotypic reaggregation of trypsin-dissociated cells requires macromolecular synthesis. Cell reaggregation stops within 1–2 h in the absence of protein synthesis, and it is suppressed within 4 h following arrest of RNA synthesis (Moscona and Moscona, 1963, 1966). There is evidence that synthesis is needed, at least in part, for replenishing cell surface macromolecules degraded by the enzyme and that some of these are required for histotypic reaggregation of the cells.

Related to this are the age-dependent changes in cell aggregability: cells dissociated from tissues of late embryos are less capable of reaggregation than cells from the same tissues of early embryos. This age-dependent decline has been demonstrated for several systems and particularly for several systems and particularly for neural retina cells (Moscona, 1962) and for cells from various parts of the embryonic brain (Garber and Moscona, 1972*a*). It has been suggested that the inability of mature cells to reaggregate after separation with trypsin may result from phenotypic specialization of their biosynthetic processes (Moscona, 1962); i.e., after morphogenesis and differentiation are completed, the need for certain products which were required earlier in development is reduced, and their synthesis is phased out. This includes molecules for cell linking, and hence their removal from the surface of mature-specialized cells is not followed by effective replacement.

Another aspect related to this age-dependent decline in cell aggregability should be mentioned. It has been established that the overall properties of the cell membrane change with differentiation. Thus in cells dissociated from early embryonic retinas the surface is very active and forms numerous filopodia and ruffled membranes; in cells from mature retinas it does not (Ben-Shaul and Moscona, 1975). Furthermore, agglutination of retina cell suspensions with concanavalin A (which binds to carbohydrate sites in the cell membrane) declines with embryonic age (Kleinschuster and Moscona, 1972; Martinozzi and Moscona, 1975). Both these changes indicate a decrease in the "fluidity" of the cell membrane in the course of cell maturation. As discussed later on, this decrease may be a contributing factor in the failure of mature cells to reestablish histotypic contacts, following their dissociation from the tissue.

3. TISSUE-SPECIFIC CELL RECOGNITION

The above discussion has dealt chiefly with reaggregation of cells in suspensions derived from a single embryonic tissue, i.e., with the abilities of the various cell types that belong in the same tissue to sort out and to reorganize their tissue pattern; as such, it dealt with *cell-type specific* cell recognition. The existence of yet another category of cell recognition—*tissue-specific* cell recognition—has been demonstrated by commingling in the same suspension cells derived from different tissues of the same embryo. When such composite cell suspensions reaggregate, the cells from the different tissues tend to sort out, to segregate into distinct groupings, and to re-form their characteristic histological pattern. For example, in mixtures of embryonic retina and scleral cartilage cells, the different cells sort out and re-form patches of retina tissue and fragments of cartilage (Moscona, 1962). It therefore appears that the surfaces of cells from various tissues (or from the major groupings of cells in the embryo) are specified by different "labels" and that these may be involved in tissue-specific cell recognition and attachment.

Similar tissue-specific sorting out of embryonic cells has been described for a variety of other binary cell combinations (Moscona, 1974). Even in mixtures of cells from different regions of the embryonic brain, cells from the same region display a preferential tendency to associate with each other (Garber and Moscona, 1972*b*). However, in brain cell combinations, cell segregation is not as complete and sharp as in mixtures of cells from entirely different tissues; depending on the embryonic age of the cells and the brain region from which they were derived, cells may sort out well or display considerable "cross-reactivity."

Accepting the assumption that tissue-specific cell recognition reflects the existence on cell surfaces of tissue-specific constituents, the question arose of whether cells from different embryonic tissues possess different cell-surface antigens. Convincing evidence for this was obtained by preparing antisera against suspensions of live cells from various tissues of the chick embryo including neural retina, liver, cerebrum, myocardium, and skeletal muscle. After thorough absorption of each antiserum with heterologous cells, the antisera were found to react preferentially with the surface of the cells used as the immunogen (Goldschneider and Moscona, 1971, 1972). Thus antiserum prepared against neural retina cells agglutinated in a wide range of titers only retina cells. Fluorescein-labeled antiserum reacted only with the surface of retina cells and stained the surfaces of all the cell types present in the neural retina. When tested *in situ*, i.e., on frozen sections of the eye, this antiserum stained intercellular contact zones only in the neural retina. It did not react with other tissues in the eye, including the pigmented epithelium. It will be recalled that, although the pigmented epithelium is developmentally related to the neural retina, it differentiates into a different type of tissue; consequently, its cell surfaces become antigenically distinct from those of neural retina cells.

A specific antiserum was obtained also against the surface of embryonic liver cells, and it did not react with the surface of retina cells of cells from other tissues. Several similarly tissue-specific antisera were obtained, attesting to the presence on embryonic cell surfaces of determinants which distinguish cells from different tissues but are shared by cells from the same tissue.

In exploratory tests with antisera against cells from various regions of the embryonic brain, the existence of antigenic differences between cell surfaces was revealed. These differences corresponded, in general, to the sorting out affinities of these cells in aggregates; however, the antigenic differences were not absolute in that cross-reactivities were noted. The cross-reactions may reflect the fact that all brain regions are developmentally related or that they all share certain similar cells. It would be of interest to determine if this cross-reactivity varies depending on embryonic stage. However, the results strongly suggest the existence also of specific antigenic differences between cell surfaces from different brain regions.

Taken as a whole, the available evidence clearly points to a consistency between *tissue-specific cell recognition* and the presence of *tissue-specific antigens* on cell surfaces. Some of these surface antigens may represent cell-surface components which make up the mechanism of cell recognition and selective cell associations. The existence of such tissue-specific cell-linking components is supported by the isolation of specific cell-aggregating factors.

4. SPECIFIC CELL-AGGREGATING FACTORS

The assumption was made that certain macromolecular components of the cell surface mediate histotypic cell recognition and association by functioning as specific *cell ligands*. The possibility arose that it might be feasible to isolate materials with tissue-specific cell-ligand effects from live cells and to assay them on aggregating cells. It was expected that the addition of such a material to a suspension of cells from the homologous tissue would enhance cell reaggregation; this would result in larger aggregates than in the controls, and in more efficient cell sorting out in mixtures of cells from different tissues.

In attempting to isolate such materials, advantage was taken of the fact that if dissociated embryonic cells are maintained dispersed in a serum-free medium they synthesize and exteriorize materials that normally are associated with the cell surface and with intercellular spaces. The initial experiments were carried out with embryonic neural retina cells in suspension cultures; subsequently, it was more convenient to work with cells in primary monolayer-type cultures, and procedures were developed for the collection and bioassay of products exteriorized into the supernatant medium (Moscona, 1962; Lilien and Moscona, 1967; Lilien, 1968; McClay and Moscona, 1974).

Electrophoresis in SDS-polyacrylamide gels of the supernatant medium from monolayer cultures of retina cells (obtained from 10-day chick embryos and maintained in culture for 24–48 h) resolved a large number of protein containing bands. Although a detailed identification of all these bands was not attempted, some probably represent products common to various embryonic tissues; others may be limited to neural cells; still others appear to be retina specific. It should be stressed that in attempting to isolate materials with retina cell-ligand activity the crucial test is tissue-specific function. As is well known, a variety of cell-derived molecules can adsorb or bind to various cell surfaces and some will cause cell clumping, clotting, flocculation, or agglutination. Such nondevelopmental effects should not be confused with morphogenetic cell association and hence with the effects postulated for specific cell–cell ligands.

In the initial experiments (Moscona, 1962; Lilien and Moscona, 1967), it was found that the supernatant collected from primary retina cell cultures contained a protein factor which when added to fresh suspensions of neural retina cells markedly enhanced cell reaggregation, resulting in significantly larger aggregates than in the controls. Most importantly, the effect of this *retina factor* was specific for neural retina cells; it did not enhance the reaggregation of cells from other embryonic tissues, including brain cells. The factor did not clot or agglutinate the cells; its effect was progressive and the reaggregated cells proceeded to reconstruct retina tissue.

The factor was rapidly removed by the cells from the medium, also at 4°C; however, response to it became noticeable only after about 2 h of cell rotation at 37°C; from then on, enhancement of aggregation progressed gradually to a maximum at about 12 h. Thus one can distinguish between binding of factor to the cells and expression of its effect on cell aggregation. Furthermore, it is known that binding of factor does not require simultaneous protein synthesis in the cells, or optimal temperatures, while expression of its effect does. This "multiple-phase" situation is undoubtedly significant with respect to the factor's mode of action. There is evidence that the initial adsorption of the retina factor to trypsin-dispersed cells may not be tissue specific, while its effect on cell aggregation is; this cautions against considering adsorption of the factor to cells as a definitive indicator of its functional characteristics or specificity.

A further important point is that active cell-aggregating factor could be obtained from retina cells only during a limited period of their ontogeny; cells from chick embryos past day 12 of development did not yield this activity under the experimental conditions used above. Corresponding to this, also the responsiveness of retina cells to the effect of the retina factor declined markedly after day 12 of development (Moscona, 1962). It is important to point out that these changes coincide with the age-dependent decline in cell aggregability (discussed earlier in this chapter); this indicates that the experimental effects of the cell-aggregating factor closely resemble the normal properties of these cells.

By use of the above procedures, cell-aggregating factors were obtained also from cerebrum and cerebellum cells of mouse embryos; these were found to enhance preferentially the reaggregation of cells from their respective brain regions but not of other neural or nonneural cells (Garber and Moscona, 1972*b*). It is particularly interesting that the mouse cerebrum factor also enhanced the reaggregation of chick cerebrum cells of the corresponding developmental age. This transspecies effect of the factor is remarkably consistent with earlier findings which have shown that the mechanisms of cell recognition can function between cells from these two species; i.e., when embryonic mouse and chick cells from a homologous tissue (e.g., cerebrum) are coaggregated, they do not segregate according to species but associate into bispecific tissue mosaics (Moscona, 1957, 1962), while in heterologous cell mixtures (e.g., cerebrum and cerebellum cells) there is cell segregation (albeit not complete) into distinct groupings (Garber and Moscona,

1972*b*). Thus the manner of cell recognition appears to correspond rather closely with the effects of cell-aggregating factors; therefore, the same functional principle may be involved in both cases. The fact that cell recognition is not strictly species specific suggests that its genetic-molecular basis may have arisen early in vertebrate phylogeny and that the fundamental features of this mechanisms are similar in higher forms.

Experiments are in progress to determine if, using the present procedures, tissue-specific cell-aggregating factors are obtainable also from cells of still other embryonic tissues. It should be borne in mind that factor effects need not always result in increases in the size of aggregates, but rather in enhancement of cell sorting out. The present procedure for collection of factors, which depends on their release from live cells into the culture medium, should not be expected to be universally effective; in some types of cells the factor may remain associated with the cell surface, or it may be inactive in the released form. It would be clearly desirable to have other methods for factor preparation and assay. Methods involving the use of isolated cell membranes (Merrell and Glaser, 1973) and of derivatized synthetic fibers (Edelman *et al.*, 1971) offer considerable promise in this direction. Fractionation of cell homogenates is another possibility (Rosen *et al.*, 1973).

5. PURIFICATION OF THE RETINA CELL-AGGREGATING FACTOR

Chromatography on Sephadex G200 of the crude preparation of the retina cell-aggregating factor (obtained fom 2-day primary monolayer cultures of 10-day chick embryo retina cells) yielded a glycoprotein fraction (Fr. II) which contained the characteristic factor activity. Electrophoresis on SDS-polyacrylamide gels separated this fraction into four protein-containing bands with approximate molecular weights of 35,000, 50,000, 65,000, and 70,000; the first three bands also stained with PAS. Further analysis showed that only the 50,000 mol wt material (Fr. IIB) contained the retina-specific cell-aggregating activity (McClay and Moscona, 1974). Occasionally activity was present also in the 65,000 mol wt fraction, possibly due to contamination with the lighter material.

Further fractionation by column chromatography and electrofocusing resulted in the isolation of a *glycoprotein* (Fr. IIB2) which possessed the typical activity of the retina cell-aggregating factor. It formed a single band on both SDS and non-SDS acrylamide gels in the region of 50,000 mol wt which stained lightly with PAS. By this stage, at least a seventy-fold purification of the cell-aggregating activity present in the crude factor preparation was achieved (Hausman and Moscona, 1975). By use of radioactive precursors (amino acids and sugars), it was demonstrated that the material in Fr. IIB2 was synthesized by the cells from which the factor was collected.

The molecular weight of this purified glycoprotein was recalculated from its sedimentation coefficient, Stoke's radius, and partial specific volume and confirmed to be 50,000 ± 5000. Its fractional ratio of 1.3 suggests, on first approximation, a particle with a long and a short axis. We were unable to detect the existence of subunits in this material by attempting to cleave interchain disulfide bonds. In fact, its biological activity is resistant to treatment with SDS, urea, and mercaptoethanol. However, this does not exclude the possibility that the biological effect of this material at the cell surface may involve a multimeric state, i.e., complexing or interactions with homologous or heterologous components. Also, we cannot rule out at present the possible presence in this glycoprotein of heterogeneities of a kind which would not be detectable by the methods used so far.

The glycoprotein nature of the purified retina cell-aggregating factor was further confirmed by analysis of its amino acid and sugar composition (Table I). It has a relatively high content of aspartic and glutamic acid residues. Its carbohydrate content is only about 10% of the total, and consists of glucosamine, mannose, galactose, and sialic acid.

The biological activity of this material is rapidly destroyed by trypsin. It is not destroyed by removal of sialic acid, by exposure to galactose oxidase (following desialation) or to β-galactosidase, or by brief treatment with periodate (Hausman and Moscona, 1975). Therefore, the intactness of the carbohydrate portion does not seem to be essential for the characteristic effect of this protein. It is, of course, possible that sugars are added to the treated material after it attaches to the cell surface. The purified retina cell-aggregating factor is not sequestered by concanavalin A, nor does this lectin prevent its effect on retina cells (Hausman and Moscona, 1973). It does not exhibit galactosyltransferase activity, nor does it serve (in the form in which it is purified) as an acceptor for UDP-galactose transferred by this enzyme (Garfield *et al.*, 1974). All this does not necessarily contradict other suggestions that carbohydrates play some role in cell recognition and cell associations. Conceivably, the mechanism of specific cell linking may require interactions of several cell surface components, and the glycoprotein isolated by us could represent only one of these elements;

Table I. Retina-Specific Cell-Aggregating Factor: Amino Acid and Sugar Composition[a]

Number of residues per mole			Number of residues per mole		
Lys	31		Val	34	
His	12		Met	3	
Arg	13		Ile	15	
			Leu	40	
		56	Tyr	10	
			Phe	10	
Asp	47		Trp	ND	
Glu	55				
					112
		102			
			Glcn	10	
Thr	23		Man	10	
Ser	22		Gal	10	
Pro	25		Sial acid	1–2	
Gly	24				
Ala	40				32
Cyst	ND				
		134			

[a] For details, see Hausman and Moscona (1975).

the others may involve carbohydrates or the activities of glycosylating enzymes (Roseman, 1970).

The tests performed so far indicate that the purified glycoprotein is homologous with constituent(s) of the embryonic retina cell surface (Hausman and Moscona, 1973, 1975). Taking all the available information, we have suggested, as an exploratory concept, that this glycoprotein is a component of the retina-specific cell-ligand mechanism (Hausman and Moscona, 1975). In the course of normal reaggregation of trypsin-dissociated cells, this mechanism is regenerated on the cell surface; if the glycoprotein is supplied exogenously to receptive cells, it binds to the cell surface, participates in the ligand mechanisms, and enhances cell reaggregation.

Also in other developmental systems, specific protein components of the cell surface are involved both in morphogenetic cell aggregation and in the effects of isolated specific cell-aggregating factors; these systems include sponge cells (Moscona, 1963, 1968; Humphreys, 1963; McClay, 1971; Henkart *et al.*, 1973; Turner and Burger, 1973), cellular slime molds (Beug *et al.*, 1973; Rosen *et al.*, 1973, 1974), and mammalian teratoma cells (Oppenheimer and Humphreys, 1971). Thus, there appears to exist a category of cell surface proteins which are constituents of cell-ligand mechanisms and whose properties determine the cognitive reactions and morphogenetic associations of cells. For brevity, the operational term "cognins" is proposed for this particular category of cell surface molecules. The isolated retina glycoprotein is considered to be a retina cognin, i.e., a specific constituent of the cell-ligand mechanisms of retina cells. "Discoidin," a specific factor isolated from *Dictyostelium discoideum* cells (Simpson *et al.*, 1974), represents a cognin of these cells. And the cell-aggregating factors obtained from sponge cells contain cognin components of sponge cell ligands. The characteristics of cognins from different tissues, cells, and organisms may be expected to differ in various ways. It is of course conceivable that more than one kind of cognin molecule may be involved in the composition of cell ligands on a given cell.

Concerning mode of action, there is considerable evidence that the primary site of action of the retina cell-aggregating glycoprotein is at the cell surface (Lilien and Moscona, 1967; Lilien, 1968; Balsamo and Lilien, 1974). A cell surface site of action has been convincingly demonstrated for the specific cell-aggregating factors isolated from sponge cells (Moscona, 1968) and slime mold cells (Rosen *et al.*, 1974). It was pointed out above that, in the case of the retina, mere binding of the glycoprotein to cells in suspension does not suffice for expression of characteristic cell-ligand activity; this expression requires ongoing protein synthesis in the test cells, and is inhibited also if the cells are maintained at 4°C. Evidently, in addition to availability of ligand molecules on the cell surface, still other processes are needed for cell reassociation.

One possibility is that the characteristic effect of the factor requires its interactions with still other cell surface components; these are also lost during cell trypsinization and cannot be replenished in the absence of protein synthesis or of active metabolism.

Another possibility has to do with the topographical organization of the cell surface. As pointed out earlier, trypsinization elicits changes in the overall configuration of the retina cell surface (Ben-Shaul and Moscona, 1975; Martinozzi and Moscona, 1975). It is conceivable that, unless and until the appropriate surface architecture is regained, cell-ligand mechanisms cannot function in a specific fashion. This implies that cell recognition depends on the ligands being arranged on the cell surface in a particular topographic pattern, a mosaic which may be characteristic for the type and the developmental stage of the cell. The reorganization of this pattern on trypsinized cells would require, in addition to metabolic processes, movements of components within the cell membrane; low temperatures, which inhibit cell aggregation, hinder such movements (Hubbel and McConnell, 1969; Nicolson, 1972; Yahara and Edelman, 1972). Accordingly, it might not be too surprising to find that cell recognition depends not just on the presence of functional cell ligands but also on their topographic arrangement on the cell surface. A similar concept was proposed for cell agglutination by lectins (Rutishauser and Sachs, 1974).

6. THE CELL-LIGAND HYPOTHESIS

At this point, it might be useful to review the hypothetical framework for the various views and interpretations expressed above. The basic concept is that morphogenetic cell recognition and selective cell contacts are mediated by interactions of cell surface components which function as specific cell ligands (Moscona, 1962, 1968, 1974). The working assumption is that the isolated retina-specific cell-aggregating glycoprotein is a constituent of the cell-ligand mechanism of these cells and that the activity of other specific cell-aggregating factors is similarly due to their content of specific cell-ligand components. How these components interact so as to link cells into tissues remains presently a speculative matter (Roseman, 1974; Moscona, 1974).

Cell recognition implies specificity of ligand interactions. In our view, the specificity of these interactions depends not only on the chemical characteristics of ligands but also on their topographic organization on the cell surface at the given stage of development. We proposed that cells of different kinds and developmental stages are distinguished by characteristic ligand "patterns." These patterns represent a combination of several conditions: the amount, composition, and diversity of the ligands on the cell surface; their topographic organization and relations with other components of the cell surface and of the intercellular space; and the changes which take place during development and differentiation. We propose that complementary ligand patterns are conducive to positive cell recognition and morphogenetic cell adhesion; absence of such complementarity results in negative recognition, nonadhesion, or transient nonspecific cell contacts. Since, in this view, cell recognition requires a matching of ligand topographies, it follows that cells with qualitatively similar ligands might not show mutual affinity in the absence of topographical–temporal correspondences in ligand displays.

Our hypothesis suggests that ligand patterns evolve, diversify, and specialize as embryo cells undergo differentiation: in the very early embryo they demarcate the cells into ectoderm, endoderm, mesoderm; then into the major tissue classes; and finally into the different cell types within each tissue. This evolving specialization and diversification is reflected in the progressive reassortments and organization of the cells into tissues during embryonic morphogenesis.

The development of ligand patterns is undoubtedly dependent, in part, on expression of genetic information for the synthesis of the cognins, i.e., the proteins of which the ligands are composed. The carbohydrate portion of cognins draws attention to the role of the Golgi system and of regulation of enzymes for carbohydrate synthesis in the modulation of the cell surface (Whaley *et al.*, 1972). However, ligand patterns may also be modulated by environmental effects (including contact with other cells), and their stability may depend on relationships of ligands with other components of the cell surface and with inframembrane structures.

Finally, it should not be overlooked that various intercellular substances, as well as junctional specializations of the cell membrane, serve in holding cells together after they have become specifically linked by their ligands. Proteoglycans and fibrous proteins which contribute to "intercellular cements" and "intercellular matrices" belong in this category of, probably, nonspecific cell adhesives.

The retina glycoprotein and other cell-aggregating factors isolated so far from embryonic tissues display tissue-specific effects and therefore belong in the category of *tissue-specific* cell-ligands. While this class of ligands can account for tissue-specific cell recognition and segregation of cells into different tissues, it does not readily explain cell-type-specific recognition, i.e., the sorting out and organization of the various kinds of cells that make up a single tissue. For example, the histological organization of neural retina, in normal development as well as in cell aggregates, depends on the correct positioning, alignment, and connections of the different retinocytes. Assuming that all retina cells are encoded with tissue-specific surface "labels" by which they are demarcated from nonretina cells, they must have still additional surface information which enables the different retinocytes to take up correct positions relative to one another. The need for such cell-type-specific surface information is obvious with respect to morphogenesis in all tissues, but especially in order to account for cell organization in the nervous system. This complex problem is further compounded by the fact that a single neuron may display different affinities in different regions of its surface; such "variegation" of affinities on the cell surface presumably enables a neuron to make appropriate contacts with other neurons and with glial cells.

The concept of *cell surface "variegation"* suggested here is not without some support. The presence of two classes of contact sites on different regions of the same cell surface has been demonstrated for aggregating slime mold cells: one of these is involved in end-to-end cell association and the other in side-by-side cell contacts (Beug *et al.*, 1973).

Analysis of the mechanisms which underlie this hierarchy of cell surface specifications (tissue specificities, cell type specificities, surface "variegations") is of prime importance for understanding morphogenesis in the nervous system. Such studies may be aided by the following speculative considerations derived from the cell-ligand hypothesis.

One possibility is that cell-type-specific (or neuron-specific) recognition results from the presence on different cells of qualitatively different cell ligands (perhaps of more than one kind). Considering the multiplicity of cell types, this possibility demands mechanisms for generating a large chemical diversity of cell ligands by the same genome. As pointed out by Edelman (1967), an attractive analogy might be the diversity of sequence and heterogeneity in antibodies generated from the products of a relatively small number of genes. (For further discussion of this attractive and heuristically useful analogy, see Edelman *et al.*, 1974.)

Another possibility is that the surface information for cell-type-specific recognition is provided by topographic-temporal differences in the distribution and organization of the cell ligands. Thus, while tissue-specific cell recognition may depend largely on qualitative differences between cell ligands, it is the differences in ligand topographies—i.e., the "variegation" of ligand patterns—that could be mostly responsible for surface specification of the diverse cells that make up a given tissue. Developmental modulation of these surface "variegation" patterns would further increase the informational versatility of such a system. If follows that a relatively limited chemical vocabulary of ligands, deployed in a variety of topographical–temporal permutations, could generate on cell surfaces a great diversity of ligand patterns; this diversity could be sufficiently large to account even for neuronal specificities in the morphogenesis of the nervous systems.

Pursuing these notions, one might venture to suggest that each of the major regions of the nervous system is characterized by qualitatively different classes of cell ligands (represented by the different cell-aggregating factors derived from these regions) and that within each region the diverse neuronal surfaces are specified by differently "variegated" and dynamic ligand patterns. This distinction is probably not absolute, and overlapping is to be expected. Also, the possibility that different kinds of ligands may be present on the same cell surface should not be excluded. The evolution and modulation of these ligand patterns during neuronal ontogeny and differentiation could, in theory, account for the progressive specification of neuronal surfaces, and consequently for cell sorting out, morphogenesis, and adjustments of cell contacts in the nervous system. Whether these predictions are fulfilled or not, at the very least their examination might generate useful insights into these problems.

Acknowledgments

The research program of my laboratory is supported by Grant HD01253 from the National Institute of Child Health and Human Development and in part by Grant 1-P01-CA 14599 to the Cancer Research Center of The University of Chicago. Personal support from the Louis Block Fund of The University of Chicago is gratefully acknowledged.

7. REFERENCES

Artzt, K., and Bennett, D., 1976, Cell surface antigens associated with cell differentiation, in: *Cell Surfaces and Malignancy* (in press).

Balsamo, J., and Lilien, J., 1974, Functional identification of three components which mediate tissue-type specific embryonic cell adhesion, *Nature (London)* **251**:522.

Barbera, A. J., Marchase, R. B., and Roth, S., 1973, Adhesive recognition and retinotectal specificity, *Proc. Natl. Acad. Sci. USA* **60**:2482.

Baylor, D. A., and Nicholls, J. G., 1971, Patterns of regeneration between individual nerve cells in the central nervous system of the leech, *Nature (London)* **232**:268.

Bennett, D., 1964, Abnormalities associated with a chromasome region in the mouse. II. Embryological effects of lethal alleles in the *t*-region, *Science* **144**:263.

Bennett, D., Boyse, E. A., and Old, L. J., 1972, Cell surface immunogenetics in the study of morphogenesis, in: *Cell Interactions: Proceedings of the Third Lepetit Colloquium, 1971* (L. G. Silvestri, ed.), pp. 247–263, North-Holland, Amsterdam.

Ben-Shaul, Y., and Moscona, A. A., 1975, Scanning electron microscopy of embryonic neural retina cell surfaces, *Dev. Biol.* **44**:386.

Beug, H., Katz, F. E., and Gerisch, G., 1973, Dynamics of antigenic membrane sites relating to cell aggregation in *Dictyostelium discoideum*, *J. Cell Biol.* **56**:647.

DeLong, G. R., 1970, Histogenesis of fetal mouse isocortex and hippocampus in reaggregating cell cultures, *Dev. Biol.* **22**:563.

DeLong, G. R., and Sidman, R. L., 1970, Alignment defect of reaggregating cells in cultures of developing brains of reeler mutant mice, *Dev. Biol.* **22**:584.

Edelman, G. M., 1967, Antibody structure and diversity: Implications for theories of antibody synthesis, in: *The Neurosciences* (G. C. Quarton, T. Melnechuk, and F. O. Schmitt, eds.), pp. 188–200, Rockefeller University Press, New York.

Edelman, G. M., Rutishauser, U., and Millette, C. F., 1971, Cell fractionation and arrangement on fibers, beads, and surfaces, *Proc. Natl. Acad. Sci. USA* **68:**2153.

Edelman, G. M., Spear, P. G., Rutishauser, U., and Yahara, I., 1974, Receptor specificity and mitogenesis in lymphocyte populations, in: *The Cell Surface in Development* (A. A. Moscona, ed.), pp. 141–164, Wiley, New York.

Garber, B. B., and Moscona, A. A., 1969, Enhancement of aggregation of embryonic brain cells by extracellular materials from cultures of brain cells, *J. Cell Biol.* **43:**41a.

Garber, B. B., and Moscona, A. A., 1972*a*, Reconstruction of brain tissue from cell suspensions. I. Aggregation patterns of cells dissociated from different regions of the developing brain, *Dev. Biol.* **27:**217.

Garber, B. B., and Moscona, A. A., 1972*b*, Reconstruction of brain tissue from cell suspensions. II. Specific enhancement of aggregation of embryonic cerebral cells by supernatant from homologous cell cultures, *Dev. Biol.* **27:**235.

Garfield, S., Hausman, R. E., and Moscona, A. A., 1974, Embryonic cell aggregation: Absence of galactosyltransferase activity in retina-specific cell-aggregating factor, *Cell Differ.* **3:**215.

Gaze, R. M., 1974, Neuronal specificity, *Br. Med. Bull.* **30:**116.

Gerisch, G., Malchow, D., Riedel, V., Muller, E., and Every, M., 1972, Cyclic AMP phosphodiesterase and its inhibitor in slime mould development, *Nature (London) New Biol.* **235(55):**90.

Goldschneider, I., and Moscona, A. A., 1971, Tissue-specific antigenic determinants on embryonic cell surfaces demonstrated by antisera prepared against suspensions of live cells, *Anat. Rec.* **169(2):**478.

Goldschneider, I., and Moscona, A. A., 1972, Tissue-specific cell surface antigens in embryonic cells, *J. Cell Biol.* **53:**435.

Hausman, R. E., and Moscona, A. A., 1973, Cell-surface interactions: Differential inhibition by proflavine of embryonic cell aggregation and production of specific cell-aggregating factor, *Proc. Natl. Acad. Sci. USA* **70(11):**3111.

Hausman, R. E., and Moscona, A. A., 1975, Purification and characterization of the retina-specific cell-aggregating factor, *Proc. Natl. Acad. Sci. USA* **72:**916.

Henkart, P., Humphreys, S., and Humphreys, T., 1973, Characterization of sponge aggregation factor: A unique proteoglycan complex, *Biochemistry* **12:**5045.

Hubbel, W. L., and McConnell, H. M, 1969, Motion of steroid spin labels in membranes, *Proc. Natl. Acad. Sci. USA* **63:**16.

Humphreys, T., 1963, Chemical dissolution and *in vitro* reconstruction of sponge cell adhesions. I. Isolation and functional demonstration of the components involved, *Dev. Biol.* **8:**27.

Hunt, R. K., and Jacobson, M., 1974, Neuronal specificity revisited, in: *Current Topics in Developmental Biology,* Vol. 8 (A. A. Moscona and A. Monroy, eds.), pp. 203–255, Academic Press, New York.

Jerne, N. K., 1967, Antibodies and learning: Selection versus instruction, in: *The Neurosciences* (G. C. Quarton, T. Melnechuk, and F. O. Schmitt, eds.), pp. 200–205, Rockefeller University Press, New York.

Kleinschuster, S. J., and Moscona, A. A., 1972, Interactions of embryonic and fetal neural retina cells with carbohydrate-binding phytoagglutinins: Cell surface changes with differentiation, *Exp. Cell Res.* **70:**397.

Lilien, J. E., 1968, Specific enhancement of cell aggregation *in vitro, Dev. Biol.* **17:**658.

Lilien, J. E., and Moscona, A. A., 1967, Cell aggregation: Its enhancement by a supernatant from cultures of homologous cells, *Science* **157:**70.

Lillie, F. R., 1913, The mechanism of fertilization, *Science* **38:**524.

Martinozzi, M., and Moscona, A. A., 1975, Binding of [^{125}I]-concanavalin A and agglutination of embryonic neural retina cells: Age-dependent and experimental changes, *Exp. Cell Res.* **94**:253.

McClay, D. R., 1971, An autoradiographic analysis of the species specificity during sponge cell reaggregation, *Biol. Bull.* **141(2)**:319.

McClay, D. R., and Moscona, A. A., 1974, Purification of the specific cell-aggregating factor from embryonic neural retina cells, *Exp. Cell Res.* **87**:438.

Merrell, R., and Glaser, L., 1973, Specific recognition of plasma membranes by embryonic cells, *Proc. Natl. Acad. Sci. USA* **70(10)**:2794.

Morris, J. E., and Moscona, A. A., 1971, The induction of glutamine synthetase in aggregates of embryonic neural retina cells: Correlations with differentiation and multicellular organization, *Dev. Biol.* **25**:420.

Moscona, A., 1952, Cell suspensions from organ rudiments of chick embryos, *Exp. Cell Res.* **3**:535.

Moscona, A., 1957, The development *in vitro* of chimeric aggregates of dissociated embryonic chick and mouse cells, *Proc. Natl. Acad. Sci. USA.* **43**:184.

Moscona, A. A., 1960, Patterns and mechanisms of tissue reconstruction from dissociated cells, in: *Developing Cell Systems and Their Control* (D. Rudnick, ed.), pp. 45–70, Ronald Press, New York.

Moscona, A., 1961, Rotation-mediated histogenetic aggregation of dissociated cells: A quantifiable approach to cell interactions *in vitro*, *Exp. Cell Res.* **22**:455.

Moscona, A. A., 1962, Analysis of cell recombinations in experimental synthesis of tissues *in vitro*, *J. Cell. Comp. Physiol.* **60**:65.

Moscona, A. A., 1963, Studies on cell aggregation: Demonstration of materials with selective cell-binding activity, *Proc. Natl. Acad. Sci. USA* **49**:742.

Moscona, A. A., 1968, Cell aggregation: Properties of specific cell ligands and their role in the formation of multicellular systems, *Dev. Biol.* **18**:250.

Moscona, A. A., 1974, Surface specification of embryonic cells: Lectin receptors, cell recognition wand specific cell ligands, in: *The Cell Surface in Development* (A. A. Moscona, ed.), pp. 67–99, Wiley, New York.

Moscona, A., and Moscona, H., 1952, The dissociation and aggregation of cells from organ rudiments of the early chick embryo, *J. Anat.* **86**:287.

Moscona, A., and Moscona, M. H., 1962, Specific inhibition of cell aggregation by antiserum to suspensions of embryonic cells, *Anat. Rec.* **142**:319.

Moscona, M. H., and Moscona, A. A., 1963, Inhibition of adhesiveness and aggregation of dissociated cells by inhibitors of protein and RNA synthesis, *Science* **142**:1070.

Moscona, M. H., and Moscona, A. A., 1966, Inhibition of cell aggregation *in vitro* by puromycin, *Exp. Cell Res.* **41**:703.

Nicolson, G. L., 1972, Topography of membrane concanavalin A sites modified by proteolysis, *Nature (London) New Biol.* **239**:193.

Oppenheimer, S. B., and Humphreys, T., 1971, Isolation of specific macromolecules required for adhesion of mouse tumor cells, *Nature (London)* **232**:125.

Roseman, S., 1970, The synthesis of complex carbohydrates by multiglycosyltransferase systems and wtheir potential function in intercellular adhesion, *Chem. Phys. Lipids* **5**:270.

Roseman, S., 1974, Complex carbohydrates and intercellular adhesion, in: *The Cell Surface in Development* (A. A. Moscona, ed.), pp. 255–271, Wiley, New York.

Rosen, S. D., Kafka, J. A., Simpson, D. L., and Barondes, S. H., 1973, Developmentally regulated, carbohydrate-binding protein in *Dictyostelium discoideum*, *Proc. Natl. Acad. Sci. USA* **70**:2554.

Rosen, S. D., Simpson, D. L., Rose, J. E., and Barondes, S. H., 1974, Carbohydrate-binding protein from *Polysphondylium pallidum* implicated in intercellular adhesion, *Nature (London)* **252(5479):**128.

Rutishauser, U., and Sachs, L., 1974, Receptor mobility and the mechanism of cell–cell binding induced by concanavalin A, *Proc. Natl. Acad. Sci. USA* **6:**2456.

Sheffield, J. B., and Moscona, A. A., 1969, Early stages in the reaggregation of embryonic chick neural retina cells, *Exp. Cell Res.* **57:**462.

Sheffield, J. B., and Moscona, A. A., 1970, Electron microscopic analysis of aggregation of embryonic cells: The structure and differentiation of aggregates of neural retina cells, *Dev. Biol.* **23:**36.

Sidman, R. L., 1968, Development of interneuronal connections in brains of mutant mice, in: *Physiological and Biochemical Aspects of Nervous Integration* (F. D. Carlson, ed.), pp. 163–193, Prentice-Hall, Englewood Cliffs, N.J.

Sidman, R. L., 1974, Contact interaction among developing mammalian brain cells, in: *The Cell Surface in Development* (A. A. Moscona, ed.), pp. 221–253, Wiley, New York.

Simpson, D. L., Rosen, S. D., and Barondes, S. H., 1974, Discoidin, a developmentally regulated carbohydrate-binding protein from *Dictyostelium discoideum:* Purification and characterization, *Biochemistry* **13:**3487.

Sperry, R. W., 1943, Visuomotor coordination in the newt *(Triturus viridescens)* after regeneration of the optic nerve, *J. Comp. Neurol.* **79:**33.

Sperry, R. W., 1951, Developmental patterning of neural circuits, *Chicago Med. School Q.* **12:**66.

Sperry, R. W., 1965, Embryogenesis of behavioral nerve nets, in: *Organogenesis* (R. L. DeHaan and H. Ursprung, eds.), pp. 161–186, Holt, Rinehart and Winston, New York.

Turner, R. S., and Burger, M. M., 1973, Involvement of a carbohydrate group in the active site for surface guided reassociation of animal cells, *Nature (London)* **244:**509.

Tyler, A., 1946, An autoantibody concept of cell structure, growth and differentiation, *Growth* **10:**7.

Weiss, P., 1947, The problem of specificity in growth and development,*Yale J. Biol. Med.* **19:**235.

Whaley, W. G., Dauwalder, M., and Kephart, J. E., 1972, Golgi apparatus: Influence on cell surfaces, *Science* **175:**596.

Yahara, I., and Edelman, G. M., 1972, Restriction of the mobility of lymphocyte immunoglobulin receptors by concanavalin A, *Proc. Natl. Acad. Sci. USA* **69:**608.

8

An *in Vitro* Assay for Retinotectal Specificity

STEPHEN ROTH and RICHARD B. MARCHASE

1. INTRODUCTION: SPECIFIC ADHESION AS A MORPHOGENETIC MECHANISM

The acquisition of form by developing organisms was one of the first biological problems considered by man. Despite this early start, however, biochemical analyses of morphogenetic movements have lagged far behind those of other biological specialities such as physiology and genetics. The lack of information about morphogenesis at the molecular level is directly attributable to the lack of information at the cell and tissue levels. It is only recently that we have come to understand that morphogenetic movements are the results of cellular properties and not of external forces of unknown origins that work their ways on cells like the wind on trees.

1.1. *Holtfreter and Selective Affinities*

The first critical experiments showing that intrinsic cellular properties could be responsible for morphogenetic activity were done by Holtfreter (1939). He dissociated cells of amphibian embryos and watched as the cells reassociated and re-formed histological structures

STEPHEN ROTH and RICHARD B. MARCHASE · Department of Biology, The Johns Hopkins University, Baltimore, Maryland.

similar to those found in the embryos before they were dissociated. Not only did cells dissociated from different germ layers give specific reassociation patterns, but cells from the same germ layer yielded different patterns if they were of different developmental ages. Holtfreter concluded that these cells possess specific affinities determined by their surface coats and that these affinities govern the way in which the cells interact with other cells. These data changed our way of thinking of morphogenesis. It was no longer necessary to look for hydrostatic pressures or block-and-tackle-like structures in the developing embryos to account for shape changes. The cells and their immediate products seemed to have the information and the force to accomplish at least a semblance of morphogenesis without obvious fluids or superstructures.

It is interesting that conceptually identical experiments were carried out in the sponge by H. V. Wilson (1907) long before Holtfreter examined the problem using amphibians. Wilson found that single cells from dissociated sponges re-form spongelike masses after hours in seawater. Furthermore, he found that if two different species of sponge cells were mixed the reaggregated sponges were of one or the other type but not mixed. These studies were not, however, accompanied by thorough histological examination as were the experiments of Holtfreter and, as Wilson himself pointed out, were subject to several alternative explanations.

1.2. *Weiss and Molecular Complementarity*

If, as the work of Holtfreter and Wilson has shown, cells have the ability to reassociate with other cells and give rise to recognizable structures, then some components of the cell surface or coat must be responsible for these abilities. One of the earliest and clearest suggestions as to how this could be accomplished was Weiss's (1947) idea of complementarity of cell surface molecules. Taking his cue from enzyme–substrate and antigen–antibody interactions, he felt that similar lock-and-key molecules could exist on the outside surfaces of various cell types. If two cells met and a sufficient number of their surface molecules were complementary, then a firm adhesion could result. Weiss speculated further that soluble, unattached molecules with similar active sites could be physiologically active either by regulating growth or by interfering with cell associations.

1.3. *Sperry and Neuronal Specificity*

Nowhere is the precision of the morphogenetic process more evident than in the development of the nervous system. Of all neural patterns in vertebrates, the most studied is the projection of the retinal ganglionic axons to the optic tecta. In the chick embryo, the first axon tips from the retina make contact with the contralateral tectum on about the sixth day of development (DeLong and Coulombre, 1965). Immediately upon reaching the tectum, the axons begin their migration across its surface until they arrive at their proper terminations. During the course of this migration, the fibers appear to move in straight lines (Goldberg, 1974), although at the level of the collateral sprouts there may well be a great deal of trial and error similar to that seen in cell culture (Bray and Bunge, 1973). By about day 9, the last of the axons have reached the tecta and by about day 12 all of the axon termini are at their physiologically correct sites on the surface of the tecta (Crossland *et al.*, 1974). In the next 1 or 2 days, all of the axons dive perpendicular to the tectal surface and proceed to the deeper tectal layers where they will eventually form synapses with the tectal neurons and complete the major communicatory pathway between the eye and the brain.

The pattern of the completed retinotectal projection is such that the dorsal retina is innervating the ventral tectum while the nasal retina is innervating the posterior tectum. Similarly, ventral and temporal retinal sections innervate dorsal and anterior tectal sections, respectively.

This pattern has been shown histologically and electrophysiologically in all of the lower vertebrates examined and is discussed in greater detail by Jacobson (1970, and this volume).

Sperry (1963) has postulated that this pattern could be explained if both the retina and the tectum had gradients of complementary molecules along both the anteroposterior and dorsoventral axes of the tectum that matched gradients along the temporonasal and ventrodorsal axes of the retina. According to this hypothesis, retinal axons would migrate along the tectal surface until they located adhesively stable positions.

Two definite predictions are made by Sperry's idea. First, axon tips from the dorsal part of the retina, for example, should be more adhesive to ventral tectal surfaces than to dorsal tectal surfaces, in accord with the ultimate physiological connection. Second, the retinal ganglion cells should have different surface moieties at their migrating tips, although this difference could be quantitative, qualitative, or both. These surface components should, of course, be involved in the recognition phenomenon that occurs between the axon tips and the tectal surface.

In order to test these predictions, an assay for intercellular adhesive specificity is required that can be adapted for use in the retinotectal system.

2. ASSAYS FOR ADHESIVE SPECIFICITY

2.1. *Sorting Out*

In experiments similar to those of Holtfreter, Moscona (this volume, and 1965) and Steinberg (1964) have extensively studied the fates of reassociating cells when they are combined with cells from various tissues and species. Concentrating on cells from the chick and mouse, both of these investigators have shown that when cells from different tissues are allowed to reaggregate together the cell types eventually segregate into groups composed almost entirely of cells from one or another of the tissue types. This separation of unlike types and reassociation of like types has been called "sorting out" and occurs in small cell aggregates over a course of 1 or 2 days in suspension cultures.

An exciting observation made with the sorting out assay (Moscona, 1965) is that tissue type dominates over species type when the two are compared. For example, cells from mouse heart will sort out with cells from chick heart rather than with cells from any other mouse tissue. The implications of these phenomena are often understated, although they clearly imply that, whatever the actual mechanism of sorting out, this specificity is accurately conserved across a wide evolutionary expanse. This language of the tissues must be important to be so rigidly conserved. Its generality should be tested further.

Unfortunately, sorting out cannot be used as an assay for adhesive specificity, for several critical reasons. First, the extent to which the sorting results from adhesive differences or migratory differences (Curtis, 1967) is unclear at the present time. Second, the degree of sorting within aggregates cannot be precisely determined. That is, there is no method for reliably distinguishing between 10% and 60% sorting out. If these two difficulties were overcome, sorting out would still be analytically cumbersome since treatment of the aggregates with drugs or other agents must be done without knowledge of the effective dose around the cells at various levels inside the aggregates.

2.2. *Collecting-Aggregate Assay*

An assay for adhesive specificity was developed that could be used for measuring the specificities of cell types that could form aggregates

(Roth and Weston, 1967; Roth, 1968). Tissues were labeled with [^{3}H]thymidine and then dissociated into their single cells. Unlabeled cell aggregates were prepared previously from any number of tissues. The unlabeled aggregates were circulated in the suspension of labeled cells. After the aggregates had been exposed to the suspension for equal periods of time, they were examined autoradiographically and the numbers of labeled cells collected by each aggregate type were compared. This protocol is shown in Fig. 1A. As long as the aggregates chosen for circulation were of equal diameters, one could assume that the numbers of collisions between the aggregates and the cells would be identical. The number of labeled cells adhering to an aggregate was, therefore, a direct estimation of the probability of adhesion between the cell type and the aggregate being examined. Within one cell suspension, the numbers of cells collected by different aggregate types could be compared and these differences were assumed to be a measure of adhesive recognition.

When a wide variety of chick and mouse embryo tissues were examined, the results were very similar to those obtained in the sorting out experiments. Labeled cells adhered preferentially to those aggre-

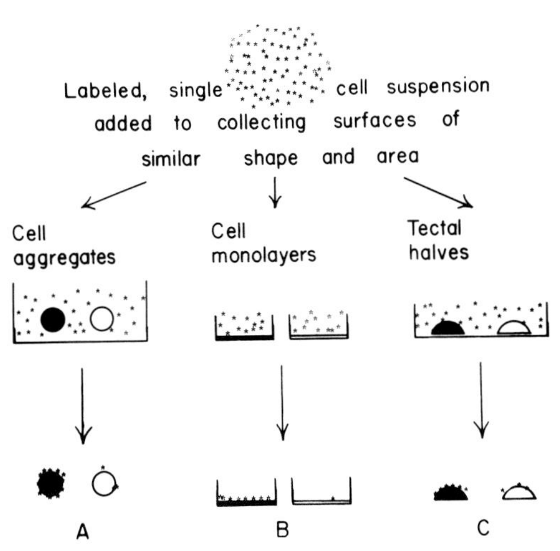

Fig. 1. Protocols for collection assays. A single-cell suspension, prepared from one tissue and labeled with a radioactive isotope, is exposed to collecting surfaces of similar size and area prepared from any number of tissues. Since the numbers of collisions between the surfaces and the labeled cells are about equal, the number of radioactive cells adhering to a surface gives a direct estimation of the relative adhesivity between the cells in suspension and the cells making up that surface. A illustrates the collecting-aggregate technique, B the cell monolayer assay, and C the retinotectal adhesion assay.

gates made from the same tissue type. This assay was able to show that, although tissue type was clearly more important for adhesive recognition than species, aggregates of identical species and tissue collected labeled cells more rapidly than aggregates of the same tissue but different species (Roth, 1968). Also, this method demonstrated adhesive recognition between living aggregates and fixed cells and *vice versa*. Taken together, these data made it clear that the collecting-aggregate assay measured some sort of adhesive phenomenon, although the relevance of that adhesive recognition to morphogenesis was unclear.

The collection assay was modified by labeling the cells with [^{32}P]phosphate to the extent that the collecting aggregates could be washed and counted in a scintillation counter (Roth *et al.*, 1971*a*; Dorsey and Roth, 1973). This avoided both autoradiography and its subsequent statistical estimates of the numbers of collected cells. The result was an assay that was much more reliable and less tedious, although the data obtained were, qualitatively at least, identical to those obtained with the autoradiographic assay.

2.3. *Variations of the Collection Assay*

a. Collecting Monolayers. Walther *et al.* (1973) have devised an adhesion assay in which the single, labeled cells are placed into small wells on the bottom of which cells in monolayer have been cultured. Since the area covered by the cultured cells is identical, the number of interactions between labeled cells and unlabeled cultured cells will be identical. After washing of the wells, the adherent cells can be counted and compared for their relative adhesive stabilities to the various cell types in the wells. This method has the definite advantage of being useful when cell types are being dealt with that do not make aggregates or make them poorly. Furthermore, the monolayers collect far larger proportions of the labeled cells than do the aggregates in the collecting-aggregate assay. The data obtained with the collecting monolayer assay are clearly representative of the majority of the cells in the labeled suspension. Using embryonic chick neural retina and liver, Walther *et al.* (1973) have obtained data essentially similar to those obtained with the collecting-aggregate assay (Roth, 1968; Roth *et al.*, 1971*a*). With cultured mouse fibroblasts, both malignant and nonmalignant, the monolayer assay (Walther *et al.*, 1973) gave results that differed slightly from the aggregate assay (Dorsey and Roth, 1973). This assay is schematically pictured in Fig. 1B.

b. Double-Labeled Collecting Aggregates. McClay and Baker (1975) have modified the original collecting-aggregate assay by using very large numbers of smaller aggregates and by differentially labeling the aggregates and the single cells. After collection occurs, the aggregates and adherent cells are separated from the nonadherent cells by filtration and the numbers of collected cells are determined by counting the ratio of the cellular isotope to the aggregate isotope. As with the monolayer assay, most of the cells of the suspension can be collected since the number of aggregates is large in comparison with the number of cells. Additionally, McClay and Baker (1975) report replicate data points with standard errors that vary less than 8%.

The data accumulated with this technique using embryonic tissues again show strong tissue preferences, although, interestingly, very little specificity is observed between the tissues of the nervous system in the chick.

c. Collecting Beads. A final variation is through the use of cells grown in monolayer but on DEAE-Sephadex beads (van Wezel, 1973). Use of these cell-covered beads to collect labeled cells has yielded very precise collection data with cultured fibroblasts (Vosbeck and Roth, 1976). The most significant advantage is the ease of preparation of the beads. They can be seeded with cells and grown in suspension cultures (van Wezel, 1973). The cultures are partially depleted to obtain hundreds of thousands of cell-coated Sephadex beads. The degree to which this method will be applicable to embryonic cell type is not known at present.

2.4. *Conclusion*

There are assays that can measure adhesive recognition between embryonic and cultured cells. These assays are versatile and precise and do not require rare or esoteric equipment or abilities. Furthermore, most of these assays are amenable to biochemical perturbation. Aggregates, monolayers, or beads can be easily treated with any water-soluble agent and their collection properties compared with those of untreated aggregates, monolayers, or beads. By use of these methods, in fact, embryonic cell collection has been dissected into two phases—nonspecific and specific (Roth, 1968; McClay and Baker, 1975). The acquisition of specificity after enzymatic dissociation of embryonic cells has been shown to require active metabolism (Roth, 1968; McClay and Baker, 1975), and the recognition between chick neural retina cells has been shown to depend, at least partially, on some cell surface carbohydrates terminating in β-galactosides (Roth *et al.*, 1971*a*).

The crucial problem with these techniques is relevance. Do the differences being measured have anything to do with morphogenesis?

It is one thing to show that retinal cells prefer to adhere to each other when given the choice between retina and liver. It is yet another thing to conclude that the retinal preference for self is the force that maintains retinal integrity *in vivo* or causes formation of the optic vesicle and cup. Not one of these assays, with the possible exception of Holtfreter's original experiments with gastrula tissues, has ever been applied to cells that interact in a meaningful way during embryogenesis.

Therefore, the application of an assay for specific adhesion to the retinotectal system, in order to test Sperry's hypothesis or anything else, is as much a test for the relevance of the assay to morphogenesis as it is for the hypothesis being tested.

3. ADHESIVE MEASUREMENTS IN THE RETINOTECTAL SYSTEM

3.1. *The Assay and Results*

The adhesive assay designed to test Sperry's hypothesis was based on the same principle as the collection assays described above: exposure of radioactively labeled single cells to different surfaces of equal size and configuration and comparison of the number of radioactive cells that adhere. The retinotectal system of the chick was chosen because of its size and because of the extensive investigations already completed on retinal cell adhesion (Moscona, 1965; Roth *et al.*, 1971*a*).

Sperry's theory predicts that axonal tips from the dorsal part of the retina should be more adhesive to ventral tectal surfaces than to dorsal tectal surfaces. In the implementation of the assay it was assumed that if molecules that were responsible for neural recognition existed on the axonal tips they would also be present on retinal cell bodies. If this were true, cell bodies from dorsal half-retinas might be expected to adhere preferentially to ventral half-tecta while cell bodies from ventral half-retinas would adhere preferentially to dorsal half-tecta.

To test this possibility, tecta were split into equal-sized dorsal and ventral halves and placed in the bottom of a petri dish (Barbera *et al.*, 1973). Cells from either the dorsal or the ventral half of the retina were labeled with [^{32}P]phosphate and dissociated with trypsin to form a single-cell suspension. The labeled retinal cells were added to the dish containing the tectal halves. After a collection period of about 30 min, the tectal halves were washed and counted individually in a liquid scintillation system. The specific activity of the retinal cells was determined, and the radioactivity present on each tectal half was converted into numbers of adhering retinal cells. This assay is schematized in Fig. 1C.

Figure 2 shows the net adhesion of dorsal retina cells to ventral and dorsal tectal halves as a function of the number of washes. Once the nonadhering cells were rinsed away, about twice as many dorsal retina cells adhered to an average ventral half-tectum as to a dorsal half-tectum. This result was reproducible with dorsal retina cells that were either freshly dissociated or preincubated in nutrient medium for up to 8 h before being exposed to the tectal halves.

When the experiment was repeated using cells from ventral half-retinas, a dependence on preincubation time was observed. Soon after trypsinization, ventral retina cells showed a slight adhesive preference for ventral tectal halves. However, when allowed to incubate for more than 3 h in a nutrient medium before exposure to the tectal halves, nearly twice as many ventral retina cells adhered to dorsal half-tecta as to ventral half-tecta after the tecta were washed identically.

This assay was later modified to provide better reliability (Mar-

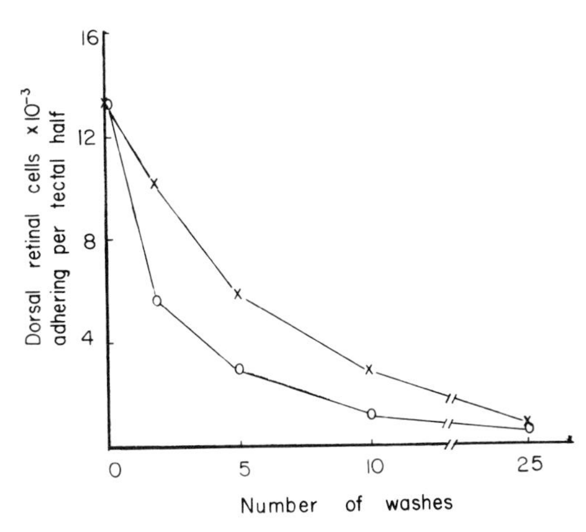

Fig. 2. Number of ^{32}P-labeled cells adhering to dorsal and ventral tectal halves as a function of the washing procedure. After a 30-min stationary incubation, the tectal halves were placed in graduated cylinders containing saline. The cylinders were rocked back and forth, each stroke counting as a wash. For each pair of data points, dorsal and ventral tectal halves were washed in the same cylinder at the same time. After the wash, they were identified as dorsal or ventral and counted. Each data point represents the average of five 12-day tectal halves. Cell suspension was made from 7-day dorsal half-retinas, 1.0×10^6 cells/ml. Symbols: ○, cells adhering to dorsal tectal halves; ×, cells adhering to ventral tectal halves.

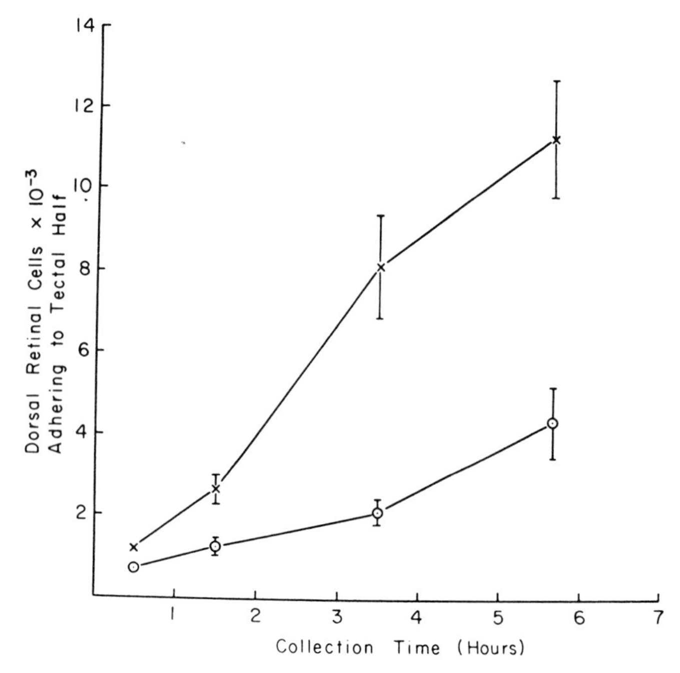

Fig. 3. Number of ^{32}P-labeled cells adhering to dorsal and ventral tectal halves as a function of collection time. Four to six tectal halves of each type from 12-day embryos were used for each data point. Standard error of the mean is shown for each point. Cell suspension was made from 7-day dorsal half-retinas, 1.0×10^6 cells/ml. Symbols: ⊙, cells adhering to dorsal tectal halves; ×, cells adhering to ventral tectal halves.

chase *et al.*, 1975; Barbera, 1975). Tectal halves were pinned to a paraffin layer in the bottom of a petri dish and reciprocated during the collection period so that the retinal cells moved back and forth across the tectal surfaces. The results obtained were very similar to those of the original assay but more consistent both within and between experiments. The results of an experiment measuring the rates of adhesion of dorsal retinal cells to dorsal and ventral tectal surfaces are shown in Fig. 3. About twice as many cells adhered to the ventral half-tecta at all times after dissociation.

When the cells in suspension were taken from ventral retina, the specificity of adhesion changed with time of collection (Fig. 4). During the first 3 h, more retinal cells adhered to ventral half-tecta, but after this time the ventral retina cells adhered preferentially to dorsal tectal halves. At 6 h, nearly twice as many retinal cells were adhering to the dorsal tectal halves. There is thus a preferential adhesion of both dorsal and ventral retina cells to the tectal half with which they normally form synapses. This selectivity is exhibited at all times after dissociation by dorsal half-retina, while ventral half-retina displays a preference for the appropriate tectal half only after 3 h of incubation.

Equivalent results with optic tecta which had never been innervated by retinal axons ensure that the observed specificity is not just a result of retinal cells adhering to their own axons on the surface of the tecta. Further controls with labeled cell suspensions prepared from liver, cerebellum, or brain stem showed no preferential adhesion to either tectal half. A typical control experiment with nonretinal cells is shown in Fig. 5.

These results thus demonstrate an adhesive selectivity between neural retina and optic tectum that mimics the innervation pattern of the tectum by the retina. This strongly supports an interpretation of

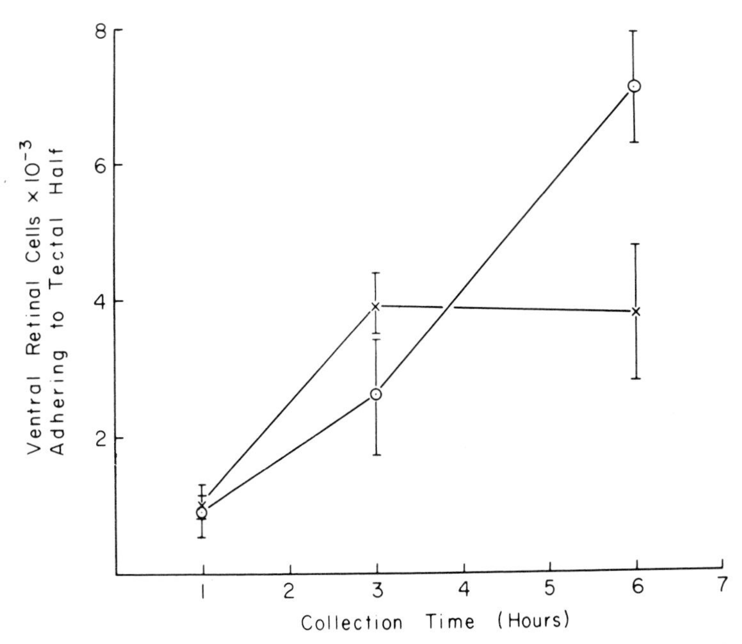

Fig. 4. Number of ^{32}P-labeled cells adhering to dorsal and ventral tectal halves as a function of collection time. Four to five tectal halves of each type from 12-day embryos were used for each data point. Standard error of the mean is shown for each point. Cell suspension was made from 7-day ventral half-retinas, 0.6×10^6 cells/ml. Symbols: ⊙, cells adhering to dorsal tectal halves; ×, cells adhering to ventral tectal halves.

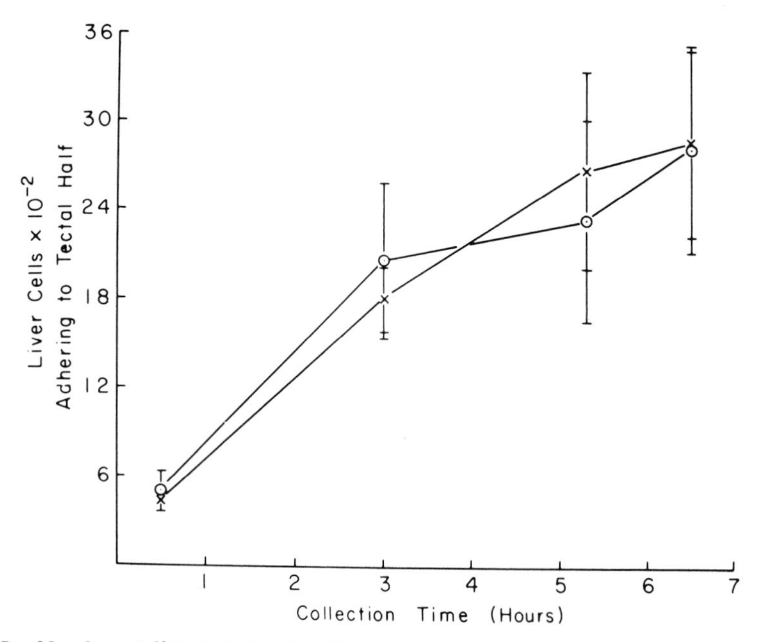

Fig. 5. Number of ^{32}P-labeled cells adhering to dorsal and ventral tectal halves as a function of collection time. Five tectal halves of each type from 12-day embryos were used for each data point. Standard error of the mean is shown for each point. Cell suspension was made from 7-day liver, 0.5×10^6 cells/ml. Symbols: ○, cells adhering to dorsal tectal halves; ×, cells adhering to ventral tectal halves.

neuronal specificity mediated by recognition between relatively permanent cell surface molecules. A simple interpretation of these data is that these surface molecules participate in a process of specific adhesion that ultimately results in the ordering of the retinotectal projection.

The success of these experiments was dependent on two unrelated premises: (1) that an *in vitro* assay is capable of detecting adhesion relevant to morphogenetic phenomena and (2) that neuronal specificity is achieved through adhesive interactions between cell surface markers, as Sperry predicted. These results thus not only recommend the study of *in vitro* adhesion as a useful tool for the understanding of morphogenesis but also support Sperry's hypothesis of neuronal specificity.

The assay is also potentially useful for unraveling the molecular mechanisms involved in neuronal specificity. Biochemical, heterochronic, and cross-species studies are all much simpler to perform *in vitro* than *in vivo*.

3.2. *Changes with Development*

Barbera (1975) has studied the appearance of selective adhesion in this system as a function of the developmental age of both the retina and the tectum. He performed these experiments by using one tissue at an age known to be capable of adhesive specificity and varying the age of the other tissue. He found that as early as dorsal and ventral could be definitively distinguished in developing tecta (about embryonic day 8) they exhibited selective adhesion that mimicked the proper projection.

Dorsal retina as young as embryonic day $2\frac{1}{2}$ was also capable of relevant adhesive selectivity. However, ventral retina exhibited preferential adhesion for dorsal half-tectum only after embryonic day 6.

The relation of these findings to other data on the time of retinal specification as well as a model encompassing these results is also discussed (Barbera, 1975).

3.3. *Which Retinal Cells Are Adhering?*

The actual retinotectal projection is created by interactions between the axonal tips of the retinal ganglion cells and the tectal surface. The success of the adhesion assay was dependent on the assumption that cell bodies from the retina would exhibit selectivity similar to that of the ganglion axonal tips. In the assay, the entire neural retina is isotopically labeled and used to make the cell suspension. Thus the suspension contains all the cells of the neural retina, not just the ganglion cells. It is therefore not known which cell types are adhering to the tectal surface. It is possible that only the ganglion cells exhibit selective adhesions with the tectal surface and that the presence of other labeled cell types only contributes to a high, nonspecific background. However, two results make this unlikely.

First, the ganglionic cells are the first cells of the retina to proceed through their final mitosis (Fujita and Horii, 1963). Younger retinas would therefore be enriched in ganglion cells compared to older retinas. However, the ratio of retinal cell adhesion to the "matching" tectal half over the "nonmatching" tectal half remains constant through this period of enrichment and beyond (Barbera, 1975).

Second, the pigmented epithelium of the retina exhibits adhesive selectivity very similar to that of the neural retina (Barbera *et al.*, 1973). The ability of the pigmented epithelium to regenerate a neural retina is well established (Alexander, 1937), and in *Triturus* (Stone, 1950; Levine and Cronly-Dillon, 1974) it has been shown that such a regenerated retina has specificities derived at least partially from the pigmented epithelium. Thus at least one cell type other than ganglion cells seems to possess the molecules responsible for selective adhesions. Furthermore, the preferences of "matching" to "nonmatching" adhesions were very comparable to those obtained using the whole neural retina. Since the pigmented epithelium is a homogeneous preparation with no labeled cells from other tissues, the magnitude of the preference exhibited is not a result of a labeled, nonpigmented contaminant. This constancy also detracts from the view that only the ganglion cells of the neural retina are capable of selective adhesion.

It seems likely, therefore, that cells in the neural retina other than the ganglion cells participate in the process of specific adhesion.

3.4. *The Tectal Surface*

Light and electron microscopic evidence collected by Barbera (1975) implies that retinal cells are not adhering to cells on the tectal surface but rather to an extracellular matrix. The retinal cells are well separated from the layers of the tectum in which they normally form synapses. This implies a distinction between positional specificity and synaptogenesis. Such a distinction has also been reported by Crossland *et al.* (1974). They observed, also in chick embryos, that incoming retinal axons arrange themselves in a proper retinotectal map while still on the exterior surface of the tectum. Only after this positioning do they dive straight down into the tectum and form synapses.

3.5. *Magnitude of the Adhesive Selectivity*

Several schemes utilizing different kinds of surface recognition molecules can be devised that account for the observed retinotectal specificity. Three general schemes can be distinguished using two criteria: qualitative vs. quantitative changes in the recognition molecules and rigorous vs. contextual specificity.

In the first general class, each retinal cell is pictured as having a unique surface molecule which other retinal cells do not possess. On the tectal surface only one site, the appropriate terminus, would possess the complementary molecule to which the retinal cell could bind. There would be no affinity for other tectal sites. Thus qualitative differences in the molecules responsible for adhesion would determine the projection. Such a scheme would provide rigorous specificity. That is, a retinal cell would adhere only to its appropriate terminus, regardless of what other areas of the tectum were unoccupied and regardless of the presence or absence of other retinal cells.

The second class of schemes could be classified as quantitative and rigorous. In this set, every retinal cell and tectal site possesses the same adhesive molecules. These molecules are distributed in such a way that the affinity between a retinal cell and its appropriate tectal target is maximal. However, because the tectal sites all utilize the same adhesive molecules, only in different numbers, a retinal cell will adhere to other tectal sites as well. The affinity a retinal cell has for other tectal loci will decrease monotonically with distance from the natural target. Since a retinal cell's adhesion is maximal at its appropriate terminus, it will prefer that point even if other areas of tectum are unoccupied. It therefore provides for rigorous specificity.

A simple model which manifests these properties is shown in Fig. 6. For simplicity, only the dorsoventral axis is shown. A similar but independent mechanism can be imagined to function in the perpendicular axis. Affinity between retinal cells and tectal sites results from the binding of complementary molecules Y and •. Linear gradients of these molecules are postulated across both retina and tectum. In both, the ventral extreme is richest in Y but has no •. The dorsal extreme is richest in • but has no Y. As a result of these gradients, every retinal cell will bind maximally to the unique point on the tectum that possesses numbers of the adhesion molecules exactly complementary to its own. Its adhesion for other tectal sites will decrease linearly with

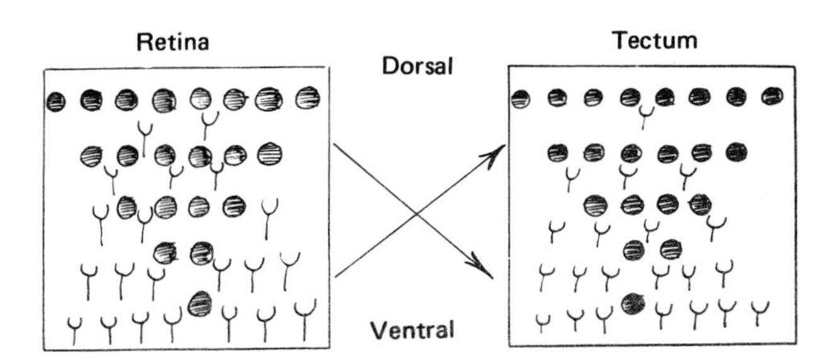

Fig. 6. A simple model of gradients of complementary molecules in the dorsoventral axis that could provide adhesive connections corresponding to the observed retinotectal map. Retinal cells from the dorsal region have many •'s but few Y's. The ventral part of the tectum has cells with few •'s but many Y's, thereby allowing a maximum number of bonds (Ẏ) to be formed. In the same fashion, ventral retina cells, rich in Y's and poor in •'s, would form the most stable connections with the dorsal tectum since it is rich in •'s but poor in Y's. The gradients are pictured as linear, with the concentration of each adhesive molecule being zero at the appropriate extremes.

separation from that point. The adhesion of other retinal cells to that point will be less strong than the adhesion of the appropriate retinal cell. Such a model would result in a rigorous, context-independent retinotectal projection.

The third class of models would also utilize quantitative variations in the binding molecules but would provide for a contextual separation of the retinal cells. One can produce such a scheme in one dimension by creating a tectal surface in which an absolute gradient of an adhesion molecule extends from one pole to the other. In the retina, a gradient of the complementary molecule is imagined. Maximal adhesion for all retinal cells would occur at the tectal extreme rich in the adhesive molecule. However, the strength of this adhesion would vary, depending on the number of complementary molecules present on the retinal cell in question. Obviously, with no other constraints all the retinal cells would pile up at the adhesive end of the tectum. If one includes an additional premise, such as a limit on the number of tectal sites available or a repulsion between retinal axons such that they tend to spread as evenly as possible, then a competitive equilibrium can create a retinotectal projection. This class of models would provide a context-dependent specificity in that the distribution of the retinal axons is dependent on what retinal cells are present and what parts of the tectal surface are available. Gaze and Keating (1972) use the term "systems matching" to refer to such a flexible projection.

These three classes of models pose very different possibilities for retinal axons on the tectal surface as they search for their proper termination site. In a strictly qualitative scheme, a retinal axon would receive only a "yes" or "no" response as its filopodia tested different tectal loci. No guidance would be provided. The retinal axon, on the average, would have to make contact with half the tectal sites before finding its appropriate terminus.

In the second class of models, this randomness could be replaced by a guidance mechanism based on movement up an adhesive gradient. Carter (1965) has observed such movement in mouse fibroblasts. As a retinal axon extended filopodia, some degree of adhesion would occur at all points on the tectal surface. However, because of increasing affinity as one neared the correct terminus, filopodia closest to the target would adhere best. If net axonal movement occurred in the direction of maximal adhesivity, a mechanism for guiding retinal axons to their respective targets would exist.

In the third class of models, all retinal axons would tend to move up the gradient to the most adhesive pole. The equilibrium between this force and the limitation of tectal sites or interaxonal repulsion would establish the correctly distributed projection.

These models would each predict different results when tested with the adhesion assay described. In the first type of model, a retinal cell binds only to its exact tectal locus. This target is very small when compared to the total surface area of the tectum (Hunt and Jacobson, 1974). In the stationary adhesion assay, the probability that a retinal cell would have the opportunity to adhere to this target is correspondingly small. Yet a relatively large fraction of the retinal cells did adhere to tecta. One can conclude that most of these cells were adhering to the tecta at positions other than the normal locus of termination of their axons. This suggests that their affinity varies quantitatively and is not dependent on individually unique adhesion molecules as suggested in the first type of model.

Any scheme belonging to the third class which utilizes actual gradients of adhesivity must cope with the result that dorsal retina binds preferentially to ventral tectum while ventral retina binds preferentially to dorsal tectum. A model dependent on adhesive gradients running completely across the retina and the tectum could account for the *in vivo* projection but would predict that in an adhesion assay all retina cells would bind preferentially to one-half of the tectum. This is not the case. A modification of such a scheme constructed of two gradients—one running from ventral extreme to midline and the other running from dorsal extreme to midline—would predict the adhesive preferences we see but would also predict much higher ratios than observed.

The second class of model seems to be most consistent with the results of the adhesion assay. The model presented in Fig. 6 utilizes gradual changes in the numbers of two complementary molecules. The affinity any retinal cell has for its correct tectal target is equal to the affinity of any other retinal cell for its correct tectal target. The affinity a retinal cell has for any other tectal site decreases linearly with separation from its correct target. The slope of this decrease is such that the dorsalmost cell exhibits half-maximal adhesion to sites at the tectal midline and no adhesion to sites at the extreme dorsal tectum. This is the maximum slope possible if the entire axis is to be determined by only these molecules.

Assuming that the probability of a retinal cell adhering to a tectal site is proportional to its strength of adhesion toward that site, the relative number of dorsal retina cells that would bind to the respective tectal halves can be calculated. The result is that 5/3 times as many dorsal retina cells bind to a ventral half-tectum than bind to a dorsal half-tectum. This is nearly identical to a grand mean of our experimental ratios, which is 1.7. This ratio of the number of retinal cells adhering to the "matching" tectal half to the number adhering to the "nonmatching" tectal half is almost always between 1.5 and 2.0. This is true of the original stationary assay and the reciprocating assay. It is true for both dorsal and ventral neural retina as well as for both halves of pigmented epithelium. It has been found across variations in retinal age and tectal age. The constancy of this result suggests that it may reflect a basic quality of the retinotectal system.

The true significance of this correspondence is unknown. It certainly lends support to quantitative models with properties similar to those of the second model, but further experimentation is required to show that the ratio is not coincidental.

This calculation does illustrate the thermodynamic properties of adhesion systems. Here is a theoretical mechanism which at equilibrium is deemed capable of assembling about 1000 retinal axons in their proper order along each axis. This implies a quantitative sensitivity of 0.1%. Yet the initial rates of adhesion to the coarsest division of the tectum possible, tectal halves, yield a preferential adhesion ratio of only 1.67. This emphasizes the caution that must be exercised when comparing kinetic rates of adhesion with equilibrium behavior.

3.6. *Biochemical Approaches*

An *in vitro* adhesion assay offers the opportunity to perturb a recognition system biochemically in a manner that would be impossible in the living animal. In the retinotectal system, the first information from such an approach was inherent in the dissociation of retinal tissue into single cells. Trypsin, a proteolytic enzyme, was used as the dissociating agent. This treatment resulted in asymmetrical adhesive behavior from the two populations of retinal cells. Dorsal retina cells displayed selective adhesion for ventral tecta both immediately after dissociation and up to 9 h later. Dissociated ventral retina cells, on the other hand, required incubation in nutrient medium before exhibiting selectivity that favored dorsal tecta. This finding suggests that molecules responsible for relevant, specific adhesion on dorsal retina are less sensitive to the dissociating agent, trypsin, than are those on ventral retina. Incubation in nutrient medium may be required by ventral retina to replace molecules removed by trypsinization.

As a complement to this finding, collection experiments were carried out in which the tectal surfaces were pretreated with crystalline

Table I. Collection by 12-Day Tecta after Trypsinization[a]

		Retinal cells adhering to tectal halves	
		Control	Trypsinization
Dorsal retina suspension[b]	Dorsal tecta	1800 ± 300	1500 ± 400
	Ventral tecta	3400 ± 700	1100 ± 300
Ventral retina suspension[b]	Dorsal tecta	4100 ± 1200	3400 ± 800
	Ventral tecta	2500 ± 600	2000 ± 500

[a] 0.02% crystalline trypsin, 15 min, 37°C.
[b] 7-day retina, 1.0×10^6 cells/ml, 4-h preincubation, 30-min collection.

trypsin. Results of such experiments are shown in Table I. Adhesion of healed ventral retina cells to either tectal half was not greatly affected by the treatment. However, the adhesion of dorsal retina cells to ventral tectal halves was reduced immensely. These results suggest that moieties located on ventral tectum and responsible for adhesion of dorsal retina cells are trypsin labile. Along with the effects of trypsin on retinal cells, these data suggest a molecular mechanism in which the adhesive unit consists of a trypsin-sensitive entity on one tissue binding to a trypsin-insensitive entity on the other tissue. In the binding of ventral retina to dorsal tectum, the trypsin-sensitive molecule is located on the retina, while in the binding of dorsal retina to ventral tectum it is located on the tectal surface.

These results suggest a double gradient scheme similar to that shown in Fig. 6 in which one of the adhesive molecules is trypsin sensitive and the other relatively trypsin insensitive.

Molecular models consistent with these data have been presented (Marchase *et al.*, 1975). In these models, the basic adhesive interaction is an intercellular binding between a cell surface glycosyltransferase on one tissue and a complementary oligosaccharide acceptor on the other. These molecules were chosen because they have been implicated as potential recognition molecules in other systems (Roseman, 1970; Roth, 1973) and because they have been shown to be present on the retinal cell surface (Roth *et al.*, 1971*b*).

3.7. *Conclusions*

The data presented above allow certain firm conclusions to be made:

1. After a preincubation period of 3–4 h, cells from the ventral half of a chick retina adhere preferentially to the dorsal half of the optic tectum.
2. Cells from the dorsal half of the retina adhere preferentially to the ventral half of the tectum at all times after dissociation of the retina.
3. Cells from the pigmented retina show this identical preference for the opposite tectal halves.
4. No other tissue type tested from these embryos shows any preference for either dorsal or ventral tectal half.
5. Retinal cells exhibit these adhesive specificities toward tectal halves that are either innervated or not innervated by the optic nerve.
6. Ventral retina cells show a preference for dorsal tectum only after the sixth day of embryonic development, whereas dorsal retina cells show an adhesive preference for ventral tectum surfaces as early as it can be dissected from the embryo, about $2\frac{1}{2}$ days.

The simplest interpretation of these experimental findings is that *in vitro* measurements of rates of adhesion of cells to biological surfaces can detect intercellular adhesive differences that are relevant to the process of embryonic morphogenetic movements. If this conclusion is valid, it follows that Sperry's hypothesis for neural specificity is clearly substantiated by these data.

An alternative explanation for the experimental results is that they result from a complex series of unknown artifacts and that the observed, apparent adhesive preferences actually have another, more trivial cause. In this event, the data obtained using the collection assay have no bearing whatsoever on any hypothesis concerning neural specificity.

Our protocol in this matter has been to rule out all known, possible artifacts in this assay and, subsequently, to proceed with the biochemical analysis of the differences being made apparent in this *in vitro* retinotectal system. The ultimate corroboration for a molecular mechanism of *in vitro* specificity is, of course, the finding of the same mechanism *in vivo*.

4. SUMMARY

This chapter advocates the position that morphogenesis is largely the result of changes in intercellular adhesive specificities. At present, there are numerous versatile and quantitative methods for the demonstration of these specificities. The most serious criticism of these methods has been the dubious relationship between the measured differences in adhesive stabilities and their roles in actual morphogenetic phenomena.

The data presented here and elsewhere (Barbera *et al.*, 1973; Marchase *et al.*, 1975; Barbera, 1975; Glaser, this volume) on the retinotectal system show a distinct correlation between kinetically measured adhesive stabilities and morphogenetic patterns. Retinal cells adhere preferentially to those areas of the tectal surface on which their retinal axons would eventually terminate. It is possible, therefore, that other morphogenetic systems can be analyzed meaningfully using *in vitro* assays for cellular recognition.

In sum, simple and reproducible assays exist that are capable of initiating objective investigations into the cellular and molecular levels of morphogenesis and, ultimately, biological form.

Acknowledgments

This work is supported by research grants from the National Institute of Child Health and Human Development of the United States Public Health Service. R. B. M. is a predoctoral fellow of the Danforth Foundation. This is contribution No. 842 from the McCollum-Pratt Institute.

5. REFERENCES

Alexander, L. E., 1937, An experimental study of the role of optic cup and overlying ectoderm in lens formation in the chick embryo, *J. Exp. Zool.* **75:**41.

Barbera, A. J., 1975, Adhesive recognition between developing retinal cells and the optic tecta of the chick embryo, *Dev. Biol.* **46:**167.

Barbera, A. J., Marchase, R. B., and Roth, S., 1973, Adhesive recognition and retinotectal specificity, *Proc. Natl. Acad. Sci. USA* **70:**2482.

Bray, D., and Bunge, M. B., 1973, The growth cone in neurite extension, in: *Locomotion of Tissue Cells* (Ciba Foundation Symposium 14), pp. 195–209, Associated Scientific Publishers, Amsterdam.

Carter, S. B., 1965, Principles of cell motility: The direction of cell movement and cancer invasion, *Nature (London)* **208:**1183.

Crossland, W. J., Cowan, W. M., and Rogers, L. A., 1975, Studies on the development of the chick optic tectum, *Brain Res.* **91:**1.

Curtis, A. S. G., 1967, *The Cell Surface: Its Molecular Role in Morphogenesis*, Logos Press, London, Academic Press, London.

DeLong, R. G., and Coulombre, A. J., 1965, Development of the retinotectal topographical projection in the chick embryo, *Exp. Neurol.* **13:**351.

Dorsey, J. K., and Roth, S., 1973, Adhesive specificity in normal and transformed mouse fibroblasts, *Dev. Biol.* **33**:249.

Fujita, S., and Horii, M., 1963, Analysis of cytogenesis in chick retina by tritiated thymidine autoradiography, *Arch. Histol. Jpn.* **23**:359.

Gaze, R. M., and Keating, M. J., 1972, The visual system and "neuronal specificity," *Nature (London)* **237**:375.

Goldberg, S., 1974, Studies on the mechanics of development of the visual pathways in the chick embryo, *Dev. Biol.* **36**:24.

Holtfreter, J., 1939, Tissue affinity, a means of embryonic morphogenesis, translated in: *Foundations of Experimental Embryology* (B. H. Willier and J. M. Oppenheimer, eds.), pp. 186–225, Prentice-Hall, Englewood Cliffs, N.J., 1964.

Hunt, R. K., and Jacobson, M., 1974, Neuronal specificity revisited, *Curr. Top. Dev. Biol.* **8**:203.

Jacobson, M., 1970, *Developmental Neurobiology,* Holt, Rinehart and Winston, New York.

Levine, R. L., and Cronly-Dillon, J. R., 1974, Specification of regenerating retinal ganglion cells in the adult newt, *Triturus cristatus, Brain Res.* **68**:319.

Marchase, R. B., Barbera, A. J., and Roth, S., 1975, A molecular approach to retinotectal specificity, in: *Cell Patterning* (Ciba Foundation Symposium 15), Associated Scientific Publishers, Amsterdam.

McClay, D. R., and Baker, S. R., 1975, A kinetic study of embryonic cell adhesion, *Dev. Biol.* **43**:109.

Moscona, A. A., 1965, Recombination of dissociated cells and the development of cell aggregates, in: *Cells and Tissues in Culture* (E. N. Willmer, ed.), pp. 489–529, Academic Press, New York.

Roseman, S., 1970, The synthesis of complex carbohydrates by multiglycosyltransferase systems and their potential function, *Chem. Phys. Lipids* **5**:270.

Roth, S., 1968, Studies on intercellular adhesive selectivity, *Dev. Biol.* **18**:602.

Roth, S., 1973, A molecular model for cell interactions, *Q. Rev. Biol.* **48**:541.

Roth, S., and Weston, J., 1967, The measurement of intercellular adhesion, *Proc. Natl. Acad. Sci. USA* **58**:974.

Roth, S., McGuire, E. J., and Roseman, S., 1971*a*, An assay for intercellular adhesive specificity, *J. Cell Biol.* **51**:525.

Roth, S., McGuire, E. J., and Roseman, S., 1971*b*, Evidence for cell-surface glycosyltransferases: Their potential role in cellular recognition, *J. Cell Biol.* **51**:536.

Sperry, R. W., 1963, Chemoaffinity in the orderly growth of nerve fiber patterns and connections, *Proc. Natl. Acad. Sci. USA* **50**:703.

Steinberg, M., 1964, The problem of adhesive selectivity in cellular interactions, in: *Cellular Membranes in Development* (M. Locke, ed.), pp. 321–366, Academic Press, New York.

Stone, L. S., 1950, Neural retina degeneration followed by regeneration from surviving retinal pigment cells in grafted adult salamander eyes, *Anat. Rec.* **106**:89.

van Wezel, A. L., 1973, Microcarrier cultures of animal cells, in: *Tissue Culture Methods and Applications* (P. F. Kruse, Jr., and M. K. Patterson, Jr., eds.), pp. 372–377, Academic Press, New York.

Vosbeck, K., and Roth, S., 1976, in preparation.

Walther, B. T., Ohman, R., and Roseman, S., 1973, A quantitative assay for intercellular adhesion, *Proc. Natl. Acad. Sci. USA* **70**:1569.

Weiss, P., 1947, The problem of specificity in growth and development, *Yale J. Biol. Med.* **19**:235.

Wilson, H. V., 1907, On some phenomena of coalescence and regeneration in sponges, *J. Exp. Zool.* **5**:245.

9

Membranes as a Tool for the Study of Cell Surface Recognition

R. MERRELL, D. I. GOTTLIEB, and L. GLASER

1. INTRODUCTION

It has been known for many years, since the early work of Townes and Holtfreter (1955) and Moscona and Moscona (1952), that embryonal tissues can be dissociated into single cells and that aggregates obtained by centrifuging single cells from different tissues into a pellet will ultimately segregate in culture so that homologous cells are in contact with one another. This phenomenon could be due to specific cell adhesion, chemotaxis, or a combination of these effects (Steinberg, 1963; Moscona, 1965).

If single cells are prepared from two different organs, mixed in a suitable medium, and allowed to aggregate, it is impossible to determine whether the initial cellular attachment occurs with organ specificity or at random, since it is usually very difficult to distinguish the different cell types in early aggregates.

Two different approaches have been used to demonstrate specific cell adhesion. Both of these measure the adhesion of radioactively

---

R. MERRELL, D. I. GOTTLIEB, and L. GLASER · Department of Biological Chemistry, Division of Biology and Biomedical Sciences, Washington University School of Medicine, St. Louis, Missouri.

labeled single cells, in one case to preformed large aggregates (Roth and Weston, 1967; Roth, 1968; Roth *et al.*, 1971) and in the second case to a monolayer of cells (Walther *et al.*, 1973).

Both of these methods are quantitative and can be used to show organ specificity of cell adhesion with cells obtained from early chicken embryos. They both have the potential disadvantage that the cells have to be kept in an *in vitro* environment for 24–48 h either to form the large aggregates or to form stable monolayers. It is quite possible that complex changes in cell surface specificity take place during this incubation period.

Cell surface specificity may be absolute—i.e., different organs may have totally different cell surface recognition systems—or the homologous cells may simply have a relatively higher affinity for each other than for heterologous cells (Steinberg, 1963). In either case, there are currently two general hypotheses of how cell recognition takes place, and these are diagrammatically shown in Fig. 1.

Model A assumes that cell surfaces contain two complementary molecules and that cell recognition is analogous to the binding of an enzyme to a substrate. A specific version of this model is that cell recognition depends on the binding of a cell surface glycosyltransferase to its substrates on an adjacent cell (Roseman, 1970; Roth and Marchase, this volume).

Model B, derived originally from work with sponges (Humphreys, 1963; Margoliash *et* al., 1965; Cauldwell *et al.*, 1973; Weinbaum and Burger, 1973), assumes the presence of an extracellular organ-specific multivalent ligand (aggregation-promoting factor) which like an antibody or a lectin will bind together cells from the same organ (see also Barondes and Rosen, this volume, for a discussion of this model in slime mold aggregation).

Model C illustrates the fact that if the multivalent ligand or aggregation factor is firmly attached to the cell membrane under *in vivo* conditions then model B converts to model A, so that these models are not really mutually exclusive in that cells prepared under certain conditions may lose specific surface proteins into the medium.

Aggregation-promoting factors have been extracted from chick retina and telencephalon either by extraction of the whole organ or by harvesting the medium in which appropriate cells have been maintained in cultures (Balsamo and Lilien, 1974*a*,*b*; Hausman and Moscona, 1973; Garber and Moscona, 1972). When these factors are added to aggregating cells, they increase the size of the homologous aggregate after 24 h of incubation. These factors are discussed in this volume by A. Moscona. It should be clear, however, that initial cell recognition

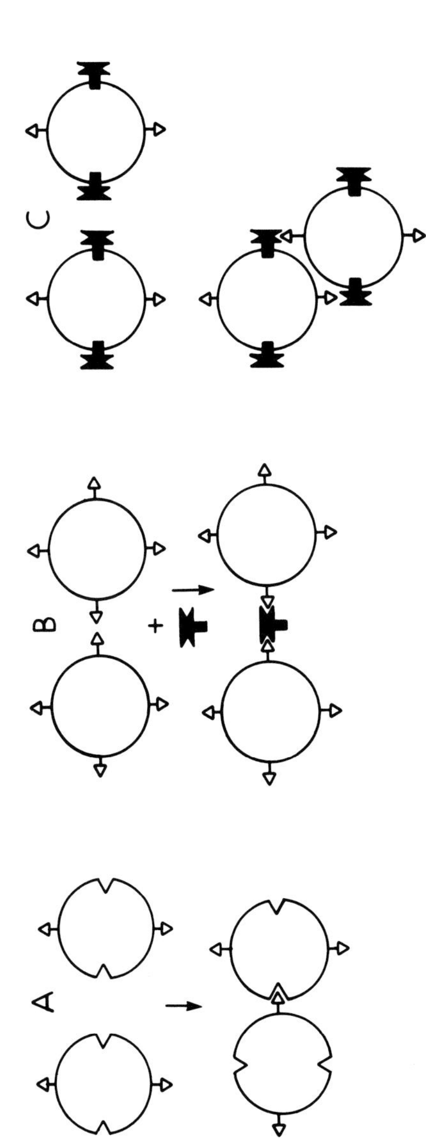

Fig. 1. Diagrammatic representation of cell adhesion models. Model A assumes direct attachment of complementary structures on the cell surface. Model B assumes that attachment occurs with the aid of a multivalent soluble ligand. Model C shows that if this soluble ligand is attached to the cell membranes then model B becomes experimentally identical to model A.

takes place in a period of 30 min and is no doubt followed by more complex cell interactions which may be reflected in the size and appearance of aggregates after 24 h. In our work, we have been concerned with the specificity of the initial recognition event.

On the assumption that model A adequately represents the events responsible for initial cell recognition, then plasma membranes isolated from embryonal cells should bind preferentially to homologous cells.

Cell aggregation is an extremely temperature-sensitive phenomenon and does not take place at 0°C. This suggests that recognition is not simply the binding of an enzyme to its substrate (which should also take place at 0°C) but also require a particular cell surface conformation or an energy-dependent change in the cell surface.

Isolated plasma membranes, under our assay conditions, inhibit cell aggregation rather than promote it. That is, they do not act as crosslinking agents (Merrell and Glaser, 1973; Gottlieb *et al.*, 1974). These observations can be explained in a number of ways. (1) The membranes retain only one of the cell surface recognition components, and cells coated with them cannot aggregate. (2) The attachment of two membrane fragments to each other is too weak to withstand the shearing forces during the aggregation assay. (3) When a membrane fragment attaches to a cell, all of its recognition sites, which are likely to be mobile in the plane of the membranes, become attached to the cell surface.

We will summarize in this chapter our studies of cell surface specificity exhibited by aggregating neuronal cells obtained from young chick embryos using plasma membranes as a tool to probe cell surface specificity. This is a first step in the isolation and characterization of the proteins responsible for cell surface recognition in this system.

It is important to point out the restrictions of this methodology. The basic experiment consists of incubating single cells in a rotating flask and measuring as a function of time the disappearance of single cells. If a cell has become attached to another cell before it is coated with membranes that prevent aggregation, it is counted as an aggregate. Therefore, the assay requires that cells be coated rapidly by an excess of membranes before they have an opportunity to interact with other cells. This means that the presence of ligands which occur only infrequently on the cell surface will go undetected. A diagrammatic example is shown in Fig. 2. Membranes which contain the ligand shown by rectangles will coat the cell surface and effectively block aggregation. Membranes that contain the ligand shown by the triangles will not

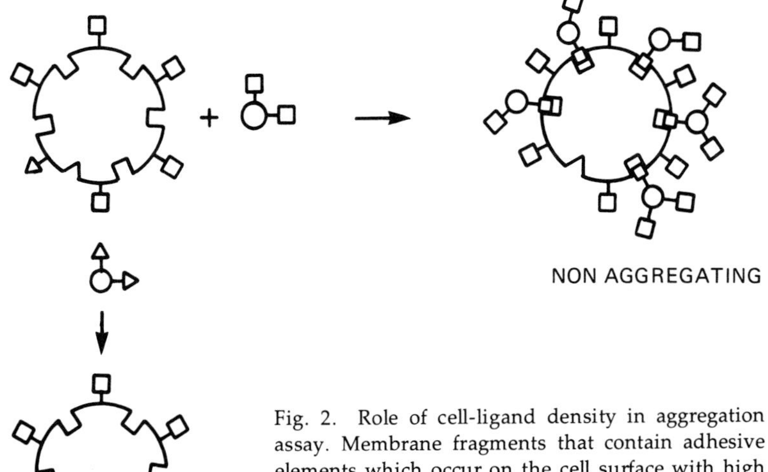

Fig. 2. Role of cell-ligand density in aggregation assay. Membrane fragments that contain adhesive elements which occur on the cell surface with high frequency will coat the cell surface and prevent aggregation. Membrane fragments that contain adhesive elements which occur infrequently on the cell surface will bind to the cell but will not prevent aggregation.

prevent homologous cell aggregation, although a specific site for this ligand is present on the cell surface.

It is clear from this simple consideration that different assays for cell surface adhesive specificity, which may not be as dependent as our assay method on the total quantity of the ligand present per unit cell surface, may indicate the presence on the cell surface of ligands other than the ones demonstrated by our assay method.

The system is very much dependent on precise methodology, and we will therefore describe the assay procedures in detail because there were some gaps in out initial description of the method (Merrell and Glaser, 1973; Gottlieb *et al.*, 1974) and also because of more recent improvements in procedure.

2. TECHNIQUES

Fertile white Leghorn chick eggs were purchased from Spafas, Inc. (Roanoke, Illinois), and stored at 10°C prior to initiating incubation at 39°C in a standard egg incubator. Under these conditions, chick em-

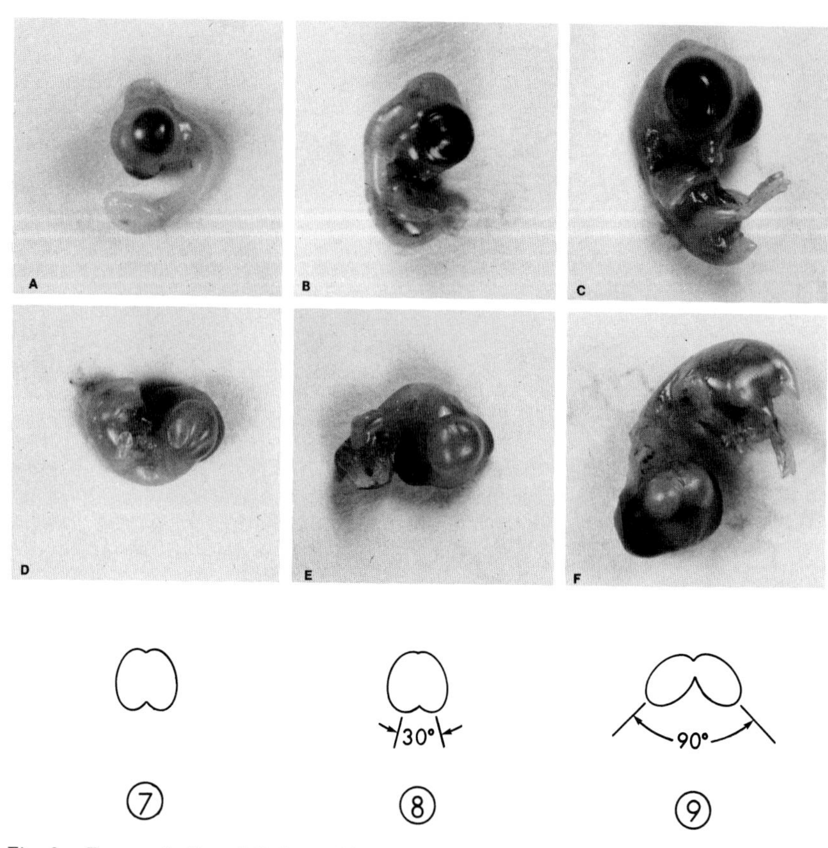

Fig. 3. Frames A, B, and C show side views of 7-, 8-, and 9-day chicks which under our incubation conditions range from Hamburger–Hamilton stage 28 to stage 33. The progressive flattening of the tecta is evident. Frames D, E, and F demonstrate the enlargement of the tecta with progressive separation of the hemispheres from close apposition to 90° as indicated diagrammatically in the bottom of the figure.

bryos developed with greater than 90% fertility and had a consistent developmental pattern. The embryos were used after 7, 8, or 9 days of incubation. Timing of the incubation was crucial and accurate developmental staging imperative. According to the Hamburger–Hamilton criteria for normal chick development (Hamburger and Hamilton, 1951), our 7-day embryos corresponded to stages 28–29, while 8-day embryos were at stages 30–31, and the 9-day embryos were scored as stages 33–34. Thus our material was 24–26 h immature compared to the standard series. We found it useful to score the development of our chicks according to the gross morphology of the optic tectum since this could be done quickly and was most pertinent to the neuroembryological state of the embryo. The 7-day tectum is translucent, protruding prominently from the head with its hemispheres tightly apposed in an anteroposterior axis. The 8-day tectum is much thicker and flatter. The hemispheres separate slightly from one another anteriorly, making an angle of no more than 30°. The 9-day tectum is quite opaque and lies smoothly in the contour of the skull. The hemispheres form an angle of greater than 90° with one another. This striking transition is shown in Fig. 3.

Dissection was done in teams with all reasonable dispatch. Two experienced people could easily harvest, for example, the neural retinas from 90 embryos in an hour. Dissection was done in Ca^{2+}-, Mg^{2+}-free Hank's solution further buffered with 0.02 M *N*-2-hydroxyethylpiperazine-*N'*-ethane sulfonic acid (HEPES), pH 7.4, and containing 10% heat-inactivated chicken serum (CMF-S). The freshly dissected tissue was stored at 0°C in CMF-S for up to 1 h. It was then gently triturated through a Pasteur pipette. If the tissue was to be dissociated to single cells, four to six tecta, retinas, etc., were pooled in 2 ml CMF-S and incubated at 37°C with gentle trituration every 5 min for 15 min. Crude trypsin (1:250 Difco) was added to a final concentration of 125–750 μg/ml and DNAse I (DN-100 Sigma) to a final concentration of 5–25 μg/ml to each tube and the digestion carried was out at room temperature for 15 min, again with trituration every 5 min. The amount of trypsin and DNAse required for an effective but gentle digestion varied with the batch of the enzymes. These conditions for dissociation are by no means harsh or thorough. Approximately 40% of the tissue remained undigested and was removed after five-fold dilution with cold CMF-S by spinning 200 rpm in a GLC-1 (Sorvall) centrifuge in an HL-4 rotor (20*g*) for 5 min. Attempts to more fully dissociate the tissue failed to provide cells maximally competent in the cell surface adhesive assay. The cells remaining in suspension after low-speed centrifugation were washed free of debris by sedimenting at 1000 rpm for 5 min in a GLC-1 centrifuge in cold CMF alone or supplemented with 0.5% fraction V bovine serum albumin, Sigma (CMF-A). The washed cells were diluted and could be kept in an ice bath for as long as 30 min prior to use in aggregation assays. More than 90% of the cells were viable as determined by trypan blue exclusion.

For aggregation assays, the cells were suspended in 0.5 ml CMF-A at a concentration of 5×10^4 cells/ml or 1×10^5 cells/ml in a 5-ml micro-Fernbach flask and rotated at 100 rpm in a New Brunswick G76 shaker bath at 37°C. After 30 min, more than 85% of the cells had formed cell–cell adhesions to produce small aggregates as determined by counting the remaining single cells in a hemocytometer. Aggregation proceeded rapidly in CMF-A, CMF alone, or in Hank's salts buffered with 0.02 M HEPES, pH 7.4. Therefore, neither cerebellum, tectum, retina, nor telencephalon demonstrated any marked requirements for divalent cation in the formation of initial adhesions. The aggregation was scored by counting the remaining single cells in a hemocytometer. Such counting gave a quantitative assessment of aggregation unlike aggregate scoring techniques. These methods for cell preparation and rotation-mediated aggregation are adapted from those previously described (Orr and Roseman, 1969). By varying these standard conditions, aggregation could be made a function of cell density, pH, ionic strength, temperature, speed of rotation, volume of cell suspension, or flask geometry (Orr and Roseman, 1969).

To isolate plasma membrane fractions, freshly dissected tissue was triturated in CMF-S and washed twice in cold CMF-A. Tissue fragments from 60–90 embryos were pooled in 7 ml CMF-A with 50 μg/ml DNAse I and homogenized in a 7-ml Dounce homogenizer, with a type B (tight) pestle (Kontes). Approximately 40 strokes were required to disrupt 95% of the cells. Because of the high nuclear/cytoplasmic ratio, breakage of the plasma membrane was almost invariably associated with breakage of the nuclear membrane. The clearance of the pestle was crucial and was checked frequently either by feel or by timing the gravity descent of the pestle through the water-filled tube. Homogenizers generally maintained their clearance for only only 3–4 weeks of active use. The homogenate was centrifuged at 50,000*g* in a RC2B Sorvall for 20 min and the supernatant was discarded. The pellet was homogenized in 60% sucrose (w/v) in CMF-A and distributed in 2-ml aliquots into cellulose nitrate ultracentrifuge tubes. This 60% base was overlaid with 1 ml each of 48% and 43% sucrose and finally 1.5 ml 37% sucrose. After 2 h centrifugation at 40,000 rpm in a SW50.1 Spinco rotor, distinct bands could be seen at each sucrose density interface and a pellet was visible. From the top of the gradient, these bands were numbered 1, 2, and 3.

The membrane fragments were freed of sucrose by dilution with CMF-A and centrifugation at 50,000*g* in glass tubes for 20 min and stored in CMF-A in 50% glycerol at −20°C. Table I summarizes the enzyme marker data which suggested that band 1 was significantly enriched in plasma membrane markers. In all subsequent references to the use of plasma membrane fractions in experiments, we refer to washed vesicles from this band 1.

On the assumption that plasma membrane markers are exclusively located on the cell surface, the recovery of plasma membrane on band 1 is no more than 20% and contamination with other cellular membranes

Table 1. Analysis of Retinal Membrane Fractions[a]

	Membrane Fraction		
	1	2	3
	Relative specific activity		
Mg^{2+} ATPase	13.4	0.47	0.7
Alkaline phosphatase	9.2	0.26	1.6
Phosphodiesterase	3.2	0.05	0.11
DPNH diaphorase	0.4	0.04	0.27
	dpm/mg protein		
[^{3}H]Glucosamine	2×10^6	4×10^5	4.0×10^5

[a] Membranes were prepared from 8-day neural retina cells which had been labeled with [^{3}H]glucosamine. The data represent the relative specific activity of the enzyme markers relative to the initial homogenate taken as unity. Radioactivity is expressed as dpm/mg protein. From Merrell and Glaser (1973).

is present in modest amounts. However, many techniques were tried for fractionating these cells and there was no superior way to isolate plasma membranes of acceptable purity which retained the adhesive cell surface determinants in an active form. A more complete discussion of this technique for membrane isolation and characterization has been reported (Merrell and Glaser, 1973).

There were two criteria for the retention of specific adhesion determinants on plasma membrane vesicles. First, isotopically labeled plasma membranes should adhere to homologous intact cells in a time-dependent and concentration-dependent fashion. Second, the adhesion of membrane vesicles to cells should affect the aggregability of these cells by either physical coverage or actual binding to adhesive sites. There were actually two possibilities for this effect. If the vesicles were multivalent with respect to adhesive determinants, they could act as ligands between cells, or, if they were effectively univalent, occupation of adhesive sites by vesicles should render the cells inaggregable. The latter case obtained. Inhibition-of-aggregation assays were done under standard conditions with the addition of small aliquots of membrane vesicle suspensions. Since the vesicles were about 1 μm in diameter, the concentration of vesicles could not exceed that which permitted easy visualization of cells and aggregates. In initial experiments we used 50–100 μg membrane protein per aggregation assay, but in later experiments this was reduced to 25–30 μg to facilitate cell counting. The vesicles tended to aggregate with one another and therefore required thorough suspension prior to use in an assay. Membranes could be stored in 50% glycerol at −20°C for about 1 week, washing twice in CMF-A before assay. Great care was taken to start the assay only after thorough mixing of cells and vesicles in the cold. In essence, three simultaneous reactions commenced after warming to 37°C: cell–cell adhesion, cell–vesicle adhesion, and vesicle–vesicle adhesion. In general, cell–cell adhesion seemed to be favored by the physical conditions of the assay, making total inhibition of aggregation difficult to achieve. All assays were stopped at 30 min, and the remaining single cells were counted in a hemocytometer. Recognition of single cells coated with membrane vesicles required some skill and the thorough agreement of all personnel involved in the study. For any given assay to be valid, no more than 20% of the control cells should be unaggregated after 30 min. For a control and experimental cell count to be considered significantly different at the 0.05% level, the number of cells counted was compared with a Poisson table. The effect of membrane vesicles on aggregation was frequently expressed as percent inhibition of aggregation.

The plasma membrane fractions were extracted to solubilize cell surface adhesive elements. The membrane vesicles from 90 embryos representing about 6–8 mg protein were suspended in 2 ml of CMF and

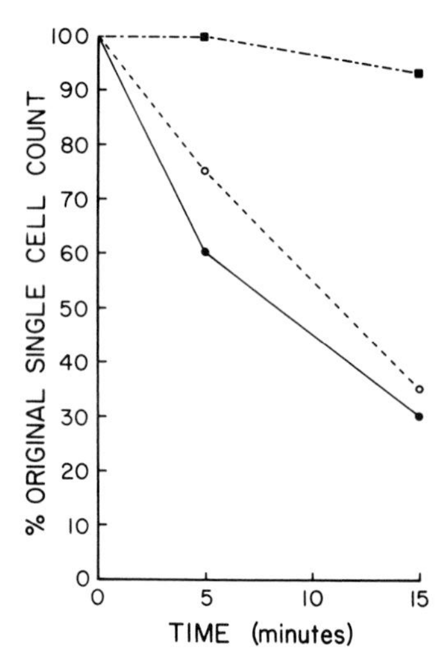

Fig. 4. Inhibition of retinal cell aggregation by membranes. Symbols: ●, 1.5 × 10^5 neural retinal cells incubated in 3 ml of medium HH with no additions; ○, 1.5 × 10^5 neural retinal cells incubated with cerebellar plasma membrane (0.1 mg of protein); ■, 1.5 × 10^5 neural retinal cells incubated with retinal plasma membrane (0.1 mg of protein). The percent remaining single cells is plotted as a function of time. Data are the average of five separate experiments with different cell suspensions and different membrane preparations; different experiments agreed within 5%. From Merrell and Glaser (1973).

pipetted into 20 ml acetone in a dry ice bath. Ice crystals were melted by careful warming and the suspension was centrifuged at 45,000*g* at −10°C for 20 min. The acetone-soluble phase was discarded and the pellet was dried under vacuum. This acetone powder could be stored at 0°C under vacuum for several weeks. The powder was extracted in 5 ml of 3×10^{-3} M lithium diiodosalicylate (LIS, prepared from twice-recrystallized diiodosalicylic acid, Eastman) at 25°C for 1 h with constant stirring. Insoluble material was removed by centrifugation in glass tubes at 45,000*g* for 20 min. The soluble extract was unstable but could generally be stored for several days at 4°C. To assay an extract, the assay medium always contained 0.5% BSA. Under these conditions, concentrations of LIS as high as 6×10^{-4} M were nontoxic and did not affect aggregation.

Protein was determined by A_{260nm}/A_{280nm} (Warburg and Christian, 1941) or a modified Biuret technique (Munkres and Richards, 1965). However, in the presence of LIS a ninhydrin (Moore and Stein, 1948) or fluorescamine assay (Nakai *et al.*, 1974) were used since LIS interfered with other protein assay methods. Wherever possible, operations with cells, membranes, or extracts were carried out in glass rather than plastic. Glassware washing was critical and generally accomplished by boiling in nitric acid or in a low-residue soap.

3. ORGAN SPECIFICITY

When membranes prepared from tissue of a given brain region are assayed, they inhibit the aggregation of cells from that same or homologous neural region, whereas membranes from a different or heterologous neural region are not recognized by intact cells and aggregation proceeds unhindered (Merrell and Glaser, 1973) (Fig. 4). When retinal membranes labeled with [^{3}H]glucosamine are incubated with retinal cells or cerebellar cells, preferential binding to the homologous cells is observed (Fig. 5).

It is important to show that under these conditions the membrane vesicles are on the cell surface and are not toxic to the cells. Figure 6 demonstrates that the adhesion of labeled membranes to intact cells is readily reversed by the addition of trypsin, showing that none of the bound radioactivity was taken up by pinocytosis. The membrane vesicles are highly sensitive to trypsin action and are inactivated by concentrations of trypsin which do not affect the aggregation of intact cells. The rate of binding of labeled vesicles to homologous cells is greatly diminished by incubation at 4°C rather than 37°C, suggesting that more is involved in adhesion than simple collision and supporting the hypothesis that adhesion requires an active cellular process or proper steric presentation of a cell surface receptor.

Neural retina/cerebellum and retina/telencephalon were mutually exclusive pairs in that aggregation of intact cells obtained from 8-day chicken embryos of either member of the pair was insensitive to the presence of plasma membrane vesicles of the other member of the pair. However, telencephalon cells tend to aggregate poorly, such that a significant fraction of the cells fail to aggregate. In the case of neural retina/optic tectum, an interesting cross-reaction was consistently observed. While tectal plasma membranes had little or no effect on the

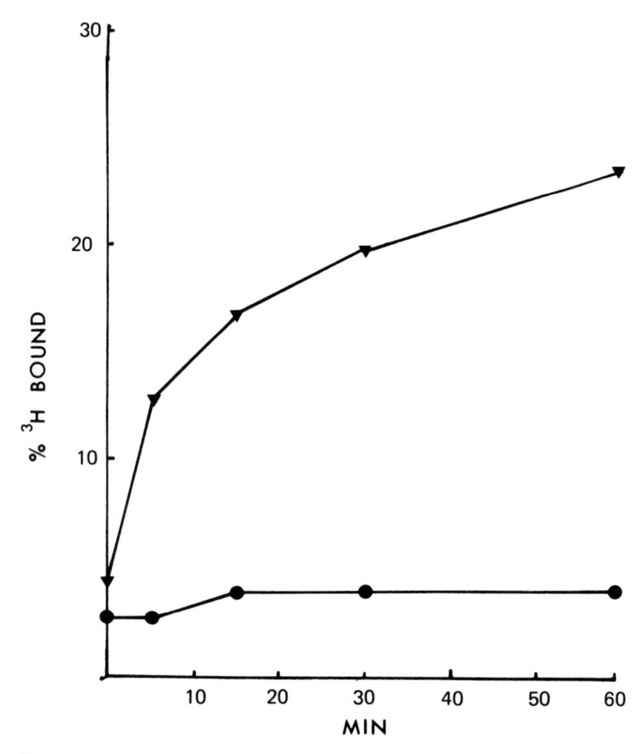

Fig. 5. Binding of radioactive membranes to cells. 6×10^5 Cells were incubated in 3 ml of medium HH with [^{3}H]glucosamine-labeled plasma membrane (20 μg of protein, 15,000 dpm). At each time point, the aggregation was stopped by threefold dilution in HH medium; cells with attached membranes were collected by centrifugation at 300*g* for 5 min. The pellet was washed with 3 ml of 5% trichloroacetic acid, dissolved in 1 ml of Protosol (New England Nuclear Corp.), and counted. Symbols: ●, cerebellar cells; ▼, retinal cells. From Merrell and Glaser (1973).

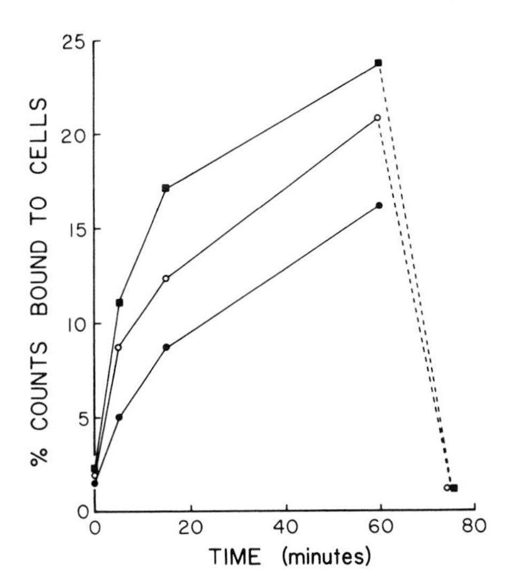

Fig. 6. Effect of cell concentration on membrane binding. The conditions are those of Fig. 3 with ^{3}H-labeled retinal plasma membranes and the following concentrations of neural retinal cells per milliliter: ●, 0.5 × 10^5; ○, 1.0 × 10^5; ■, 1.5 × 10^5. After 60 min, duplicate flasks were made 0.125% in crystalline trypsin and incubated for 15 min at 37°C, and cells were prepared for counting in the usual way. These cells are indicated by the 75-min point. The trypsinized cells were intact, as judged by the trypan blue exclusion. Data are the average of five different membrane preparations mixed with different cell suspensions. The standard error was 15%. From Merrell and Glaser (1973).

aggregation of retina cells, retinal plasma membranes inhibited the aggregation of tectal cells, although less effectively than the homologous tectal plasma membranes (Fig. 7).

The unidirectional inhibition of tectal cell aggregation by retinal membranes may reflect either the presence of a tectal-specific ligand on the retinal membrane or the fact that the recognition protein(s) on the tectal cell surface has a broad enough specificity that it can recognize the retinal-specific ligand on retinal membranes responsible for tectal specificity. Studies with soluble ligands discussed in Section 5 favor the latter possibility.

4. TEMPORAL SPECIFICITY

The techniques described in the previous sections can be successfully used only over a limited period in the development of the nervous system of the chicken. At very early developmental stages, it is very

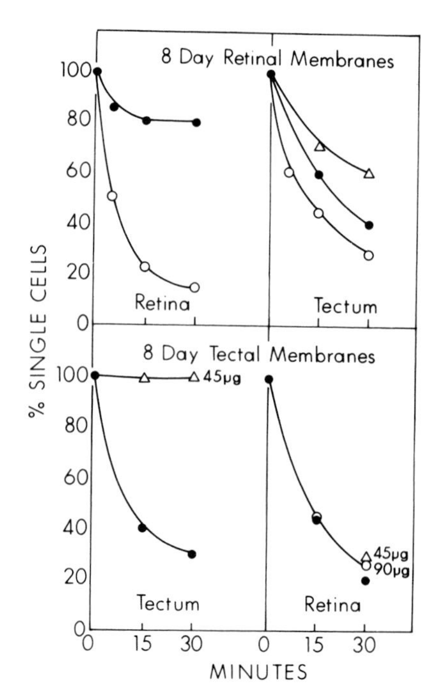

Fig. 7. Time course of cell aggregation in the presence of membranes. The top portion of the figure shows the time course of cell aggregation of 8-day retinal and tectal cells in the presence of retinal membranes. Symbols: ○, no added membranes; ●, 31 μg of retinal membrane protein; Δ, 62 μg of retinal membrane protein. The bottom of the figure shows similar data with tectal membranes with the quantity of membrane protein indicated. From Gottlieb *et al.* (1974).

difficult to obtain enough material to prepare membranes, and in our experience neuronal cells from chickens older than 9 days aggregate very slowly and show clear signs of heterogeneity. Within this limited time period, we have examined whether cell surface specificity changes with time of development within a given organ (Gottlieb *et al.*, 1974). Figure 8 summarizes experiments with plasma membrane vesicles from 8-day neural retina showing a striking temporal specificity for the inhibition of 8-day cell aggregation. Temporal specificity was expressed in the interaction of intact cells and membranes over days 7 through 9 of development both in the retina and in the optic tectum (Fig. 9). The retinotectal cross-reaction is synchronous with this temporal pattern (Fig. 8).

The change in cell surface specificity during development could be due to changes in the cell surface of preexisting cells or the appearance of new cell types. The elegant birthday studies of tectal neurons by LaVail and Cowan (1971*a*,*b*) and neural retina cells (Kahn, 1973, 1974) place the last neuronal division approximately at stage 30, or 8 days of development with our embryos. Thus the changes in cell surface specificity between days 7 and 8 could be due to cell proliferation while those between days 8 and 9 could represent continued cell membrane proliferation and maturation. However, this picture is a gross oversimplification. The cell birthday studies also show in each organ a 2- or 3-day gradient of morphological development, which would have predicted an overlap of specificity in our assay, if the adhesive potential is an integral part of the observed morphological maturation.

There are two other instances in which the morphological changes do not predict cell surface properties. Goldschneider and Moscona (1972) have shown that neural retinas at all ages share common antigenic determinants. Barbera *et al.* (1973) have shown that retinal cells will adhere to optic tectum, obtained both from normal embryos and

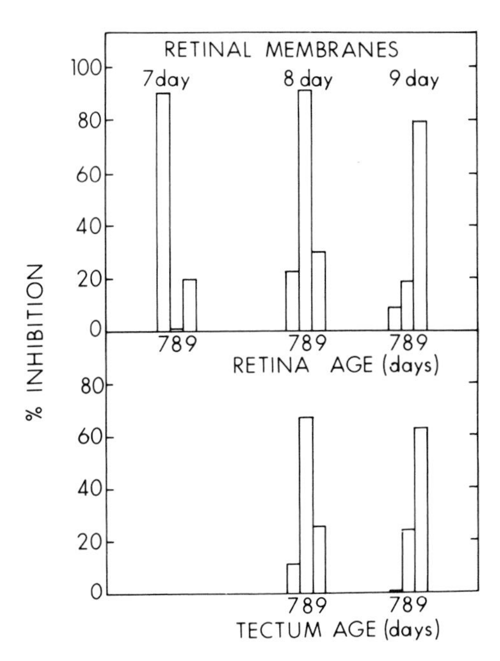

Fig. 8. Effect of embryonal age on membrane specificity. All experiments were carried out with membranes prepared from whole neural retina, obtained from 7-, 8-, or 9-day embryos. The membranes were used to inhibit aggregation of either retinal or tectal cells of the ages indicated. One-hundred micrograms of membrane protein was used in each assay to maximize cross-reactivity. From Gottlieb *et al.* (1974).

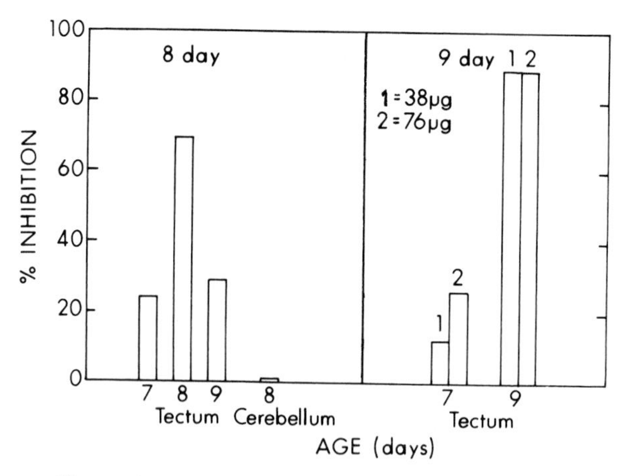

Fig. 9. Age specificity of tectal cells. In the left panel is shown the inhibition of cell aggregation obtained with 8-day tectal membranes (90 μg of protein) using 7-, 8-, and 9-day tectal cells and 8-day cerebellar cells. In the right panel is shown the inhibition obtained at two different protein concentrations when 9-day tectal membranes were tested with 7- and 9-day tectal cells. From Gottlieb *et al.* (1974).

from embryos subjected to early optic ablation to block retinal fiber projections to the tectum. Barbera *et al.* also found that cells obtained from the ventral region of the retina bind preferentially to the dorsal region of the tectum and conversely cells from the dorsal region of the retina bind preferentially to the ventral region of the tectum. This also would not have been predicted by the distribution of differentiationally distinct cells because the sequence of differentiation in retina is from optic disc region to ora serrata and was not complete at the earliest time period examined by Barbera *et al.* The selective adhesion observed by Barbera *et al.* is, however, in aggreement with the normal retina tectal projection and is discussed in detail Roth and Marchase in this volume.

It would be possible in our cell preparation to select only a few cell types which shared a common differentiational profile so that different cells would be selected at each stage of development. However, that would imply enrichment of the tissue not dissociated by trypsin for cells at other levels of development and potentially other adhesive specificities. To test this possibility, we prepared plasma membranes from freshly dissociated single cells and from that tissue left intact by the gentle trypsinization. Plasma membranes from either tissue fraction showed precisely the same adhesive and inhibitory activity, suggesting that the specific adhesive properties we study are shared by the preponderance of the cells in a given region.

5. SOLUBILIZATION OF AGGREGATION INHIBITORY FACTORS

A better understanding of these adhesive events followed solubilization of the specific aggregation inhibitory activity in plasma membrane vesicles. The LIS-soluble extract displayed all the regional and temporal specificity of the membrane vesicles as an inhibitor of cell aggregation as described in Fig. 10 for 8-day retina extract and 8-day tectum extract. The aggregation inhibitory factor of the extract was highly trypsin sensitive but survived boiling for 3 min. The concentration dependence of the inhibition of aggregation on factor concentration is shown in Fig. 11 and the tightness of the retinotectal cross-reaction can again be seen. The highly cooperative inhibition of aggregation seen in Fig. 11 is subject to a number of interpretations, the simplest of which is that a large majority of the sites on the cell surface need to be occupied by the factor before a significant decrease in the rate of cell aggregation is observed.

The factor(s) in the LIS extract seems to exert its effect at the cell surface in that the inaggregable cells could be liberated from inhibition by brief exposure to trypsin, after which the cells aggregated normally. The reversibility of inhibition by gently trypsinization suggests a very important aspect of the initial adhesive event. An inhibited cell appar-

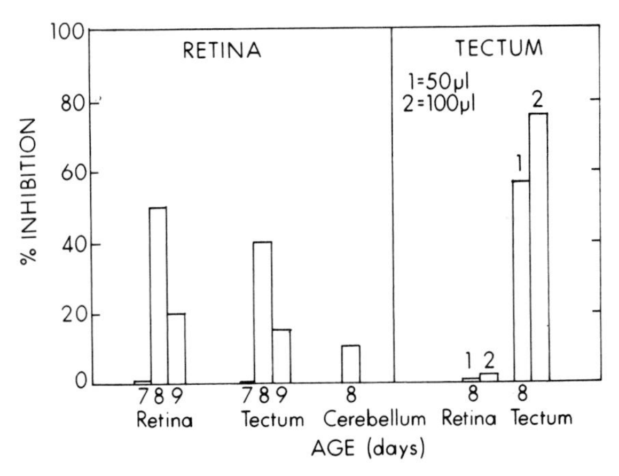

Fig. 10. Specificity of LIS extracts. The left panel shows the inhibition of 8-day retinal plasma membrane extract of the aggregation of tectal, retinal, and cerebellar cells obtained from embryos of the ages indicated at the bottom of the figure. The right panel shows the inhibition by 8-day tectal plasma membrane extract of the aggregation of retinal cells as well as the aggregation of tectal cells from 8-day embryos. Note that even at twice the concentration of tectal extract that maximally inhibits homotypic cell aggregation there is no inhibition of retinal cell aggregation.

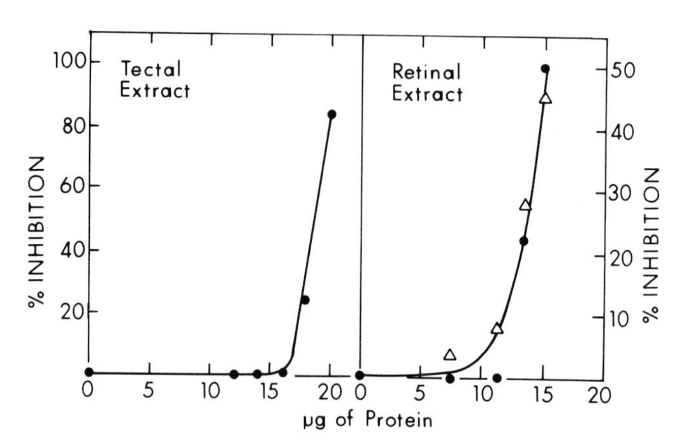

Fig. 11. Concentration dependence of inhibition of cell aggregation. The data show the inhibition of cell aggregation as a function of extract concentration. Symbols: ●, tectal cells; Δ, retinal cells. Extract and cells were obtained from 8-day embryos.

ently has most of its recognition sites occupied by the solubilized factor. We know that the soluble factor loses its recognition or binding activity in the presence of trypsin. However, in these experiments we see the soluble factor dissociating from the receptor site in the presence of trypsin, leaving the site active and apparently intact. If recognition involves sugar residues on a glycoprotein as is the case for sponges, a drastic change in K_m by removal or denaturation of the protein is not surprising and trypsinization of the soluble factor would lead to release of the recognized or bound portion of the molecule. These observations favor an enzyme–substrate model for cell adhesion and give no indication that binding and initial recognition lead (at least immediately) to a covalent linkage between cell surfaces.

The soluble extract provided an opportunity to test whether the same component in the retinal extract was recognized by both retinal and tectal cells. If the cells were recognizing different molecules which subsequently bound to very different adhesive sites on the respective cells, the adsorption of a retina extract with retina cells should not diminish the inhibitory activity of that extract toward tectal aggregation and *vice versa*. The results depicted in Fig. 12 indicate that retina and tectal cells bind the same molecule in a retina extract in that adsorption of a retinal extract with retinal cells removes inhibitory activity against the aggregation of both retinal and tectal cells. Note that adsorption with cells from the telencephalon, a truly heterologous region, has no effect on the inhibitory activity of the retinal extract.

Efforts to further purify and characterize the extracts were largely thwarted by the very small mass of material available. Also, repeated attempts at *in vivo, in vitro,* and chemical isotopic labeling failed to achieve suitable specific activity of the inhibitory molecules to permit binding studies. Ninety to one hundred twenty 8-day embryos yielded 6–8 mg of membrane protein. This was reduced to about 0.2 mg soluble protein after delipidation and LIS extraction. This extract was unstable under a variety of conditions including dialysis removal of LIS, freezing, or storage at 4°C for more than 72 h. If the LIS extract was filtered through a PM-10 Amicon membrane, the activity quantitatively passed through the membrane and could then be concentrated on a UM-2 filter (Amicon), suggesting that a molecule of 10,000 mol wt is responsible for the inhibition. About 2 μg protein of this final solution would inhibit the aggregation of 5×10^4 cells in 0.5 ml, representing a tenfold purification over the crude LIS extract. A single major protein band was obtained when this material was examined by sodium dodecylsulfate electrophoresis. This protein stained as a glycoprotein and represented less than 1% of the total membrane protein. However, ion exchange

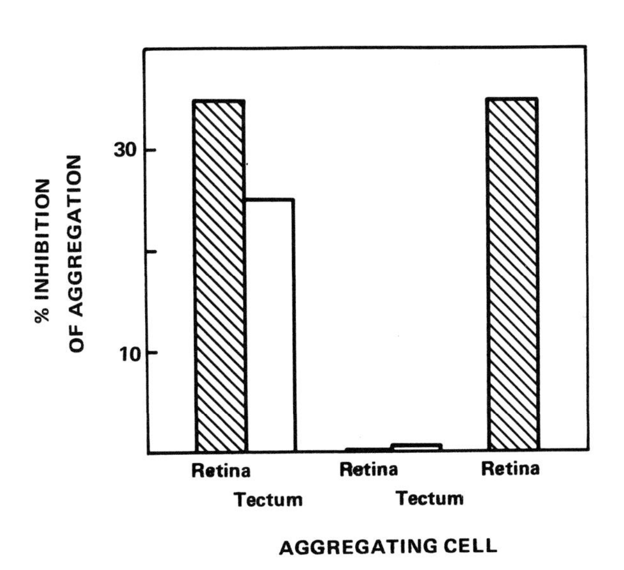

Fig. 12. Adsorption of retinal extract. Retinal LIS extract was incubated for 20 min at 37°C in 1 ml of CMF-A either with no addition or with 1×10^5 retinal cells or telencephalic cells. The cells were removed by centrifugation at 300*g* for 5 min, and 0.45 ml of the supernatant fluid was used in a standard aggregation assay. The first two bars represent the original extract, the second two bars the same material after adsorption with retina cells, and the last bar the same material after adsorption with telencephalic cells.

chromatography on DEAE-cellulose resulted in a partial separation of this major protein from the cell aggregation inhibitory activity. Thus the major glycoprotein seen on gel electrophoresis of the UM-2 retentate probably does not represent the cell surface recognition component. The active element thus seems to be a minor component of the UM-2 retentate.

Quantitatively this is not particularly surprising, if we assume that a cell contains on its surface 10^5 adhesive sites, and, assuming a very high affinity, 10^{10} molecules of inhibitor per milliliter should inhibit the aggregation of 10^5 cells. If the inhibitor has a molecule weight of 10,000, this corresponds to 0.2 μg/ml, while with the most purified preparation available to us 4 μg/ml of total protein is required to inhibit cell aggregation.

6. TROPHIC EFFECT OF NERVE GROWTH FACTOR ON TEMPORAL DIFFERENTIATION

In an effort to obtain radioactively labeled adhesive determinants, we have cultured early neuronal cells to determine the *in vitro* requirements for a change in cell surface specificity *in vitro*. Our requirements for this system were that under the *in vitro* conditions the cells should remain competent to aggregate and show specific membrane recognition and maximal survival.

Our attempts to culture retinal cells have failed to date in that cell survival was invariably very poor. However, 7-day optic tectum cells survived culture conditions quite well such that 70–80% of the cells could be recovered after 24 h incubation. Initial experiments were carried out by suspending 5×10^6 7-day tectum cells in 10 ml of medium 199 (Gibco) with glutamine, further buffered with 0.02 M Hepes, pH 7.4, and supplemented with 1000 units/ml of penicillin G, 100 μg/ml of streptomycin sulfate (Sigma), and 1% heat-inactivated chicken serum (Gibco). The suspension was rotated at 37°C in a 25-ml capped Erlenmeyer flask at 130 rpm in a G76 New Brunswick water bath shaker. Tight aggregates formed promptly and were dissociated after 24 h in culture in the same manner as fresh tissue. The resultant single cells were assayed as before for their recognition of the 7-day or 8-day optic tectum membranes by inhibition of aggregation. Under these circumstances, the acquisition of 8-day cell surface determinants or specificity was a very rare event depending in a rough way on serum

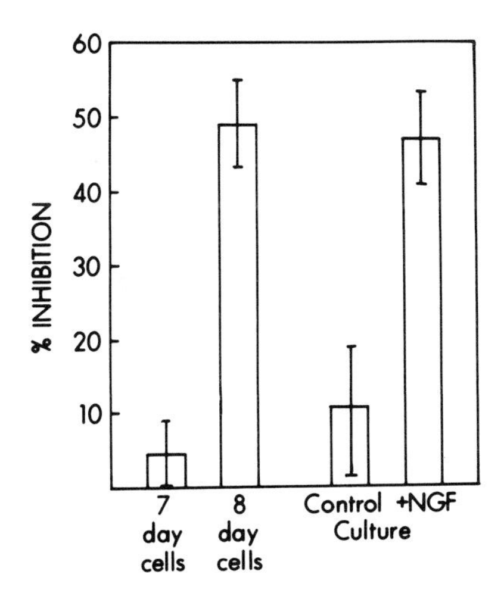

Fig. 13. Effect of NGF on cell surface specificity. Seven-day optic tectum cells were incubated for 24 h *in vitro* as described in the text. Single cells were prepared from the aggregates, and the ability of plasma membranes prepared from 8-day optic tectum to inhibit the aggregation of these cells was determined. The bar graphs from left to right show the inhibition of 7-day cell aggregation by these membranes, 8-day cell aggregation, aggregation of 7-day cells maintained for an additional day in culture, and 7-day cells maintained for 1 day in culture with NGF. The data are the average of 13 independent experiments.

lot and serum concentration. Supplementation with 8-day chick embryo extract generally supported the transition, but not with the consistency desired.

Of many agents tested for the capacity to support the transition from 7-day tectum to 8-day tectum *in vitro*, one compound was consistently effective. Nerve Growth Factor (NGF) prepared from mouse submaxillary gland (Bocchini and Angeletti, 1969) at 1×10^{-7} M in medium 199 with 1% chicken serum seemed an absolute requirement for the acquisition of 8-day determinants by 7-day tectum cells in culture (Fig. 13). The addition of NGF did not simply fulfill some gross metabolic requirement in that in parallel studies of cells in 1% chicken serum or 1% serum plus NGF, total cell recovery, aggregability, [^{3}H]leucine incorporation, and [^{3}H]thymidine incorporation were not significantly different. Neither NGF nor serum could be replaced by a mixture of dexamethasone (10^{-6} M) and insulin (2×10^{-8} M). Therefore, a somewhat specific trophic effect of NGF was suggested. [^{3}H]Thymidine incorporation studies indicated insufficient DNA synthesis in the cultured cells to account for the appearance of a new differentiated property as a result of cell division. This was borne out by the failure of cytosine arabinoside to block the NGF-induced differentiation. We had difficulties similar to those of Schwartz (1973) in using actinomycin D as an inhibitor of transcription in these neural cells in that even minimal concentrations were toxic and we were unable to thoroughly adhere to the medium formulation recommended by Jones and Moscona (1974) to correct this problem.

In collaboration with Dr. Ralph Bradshaw and his colleagues at Washington University, a number of further studies were undertaken (Merrell *et al.*, 1975). Intact 7-day tectal cells possess 3–6 × 10^3 specific binding sites for NGF (Frazier *et al.*, 1974). The mere presence of NGF did not confer 8-day specificity, as demonstrated by assay of fresh 7-day cells in the presence of 10^{-7} M NGF. Pulses of 2 and 4 h with NGF at the outset of the 24-h culture period also did not elicit the expression of 8-day cell surface specificity. Apparently, persistent occupation of the NGF receptor site is obligatory.

Oxidation of tryptophan residues 21 and 99 in NGF by bromosuccinamide results in an NGF molecule which is inactive in the dorsal root ganglion bioassay without leading to dimer dissociation or gross alteration of tertiary structure (Frazier *et al.*, 1973*b*). However, in this culture system oxidized NGF gave a weaker but significant effect while appearing totally inert in parallel dorsal root ganglion bioassays. NGF isolated from the venom of *Naja naja* (kindly provided by Dr. Ruth Hogue-Angeletti) gives a somewhat blunted effect in bioassay with dorsal root ganglia, but showed absolutely no activity in out culture system. NGF bears structural similarities ito insulin (Frazier *et al.*, 1973*b*) and belongs to a family of low molecular weight proteins called pleiotypic agents which have the common property of promoting growth in specific target tissues. In the case of cell surface transition, whether NGF has a role in this transition *in vivo*, or is acting under the *in vitro* conditions as an analogue of one of the other known or possibly unknown pleiotypic agents, remains to be established. A great deal has been reported about NGF activity in dorsal rool ganglion and other neural tissues (Levi-Montalcini *et al.*, 1973; Stenevi *et al.*, 1974; Frazier *et al.*, 1973*a*,*b*), and further experiments to examine the relationship between tectal cell surface maturation and other proposed NGF functions are in progress in our laboratory.

When maximal concentrations of NGF were used, the cells totally lost the ability to recognize 7-day membranes. Surprisingly, at maximal NGF concentration the transition was cycloheximide resistant, suggesting that new protein synthesis is not required for the transition and that it represents either a change in a preexisting surface component or the externalization of a preexisting molecule; thus in a sense the cells are preprogrammed for this developmental change and NGF only allows this program to be expressed. Whether this change in cell surface adhesion specificity involves new carbohydrate components (Roseman, 1970; Oppenheimer *et al.*, 1969) is not known.

7. SUMMARY

Complex events at the cell surface accomplish the crucial purpose of cellular communication with the external milieu. Among these events are cell–cell interactions by which the cell discovers and imparts largely unknown information in congress with its fellow cells. We have examined one potential area of cell–cell interaction in studying cell surface recognition in developing neural tissue. Such recognition could play a crucial role in the elegant ordering of this, the most intricate of all tissues. We find a system of adhesive determinants on the surface of retinal, tectal, and other cells which define a regional identity and suggest the rapid alteration of cell surface affinities from day to day during embryonic development. Attention has been focused on the cell surface itself by isolation of the plasma membranes of these cells and chemical extraction and partial purification of at least one component of the adhesive system. Studies with the extracted determinant are compatible with an enzyme–substrate model for cell surface adhesion. However, no enzymatic activity has been clearly described for either component. At least one known embryonic trophic agent, nerve growth factor, has been strongly implicated in the trasition from 7-day to 8-day cell surface specificity for optic tectum cells.

The inductive or regulatory events of growth and development have in the past been neatly partitioned into cell surface events and humoral and/or chemotactic effects. Should the *in vitro* effect of NGF in promoting the maturation of cell surface adhesive determinants prove physiologically important, a close relationship between these two modes of developmental control will be confirmed.

ACKNOWLEDGMENTS

Work in the author's laboratory was supported by Research Grants NIH GM 18405 and NSF BMS-74-22638. R. M. was supported by Grant 5T01-GM 60371 and D. I. G. by Grant F22-NS166.

8. REFERENCES

Balsamo, J., and Lilien, J., 1974*a*, Embryonic cell aggregation: Kinetics and specificity of binding of enhancing factors, *Proc. Natl. Acad. Sci. USA* **71**:727.

Balsamo, J., and Lilien, J., 1974*b*, Functional identification of three components which mediate tissue-type specific embryonic cell adhesion, *Nature (London)* **251**:522.

Barbera, A. J., Marchase, R. B., and Roth, S., 1973, Adhesive recognition and retinotectal specificity, *Proc. Natl. Acad. Sci. USA* **70**:2482.

Bocchini, J., and Angeletti, P. U., 1969, Nerve growth factor purification as a 30,000 molecular weight protein, *Proc. Natl. Acad. Sci. USA* **64**:787.

Cauldwell, C. B., Henkart, P., and Humphreys, T., 1973, Physical properties of sponge aggregation factor, a unique proteoglycan complex, *Biochemistry* **12**:3051.

Frazier, W. A., Ohlendorf, C. E., Boyd, L. F., Aloe, L., Johnson, E. M., Ferendelli, J. A., and Bradshaw, R. A., 1973*a*, Mechanism of action of nerve growth factor and cyclic AMP on neurite outgrowth of embryonic chick sensory ganglia: Demonstration of independent pathway of stimulation, *Proc. Natl. Acad. Sci. USA* **70**:2448.

Frazier, W. A., Hogue-Angeletti, R. A., Sherman, R., and Bradshaw, R. A., 1973*b*, Topography of mouse 2.5 S nerve growth factor: Reactivity of tyrosine and tryoptoophan, *Biochemistry* **12**:3281.

Frazier, W. A., Boyd, L. F., and Bradshaw, R. A., 1974, Properties of the specific binding of ^{125}I-nerve growth factor to responsive peripheral neurons, *J. Biol. Chem.* **249**:5513.

Garber, B. B., and Moscona, H. A., 1972, Reconstruction of brain tissue from cell suspensions: Specific enhancement of aggregation of embryonic cerebral cells by medium from homologous cell cultures, *Dev. Biol.* **27**:235.

Goldschneider, I., and Moscona, H. A., 1972, Tissue specific cell surface antigens in embryonic cells, *J. Cell Biol.* **53**:435.

Gottlieb, D. I., Merrell, R., and Glaser, L., 1974, Temporal changes in embryonal cell surface recognition, *Proc. Natl. Acad. Sci. USA* **71**:1800.

Hamburger, V., and Hamilton, H. L., 1951, A series of normal stages in the development of the chick embryo, *J. Morphol.* **88**:49.

Hausman, R. E., and Moscona, A. A., 1973, Cell surface interactions: Differential inhibition by proflavin of embryonic cell aggregation and production of specific cell aggregating factor, *Proc. Natl. Acad. Sci. USA* **70**:3111.

Humphreys, J., 1963, Chemical dissolution and *in vitro* reconstruction of sponge cell adhesions. I. Isolation and functional demonstration of the components involved, *Dev. Biol.* **8**:27.

Jones, R. E., Moscona, M. H., and Moscona, A. A., 1974, Induction of glutamine synthetase in cultures of embryonic neural retina, *J. Biol. Chem.* **249**:6021.

Kahn, A. J., 1973, Ganglion cell formation in the chick neural retina, *Brain Res.* **63**:285.

Kahn, A. J., 1974, An autoradiographic analysis of the time of appearance of neurons in the developing chick neural retina, *Dev. Biol.* **38**:30.

LaVail, J. H., and Cowan, W. H., 1971*a*, The development of the chick optic tectum. I. Normal morphology and cytoarchetectonic development, *Brain Res.* **28**:421.

LaVail, J. H., and Cowan, W. H., 1971*b*, The development of the chick optic tectum. II. Autoradiographic studies, *Brain Res.* **28**:391.

Levi-Montalcini, R., Aloe, L., and Johnson, E. M., 1973, Interaction between nerve growth factor and adrenergic blocking agents in sympathetic neurons, in: *Biochemical Pharmacology Supplement I* (E. Usdin and S. H. Snyder eds.), pp. 199–208, Pergamon Press, New York.

Margoliash, E., Schenk, J. R., Hargie, M. P., Burokds, S., Richter, W. R., Barlow, G. A., and Moscona, A. A., 1965, Characterization of specific cell aggregating materials from sponge cells, *Biochem. Biophys. Res. Commun.* **20:**353.
Merrell, R., and Glaser, L., 1973, Specific recognition of plasma membranes by embryonic cells, *Proc. Natl. Acad. Sci. USA* **70:**2794.
Merrell, R., Pulliam, M. W., Randono, L., Boyd, L. F., Bradshaw, R. A., and Glaser, L., 1975, Temporal changes in tectal surface specificity induced by nerve growth factor, *Proc. Natl Acad. Sc. USA* **72:**4270–4274.
Moore, S., and Stein, W., 1948, Photometric ninhydrin method for use in the chromatography of amino acids, *J. Biol. Chem.* **179:**367.
Moscona, A. A., 1965, Recombination of dissociated cells and the development of cell aggregates, in: *Cells and Tissues in Culture* (E. N. Willmer, ed.), pp. 489–529, Academic Press, New York.
Moscona, A., and Moscona, H., 1952, The dissociation and aggregation of cells from organ rudiments of the early chick embryo, *J. Anal.* **86:**287.
Munkres, K. D., and Richards, F. M., 1965, The purification and properties of *Neurospora* malate dehydrogenase, *Arch. Biochem. Biophys.* **109:**466.
Nakai, W., Lai, C. Y., and Horecker, B. L., 1974, Use of fluorescamine in the chromatographic analysis of peptides from proteins, *Anal. Biochem.* **58:**563.
Oppenheimer, S. B., Edidin, M., Orr, C. W. E., and Roseman, S., 1969, An L-glutamine requirement for intercellular adhesion, *Proc. Natl. Acad. Sci. USA* **63:**1395
Orr, C. W., and Roseman, S., 1969, Intercellular adhesion: A quantitative assay for measuring the rate of adhesion, *J. Membr. Biol.* **1:**110.
Roseman, S., 1970, The synthesis of complex carbohydrates by multiglycosltransferase systems and their potential function in intercellular adhesions, *Chem. Phys. Lipids* **5:**270.
Roth, S., 1968, Studies on intercellular adhesive selectivity, *Dev. Biol.* **18:**602.
Roth, S., and Weston, J. A., 1967, The measurement of intercellular adhesion, *Proc. Natl. Acad. Sci. USA* **58:**974.
Roth, S., McGuire, E. J., and Roseman, S., 1971, An assay for intercellular adhesive specificity, *J. Cell Biol.* **51:**525.
Schwartz, R. J., 1973, Control of glutamine synthetase synthesis in the embryonic chick neural retina, *J. Biol. Chem.* **248:**6426.
Steinberg, M. S., 1963, Tissue reconstruction by dissociated cells, *Science* **141:**401.
Stenevi, U., Bjerre, B., Bjorklund, A., and Mobley, W., 1974, Effects of localized intracerebral injections of nerve growth factor on the regenerative growth of lesioned central noradrenergic neurons, *Brain Res.* **69:**217.
Townes, P. L., and Holtfreter, J., 1955, Directed movements and selective adhesion of embryonic amphibian cells, *J. Exp. Zool.* **128:**53.
Walther, B. T., Ohman, R., and Roseman, S., 1973, A quantitative assay for intercellular adhesion, *Proc. Natl. Acad. Sci. USA* **70:**1569.
Warburg, O., and Christian, W., 1941, Isolierung und Kristallisation des Garungsferments Enolase, *Biochem. Z.* **310:**384.
Weinbaum, C., and Burger, M. M., 1973, Two component systems for surface guided reassociation of animal cells, *Nature (London* **244:**510.

10

Morphogenetic Role of Glycosaminoglycans (Acid Mucopolysaccharides) in Brain and Other Tissues

BRYAN P. TOOLE

1. INTRODUCTION

Morphogenesis of an organ or tissue is a complex and specific series of events leading to a unique, functional organization of cellular and extracellular elements. Underlying these events are several types of cell behavior such as cell movement, recognition and adhesion, shape changes, mitosis, and death. Morphogenesis is usually accompanied by and necessary for cellular differentiation, in which the component cells undergo a series of distinctive metabolic and cytological changes that result in the synthesis or secretion of characteristic and functional cell products. Although *morphogenesis* and *differentiation* are not exact or discrete terms (they both involve metabolic, cytological, and behavioral changes), it is the behavioral aspects that are emphasized when morphogenesis is spoken of and the metabolic and cytological changes that are emphasized when differentiation is spoken of.

BRYAN P. TOOLE · Developmental Biology Laboratory, Departments of Medicine and Biological Chemistry, Harvard Medical School at Massachusetts General Hospital, Boston, Massachusetts.

Investigation of the potential role of the extracellular macromolecules in morphogenesis and differentiation has burgeoned over the past decade, and the involvement of these macromolecules in the control of several important facets of cell behavior and metabolism has been suggested. The intellectual framework for modern concepts of extracellular matrix function in morphogenesis finds it origins in several experimental approaches, among which are manipulation of the interface between epithelium and mesenchyme (Grobstein, 1967), effects of conditioned medium on cell differentiation (Konigsberg, 1970), specific adhesion of cells (Moscona, 1960), and contact guidance of cell movement (Weiss, 1958; Hay, 1973). In all of these systems a role for extracellular substances in influencing cell behavior and metabolism has been invoked, but their mechanism of action has remained unresolved. Indeed, little evidence exists which clearly illustrates direct involvement of specific extracellular matrix macromolecules in the control of morphogenesis. Nevertheless, the number of correlations and circumstantial indications has reached a sufficient level to warrant serious consideration of this possibility.

Glycosaminoglycans (GAG) (formerly termed "acid mucopolysaccharides") and their protein complexes or proteoglycans are ubiquitous and characteristic components of the extracellular matrix or intercellular space, a compartment whose very existence in central nervous tissue was questioned for some time. However, measurements made by electron microscopy (Van Harreveld *et al.*, 1965; Bondareff and Pysh, 1968), by distribution of inulin and other extracellular markers (Bradbury *et al.*, 1968), and by electrical impedance (Nevis and Collins, 1967) indicate that about 20% of mature brain is intercellular space. In the immature animal this space can occupy as much as 40% of the tissue (Bondareff and Pysh, 1968) (Fig. 1), whereas in senescence it is reduced to about 10% (Bondareff and Narotzky, 1972). Histochemical and biochemical measurements have shown that this space contains GAG but not collagen, the other characteristic component of extracellular matrices. The GAG, whose presence in brain are established, are hyaluronate, chondroitin sulfate, and heparan sulfate (Szabo and Roboz-Einstein, 1962; Margolis, 1967; Singh *et al.*, 1969; Margolis and Margolis, 1974).

Several functions have been ascribed to the GAG of the nervous system, notably (1) in determining the status of divalent cations, electrical conductivity, and impedance changes (Wang and Adey, 1969; Van Harreveld *et al.*, 1971; Custod and Young, 1968) and (2) as structural components of cellular sheaths, meninges, etc. (Ashhurst and Costin, 1971; Mustafa and Kamat, 1973). In connective tissues, the traditional and most convenient site of study of extracellular macromolecules, GAG

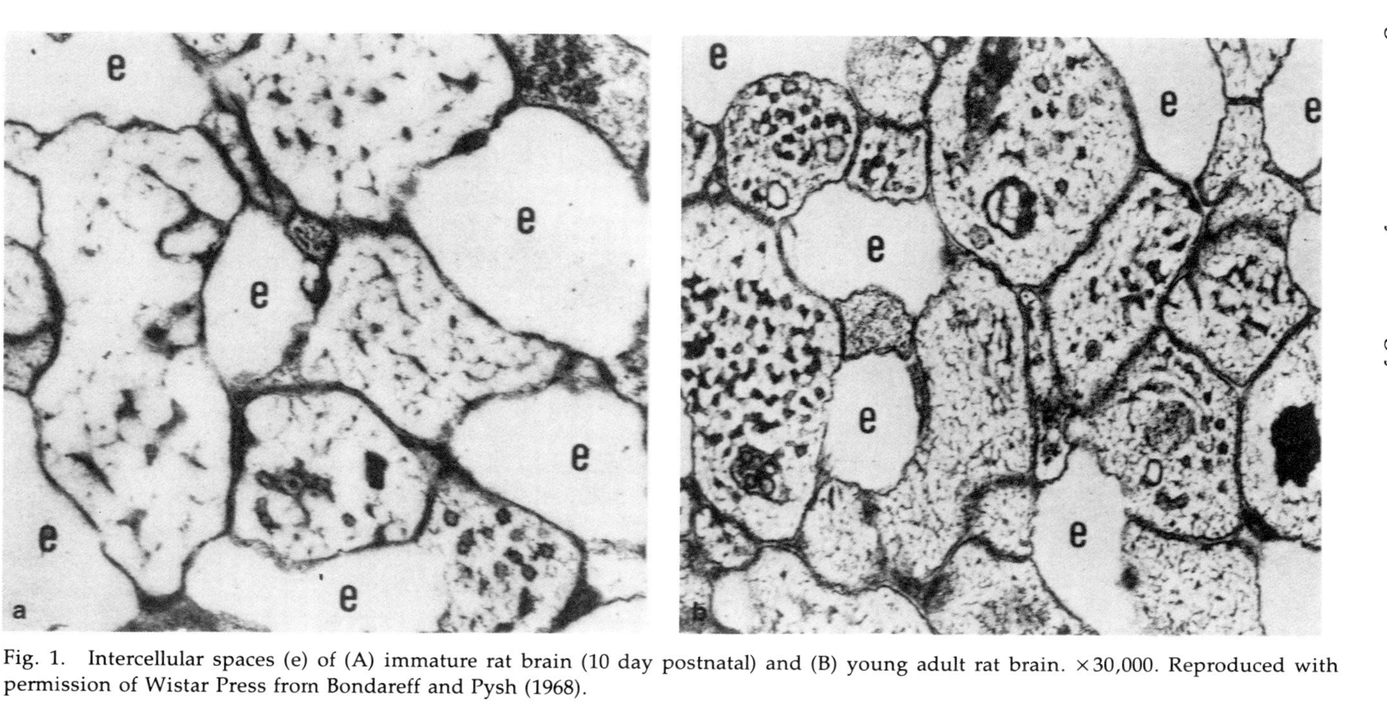

Fig. 1. Intercellular spaces (e) of (A) immature rat brain (10 day postnatal) and (B) young adult rat brain. ×30,000. Reproduced with permission of Wistar Press from Bondareff and Pysh (1968).

are thought to play a structural role. Many elegant biophysical measurements have indeed demonstrated the ability of hyaluronate or cartilage chondroitin sulfate-proteoglycan to form entangled molecular networks which resist solvent flow, restrict solute diffusion, and exclude other macromolecules from their molecular domains (Laurent, 1966; Schubert and Hamerman, 1968; Ogston, 1970).

In this chapter I wish to focus on the evidence that hyaluronate is an influential participant in *morphogenesis* and to present a hypothesis regarding the possible function of this macromolecule in the embryonic development of several tissues including brain. To facilitate a meaningful presentation I will also review briefly salient features of the structure and possible morphogenetic roles of the sulfated proteoglycans, the closest molecular relatives of hyaluronate. The bulk of evidence suggesting a morphogenetic role for GAG comes from studies of biological systems other than the central nervous system. The relevant details of this work, and the biochemistry necessary to appreciate the properties of GAG and proteoglycans, are gathered together in the next three sections. In the final section the major conclusions of these endeavors are applied to brain morphogenesis to the extent that it seems reasonable at this stage of our knowledge.

2. GLYCOSAMINOGLYCANS AND PROTEOGLYCANS

A distinctive aspect to recent studies of the role of extracellular materials in morphogenesis and differentiation is the attempt to deal with known GAG, proteoglycans, and collagens of established structure rather than simply extracellular "factors" produced in culture or *in vivo*. There is a growing awareness of the reactivity and heterogeneity of these macromolecules and the consequent potential for their involvement in differential and specific interactions. In addition, the increase in knowledge and techniques relating to these compounds now makes it possible to examine and manipulate them in developmental systems. It is thus appropriate to review briefly some of the relevant aspects of GAG and proteoglycan structure, properties, and metabolism.

2.1. *Structure*

Glycosaminoglycans are highly charged polyanions composed of repeating dissaccharide units which bear either a carboxyl or sulfate group or both and which, except for keratan sulfate, consist of uronic acid and hexosamine moieties. Eight GAG have been studied in detail:

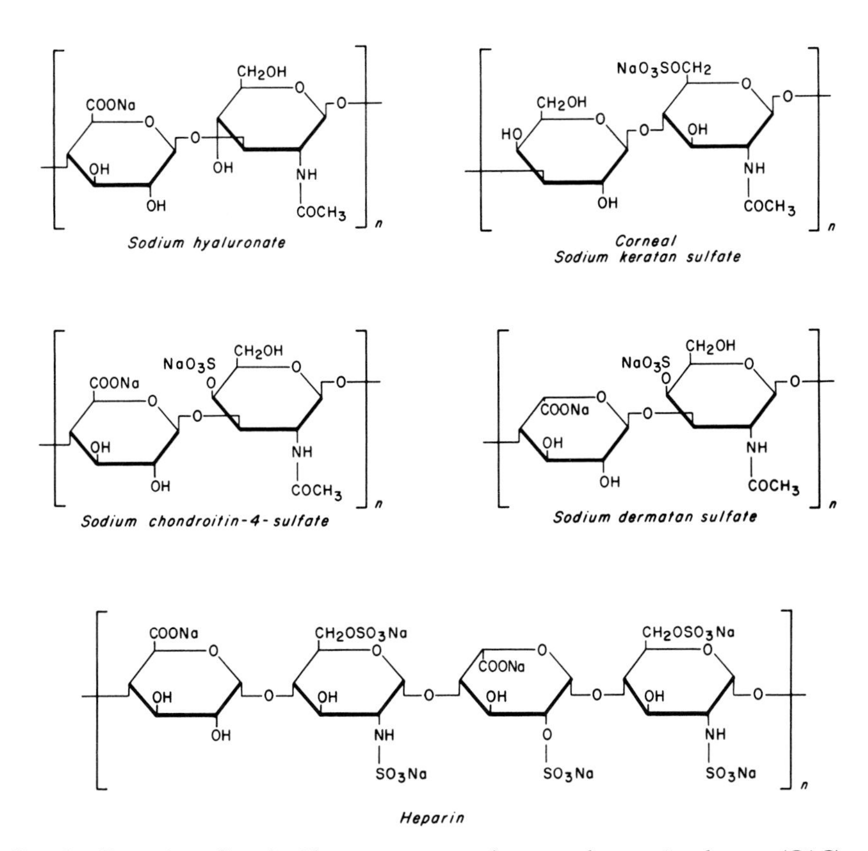

Fig. 2. Repeating disaccharide components of some glycosaminoglycans (GAG). Hyaluronate is composed of glucuronic acid and *N*-acetylglucosamine; keratan sulfate of galactose and *N*-acetylglucosamine-6-sulfate; chondroitin sulfates of glucuronic acid and *N*-acetylgalactosamine-4- or -6-sulfate; dermatan sulfate of iduronic acid and *N*-acetylgalactosamine-4-sulfate (with small amounts of glucuronic acid); heparin of glucuronic acid or iduronic acid-2-sulfate and glucosamine with both *N*-sulfate and *O*-sulfate groups (iduronic acid comprises about 70% of the uronic acid moieties). In heparan sulfate the uronic acid is chiefly glucuronic acid and some of the glucosamine moieties are *N*-sulfated while others are *N*-acetylated. Heparan sulfate contains about 1 mole of sulfate per disaccharide whereas heparin contains 2–3 moles.

hyaluronate, chondroitin, chondroitin-4- and -6-sulfates, dermatan sulfate, keratan sulfate, heparan sulfate, and heparin. The structures of the component dissaccharides of several of these are given in Fig. 2. The molecular weights of the sulfated GAG are usually in the range $1–2 \times 10^4$ but hyaluronate can be as large as $1–2 \times 10^7$.

The studies of morphogenesis to be described in the following

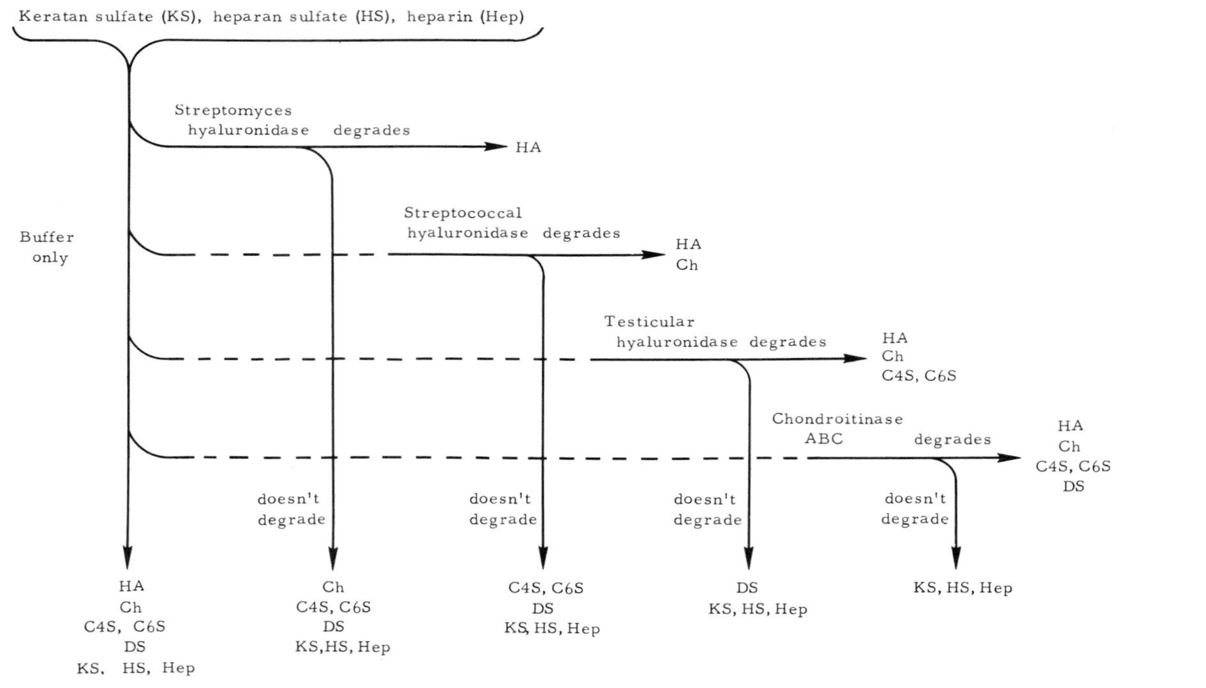

Fig. 3. Method of identification of types of GAG by their differential susceptibility to polysaccharidases. Aliquots of labeled GAG along with carrier are precipitated after treatment with the enzymes, and radioactivity of the enzyme-resistant GAG is measured. Controls incubated in buffer only are used to estimate total GAG. The label in each individual enzyme-susceptible GAG can be calculated by subtraction of label resistant to one enzyme from that resistant to another with broader specificity—e.g., hyaluronate label is the difference between total and streptomyces hyaluronidase-resistant label and that in the chondroitin sulfates is the difference between streptococcal and testicular hyaluronidase-resistant labels.

sections of this chapter have depended greatly on the development of simple and sensitive techniques for measurement and identification of the various GAG after labeling with isotopic precursors administered to the animals, tissues, or cells being studied. The approach used by us (Toole and Gross, 1971) (Fig. 3) has been to depolymerize individual or defined groups of GAG with specific polysaccharidases and then recover and measure the radioactivity of the enzyme-resistant GAG. By subtraction, the radioactivity of enzyme-susceptible GAG can then be calculated. This method is very sensitive, specific, reproducible, and simple and could potentially be used to identify all the known GAG (Table I).

With the exception of hyaluronate, GAG do not usually exist *in vivo* as isolated polysaccharide chains, but as proteoglycans where the GAG chains are linked covalently to protein (Schubert and Hamerman, 1968; Muir, 1969; Sajdera *et al.*, 1970; Roden, 1970*a*,*b*; Hascall and Heinegard, 1975). In most cases studied (Roden, 1970*a*,*b*; Fransson, 1970; Lindahl, 1970), the covalent linkage region between protein and each GAG chain has been shown to have the sequence

serine—xylose—galactose—galactose—uronic acid—hexosamine—

However, skeletal keratan sulfate is linked to serine and threonine via

Table I. Polysaccharidase Specificities[a]

Enzyme	Sensitive GAG	References
Streptomyces hyaluronidase	HA	Ohya and Kaneko (1970)
Leech hyaluronidase	HA	Meyer *et al.* (1960)
Streptococcal hyaluronidase	HA, Ch	Meyer *et al.* (1960)
Tadpole tail hyaluronidase	HA, Ch, C4S	Silbert and DeLuca (1970)
Testicular hyaluronidase	HA, Ch, C4S, C6S	Meyer *et al.* (1960)
Chondroitinase AC	HA, Ch, C4S, C6S	Saito *et al.* (1968)
Chondroitinase ABC	HA, Ch, C4S, C6S, DS	Saito *et al.* (1968)
Heparitinase	HS	Hovingh and Linker (1970)
Heparinase	HS, Hep	Hovingh and Linker (1970)
Keratanase	KS	Hirano and Meyer (1971)

[a] Use of the above enzymes would allow identification of all the known vertebrate GAG in the scheme presented in Fig. 3. For abbreviations, see Fig. 3.

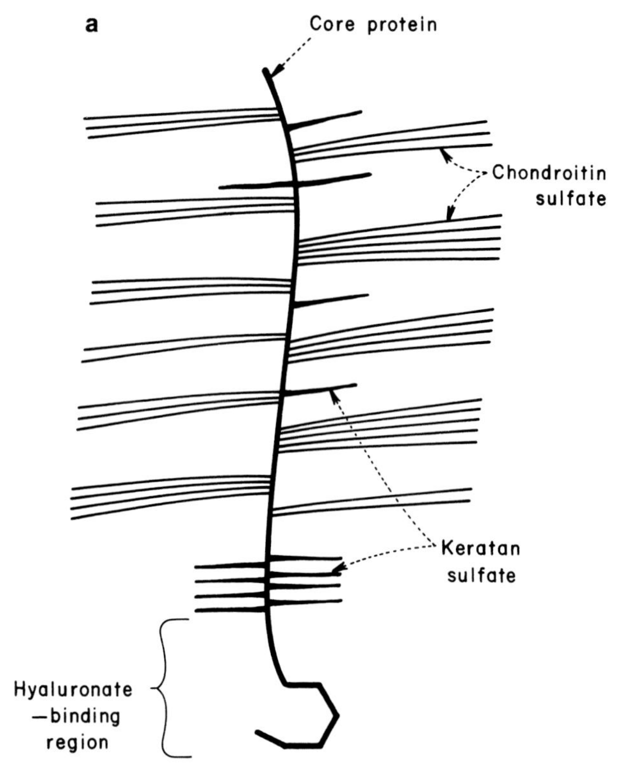

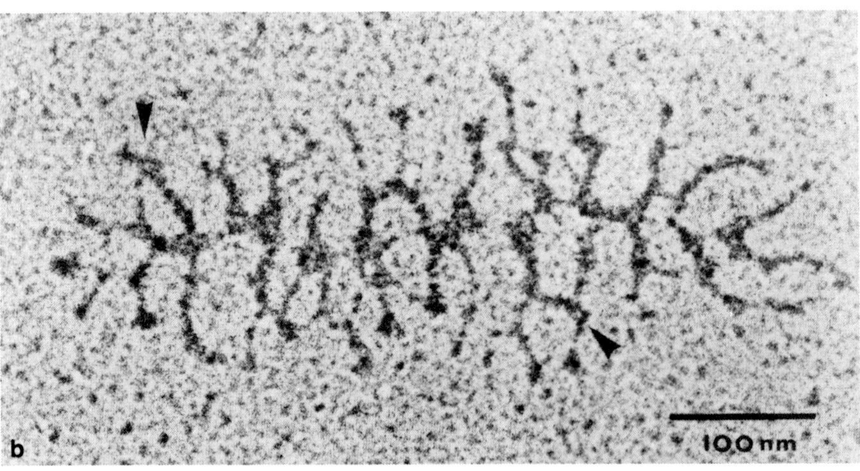

Fig. 4. Cartilage proteoglycan subunit. A: Diagrammatic representation of arrangement of constituents. Adapted from a diagram kindly supplied by Dr. Vincent Hascall (see Hascall and Heinegard, 1975). B: Electron micrograph of a single molecule of proteoglycan subunit. ×216,000. Reproduced with permission of the American Society of Biological Chemists from Rosenberg *et al*. (1970).

an *N*-acetylgalactosamine derivative and corneal keratan sulfate to asparagine via *N*-acetylglucosamine (Roden, 1970*a,b*).

The chemical structure of the proteoglycan from cartilage has been studied carefully, and it is comprised of a "core protein" from which radiate numerous chondroitin sulfate and keratan sulfate chains. Extracts of cartilage usually contain protein–polysaccharide complexes of variable size and composition. These complexes can be dissociated by treatment with, or extraction in, 4 M guanidinium chloride or 3 M $MgCl_2$ and the components purified by centrifugation in a CsCl density gradient containing the same dissociating agent (Sajdera and Hascall, 1969). The resulting proteoglycan subunit preparation still contains molecules of various sizes which, for bovine nasal cartilage, are in the range $1–4 \times 10^6$ mol wt and are comprised of 40–150 chondroitin sulfate chains of about 20,000 mol wt and 30–60 keratan sulfate chains of 4000–8000 mol wt linked to a protein of 200,000 mol wt (Hascall and Sajdera, 1970; Hascall and Heinegard, 1975). It is not yet established whether the variable size of proteoglycan subunits is due to a true heterogeneity where several different protein core molecules are present or to a heterodispersity arising from variable rates of GAG chain initiation or elongation or variable partial degradation.

Figure 4 shows an electron micrograph and a recent diagrammatic version of the arrangement of GAG chains in this complex molecule, showing other features such as the asymmetrical distribution of chondroitin sulfate and keratan sulfate chains, the clustering of the former, and the polysaccharide-free portion of the core protein which is thought to be responsible for aggregation of the subunits (Hascall and Heinegard, 1975).

The subunits interact to form very large aggregates, a process mediated by several "link" factors which include at least two proteins and, most importantly, hyaluronate (Hardingham and Muir, 1972, 1974; Gregory, 1973; Hascall and Heinegard, 1974*a,b;* Heinegard and Hascall, 1974). As shown in Fig. 5, hyaluronate reacts specifically with the GAG-free end of the core protein bridging several proteoglycan subunits at a frequency of once every 30–60 hyaluronate disaccharides to produce a "megacomplex" which contains approximately 0.5% (w/w) hyaluronate and whose size depends on the length of the hyaluronate chains present (Hascall and Heinegard, 1975). Ultrastructural evidence confirms the presence of these complexes *in vivo* and the importance of hyaluronate in their integrity (Quintarelli *et al.*, 1974).

The structures of other proteoglycans, including those present in nervous tissue, have not been studied in detail. However, from those limited investigations performed so far, two important conclusions can be made with a reasonable degree of certainty. First, all of the sulfated

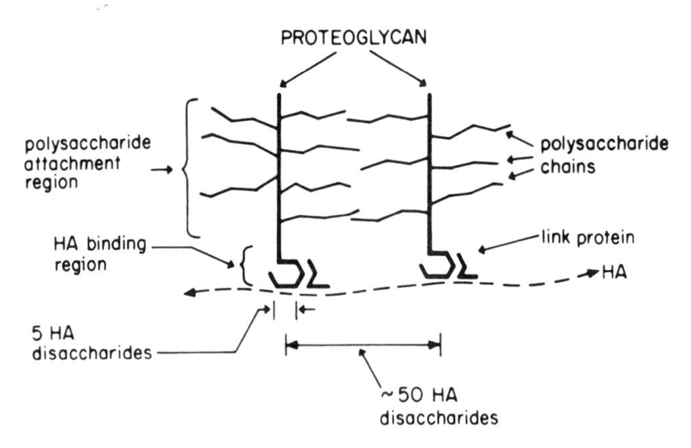

Fig. 5. Diagram of cartilage proteoglycan aggregate illustrating relationship of the subunits, link protein, and hyaluronate (HA). Reproduced with permission of the American Society of Biological Chemists from Heinegard and Hascall (1974).

GAG exist *in vivo* as proteoglycans with the protein-to-GAG linkages discussed above. Second, the structural arrangement of core protein and GAG chains varies among the different GAG and also with the same GAG in different tissues. For example, the chondroitin sulfate-proteoglycan of cartilage is clearly different from that of heart valves, which contains no keratan sulfate and whose molecular weight would allow for only two chondroitin sulfate chains to one protein core (Lowther *et al.*, 1970). There is also evidence that more than one type of chondroitin sulfate proteoglycan exists in a single tissue—e.g., cartilage (Tsiganos *et al.*, 1971; Palmoski *et al.*, 1974) and heart valves (Lowther, *et al.*, 1970). Similarly, dermatan sulfate-protein has a lower molecular weight, different physicochemical properties, and different tissue distribution when compared with cartilage proteoglycan (Toole and Lowther, 1968*a*; Obrink, 1972). An interesting illustration of the possible developmental importance of these types of heterogeneity comes from recent work on proteoglycan synthesis by mesenchymal cells and chondrocytes during chick limb development. Palmoski and Goetinck (1972) demonstrated that the bulk of chondroitin sulfate-proteoglycan produced by nanomelic or 5-bromodeoxyuridine-treated chondrocytes was considerably lower in molecular weight than the normal chondrocyte product. Levitt and Dorfman (1974) and Goetinck *et al.* (1974) extended this observation to show that limb bud mesoderm synthesizes a proteoglycan similar to that from the treated chondrocytes and that, as the mesodermal cells differentiate to chondrocytes *in vivo* or *in vitro*, this is replaced by a high molecular weight proteoglycan which is possibly cartilage specific.

In contrast to the sulfated GAG, hyaluronate may not exist as a proteoglycan but as individual GAG chains in the extracellular matrix (Varma *et al.*, 1974). The molecular weight of hyaluronate far exceeds that of the sulfated GAG chains but varies extensively from source to source. Molecules as large as 2–20 × 10^6 mol wt have been isolated from synovial fluid (Silpananta *et al.*, 1968), heart valves (Meyer *et al.*, 1969), and rooster comb (Swann, 1969), whereas preparations from vitreous humor are usually smaller and very polydisperse (Laurent *et al.*, 1960; Cleland, 1970).

2.2. *Properties*

The unusual physicochemical behavior of hyaluronate molecules in solution has led to extensive study and the emergence of interesting models and concepts relating to the possible structural role of this polymer in connective tissue matrices (Ogston, 1970; Laurent, 1966, 1970). Because of its immense size and negative charge, each molecule, in dilute solution, forms an extended, somewhat stiff random coil which occupies a vast molecular "domain" as much as 10^4 times larger than the space occupied by the same amount of molecular matter in

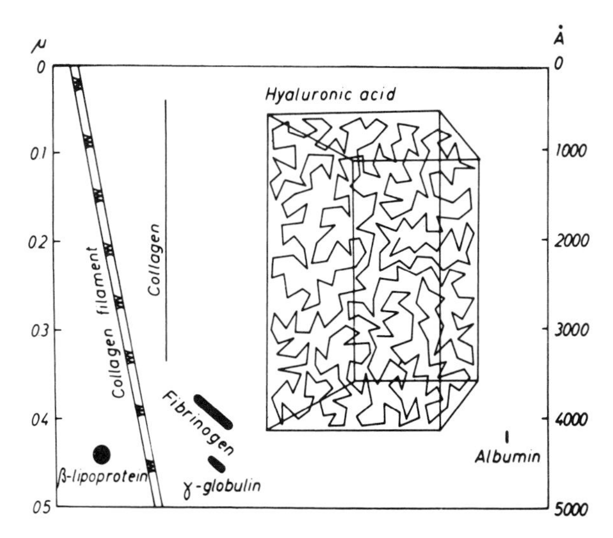

Fig. 6. Schematic comparison of the "molecular domains" of hyaluronate and some compact macromolecules. Reproduced with permission of Academic Press from Laurent (1970).

compact form. A comparative diagram of hyaluronate and other more compact macromolecules is shown in Fig. 6. In somewhat more concentrated solutions—e.g., approximately 0.5 mg/ml for molecules of 2×10^6 mol wt—the individual molecular domains begin to overlap and thus the molecules entangle. At physiological concentrations, a dense meshwork would result which has the capacity (1) to "trap" or restrict the flow of water (Fessler, 1960; Ogston and Sherman, 1961; Preston *et al.*, 1965; Ogston, 1966), (2) to interfere with diffusion and transport of solutes, especially macromolecules (Ogston and Sherman, 1961; Laurent *et al.*, 1963), (3) to exclude macromolecules or particles from the meshwork (Ogston and Phelps, 1961; Laurent, 1964), and (4) to exert an osmotic pressure (Laurent and Ogston, 1963; Ogston, 1966). Similar but somewhat less dramatic physicochemical properties have been demonstrated for the proteoglycan of cartilage (Weinstein *et al.*, 1963; Gerber and Schubert, 1964; Disalvo and Schubert, 1966; Hascall and Sajdera, 1970), while some other proteoglycans probably do not exhibit these properties (Toole and Lowther, 1968*a*; Lowther *et al.*, 1970). The above characteristics of hyaluronate and cartilage proteoglycan have been related to various features of connective tissue—e.g., high water content (Ogston, 1966; Rienits, 1960; Szirmai, 1966; Watson and Pierce, 1949; Comper and Preston, 1974), low serum protein content (Schubert and Hamerman, 1964, 1968), and turgidity, viscosity, and other mechanical properties (Ogston and Stanier, 1953; Kempson *et al.*, 1970; Meyer *et al.*, 1971; Minns *et al.*, 1973). In addition, the highly charged nature of these polyanions has suggested a potential role as ion exchangers involved in the regulation of the ionic composition of tissues, particularly in regard to the calcium ion (Dunstone, 1962; Katchalsky, 1964; Scott, 1968; Schubert and Hamerman, 1968; Preston and Snowden, 1972). Moreover, it has been shown that GAG can adopt helical configurations in stretched films (Arnott *et al.*, 1973; Dea *et al.*, 1973; Atkins *et al.*, 1972, 1974), but the relationship of this property to physiological conditions is not yet clear.

An additional area of active investigation, which again highlights the reactivity and heterogeneous nature of proteoglycans, is the capacity of certain proteoglycans to interact with collagen. Briefly (1) electron microscopy has revealed a regular pattern in the relationship of proteoglycan to collagen fibrils in several tissues (Smith and Frame, 1969, 1969; Serafini-Fracassini *et al.*, 1970; Nakao and Bashey, 1972; Eisenstein *et al.*, 1973), as is shown in Fig. 7 for the collagenous basement membrane and stroma of the embryonic chick cornea (Trelstad *et al.*, 1974), and (2) biochemical studies *in vitro* have demonstrated the capacity of dermatan sulfate-proteoglycan and cartilage chondroitin sulfate-proteoglycan to react with collagen molecules in solution in a specific

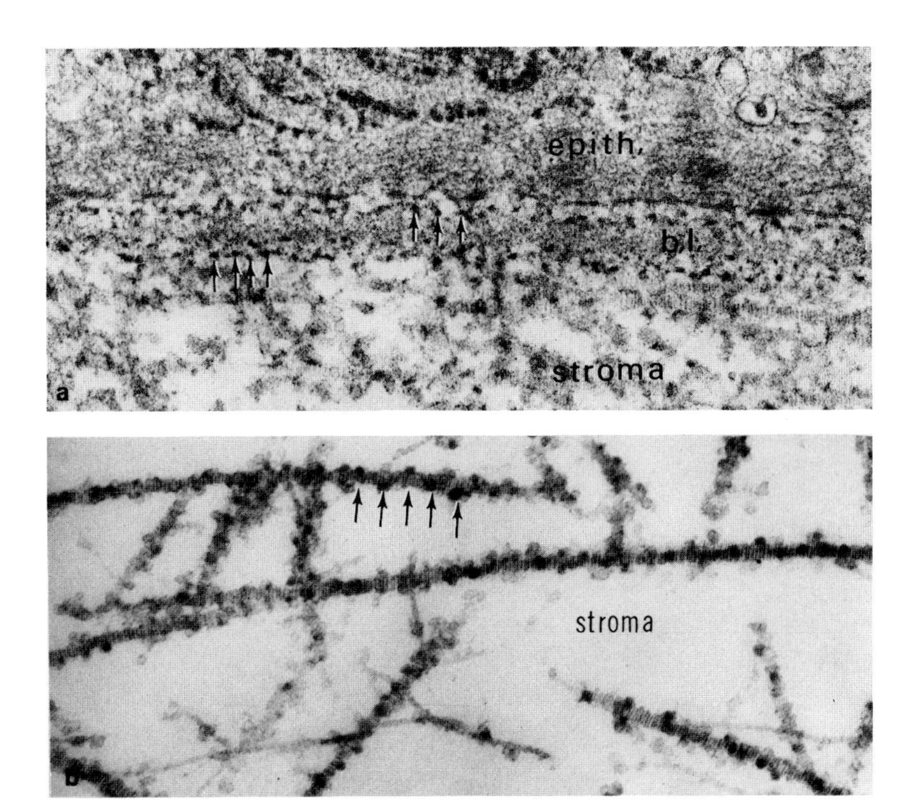

Fig. 7. Relationship of proteoglycan to (A) basement membrane (×85,000) and (B) stromal (×52,000) collagens in the chick embryo cornea. In the basement membrane, ruthenium red stained chondroitin sulfate-proteoglycan (arrows) is arranged in a regular pattern on either side of the electron-dense basal lamina (b.l.) (Trelstad *et al.*, 1974). Proteoglycan is also associated with the stromal fibrils in a specific repeated location. Micrographs kindly supplied by Dr. R. L. Trelstad.

manner to form aggregates which appear to be precursors of collagen fibrils (Mathews, 1965; Mathews and Decker, 1968; Toole and Lowther, 1967, 1968*a*,*b*; Podrazky *et al.*, 1971; Obrink, 1973*a*,*b*; Trelstad, 1975). Hyaluronate and chondroitin sulfate-proteoglycan from noncartilagenous tissues do not appear to exhibit these interactive properties with collagen (Toole and Lowther, 1968*a*,*b*; Obrink, 1973*a*).

2.3. *Metabolism*

GAG synthesis is catalyzed by coordinated groups of glycosyltransferases. In the case of chondroitin sulfate, summarized in Fig. 8, the

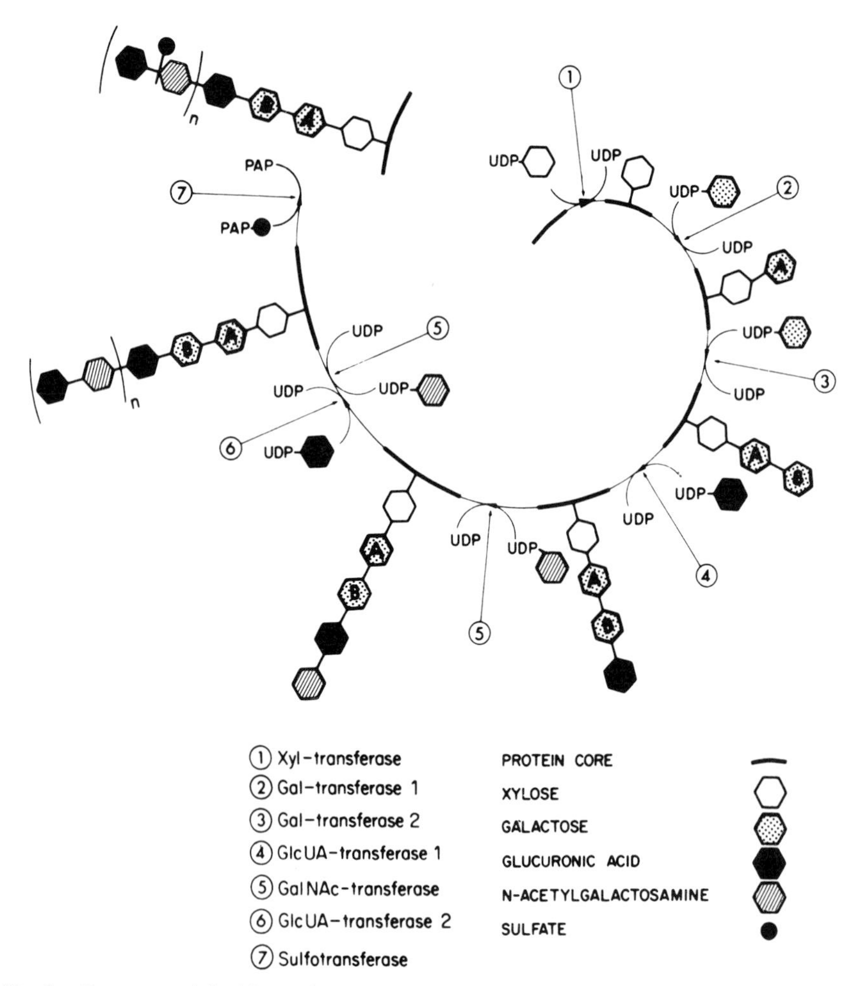

Fig. 8. Summary of the biosynthetic steps involved in chondroitin sulfate-proteoglycan synthesis. Reproduced with permission of Academic Press from Roden (1970*a*).

action of six glycosyltransferases and a sulfotransferase is required and at least some of these enzymes may exist *in vivo* as a membrane-bound multienzyme complex (Schwartz *et al.*, 1974, 1975). The initial step is the transfer of xylose from UDPxylose to a serine moiety in the core protein acceptor (Robinson *et al.*, 1966) followed by serial addition at the nonreducing end of two galactose residues, glucuronic acid, *N*-acetylgalactosamine and glucuronic acid, each step requiring a separate transferase (Telser *et al.*, 1966; Roden, 1970*a,b*). *N*-acetylgalactosamine and glucuronic acid are then added repetitively by the last two transferases. Sulfation probably occurs as the chain elongates (DeLuca *et al.*, 1973) but with sufficient lag to allow continued chain elongation, which would be inhibited if a terminal *N*-acetylgalactosamine became sulfated (Telser *et al.*, 1966).

In the case of hyaluronate, chain elongation proceeds in a similar fashion to chondroitin sulfate with alternating additions of glucuronic acid and *N*-acetylglucosamine from sugar nucleotides to the nonreducing end of nascent oligosaccharides (Hopwood *et al.*, 1974). However, chain initiation remains a mystery, although it does not appear to require the concomitant synthesis of a "core protein" as is the case for chondroitin sulfate synthesis (Stoolmiller and Dorfman, 1969).

Degradation of GAG has not been extensively studied despite the defects in GAG catabolism thought to occur in the many different types of mucopolysaccharidoses (Neufeld, 1974). Hyaluronidase has been isolated from liver and bone lysosomes and shown to act as an endoglycosidase capable of depolymerizing hyaluronate and chondroitin sulfates to form a mixture of oligosaccharides ranging from tetra- to decasaccharides (Vaes, 1967; Aronson and Davidson, 1967*a*, *b*). These oligosaccarides can in turn be reduced to smaller units by the action of the exoglycosidases: β-glucuronidase, β-*N*-acetylglucosaminidase, and β-*N*-acetylgalactosaminidase (Davidson, 1970). However, in the case of chondroitin sulfate-proteoglycan, the initial degradative step is most likely proteolytic and possibly mediated by cathepsin D or neutral proteases (Dingle *et al.*, 1971; Wasteson *et al.*, 1972; Sapolsky *et al.*, 1974), giving rise to relatively intact but free chondroitin sulfate chains which may then be released from the tissue into the circulatory system or degraded further to smaller products within the tissue.

3. MORPHOGENETIC ROLE OF HYALURONATE

The variable structure, tissue distribution, reactivity, and metabolism of GAG and proteoglycans have led many investigators to believe that each of these macromolecules has a discrete biological role. The changes in concentration and type of GAG that occur during physiological and developmental events have in turn led to consideration of possible morphogenetic roles for these molecules. In the case of hyaluronate in particular, considerable evidence has now been obtained to suggest a role in development and possibly also in physiological repair and remodeling.

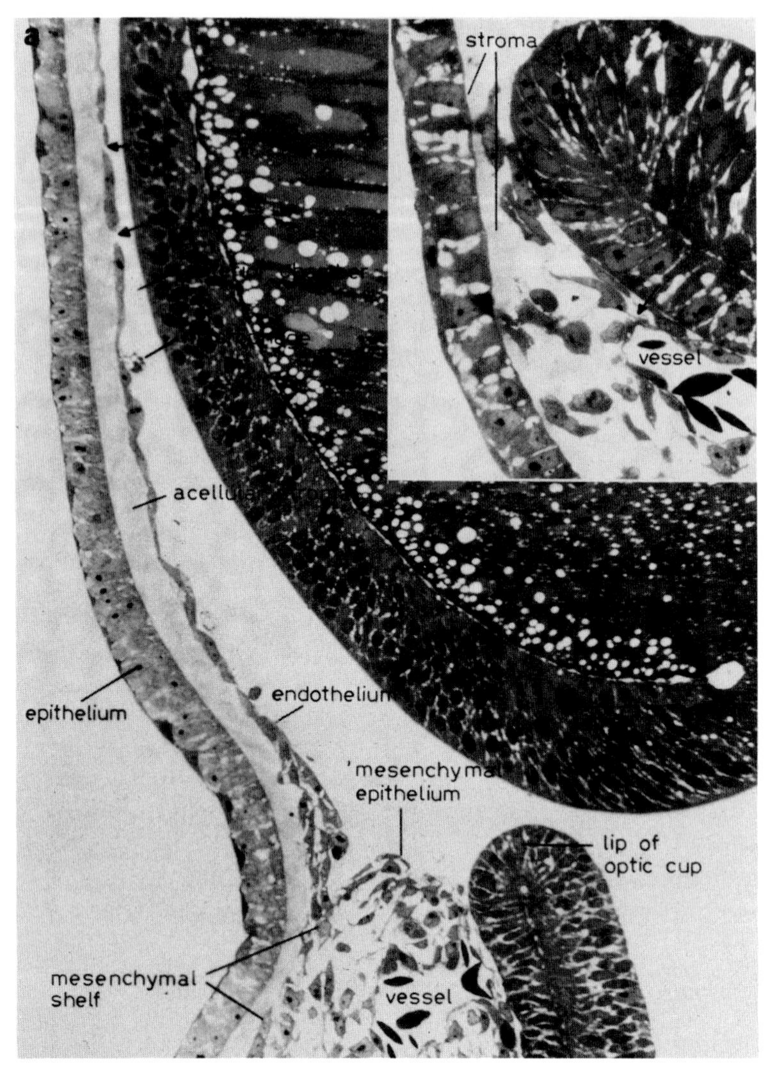

Fig. 9. Embryonic chick cornea at (a) early stage 27 (inset: stage 22) and (b) stage 28 (inset: late stage 27) of development. Note the acellular stroma at early stage 27(a) and the beginnings of mesenchymal cell migration into the

3.1. *Correlation of Hyaluronate Synthesis and Hyaluronidase Activity with Morphogenetic Events in Vivo*

The observation that hyaluronate is relatively enriched in embryonic or young tissues in comparison to adult tissues, including brain, has been made by several investigators over the past two decades

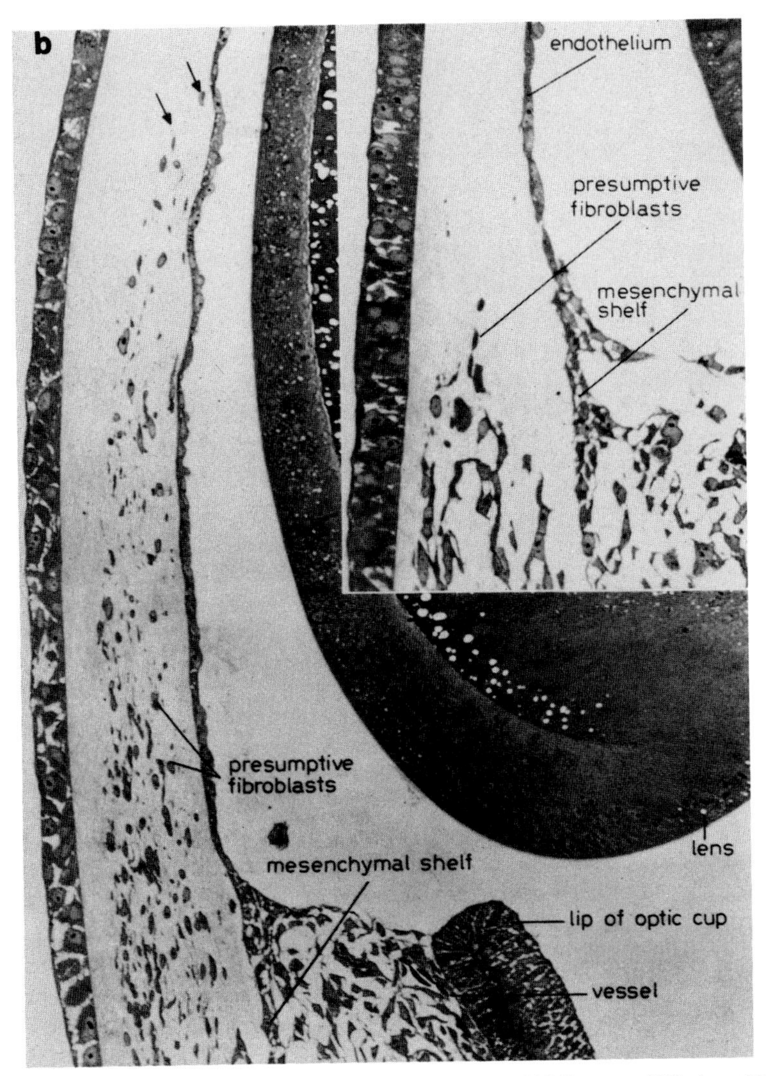

swollen late stage 27 and stage 28 stroma (b). a, ×520 (inset, ×820); b, ×350 (inset, ×755). Reproduced with permission of Karger from Hay and Revel (1969).

(Loewi and Meyer, 1958; Zugibe, 1962; Singh and Bacchawat, 1965, 1968; Young and Custod, 1972; Margolis *et al.*, 1975; Lipson *et al.*, 1971; Nanto, 1969; Breen *et al.*, 1970, 1972). However, these studies have usually been addressed to general maturational or aging trends in tissue structure rather than to morphogenetic sequences. The results have usually been correlated with changes in tissue water content rather than with cellular events.

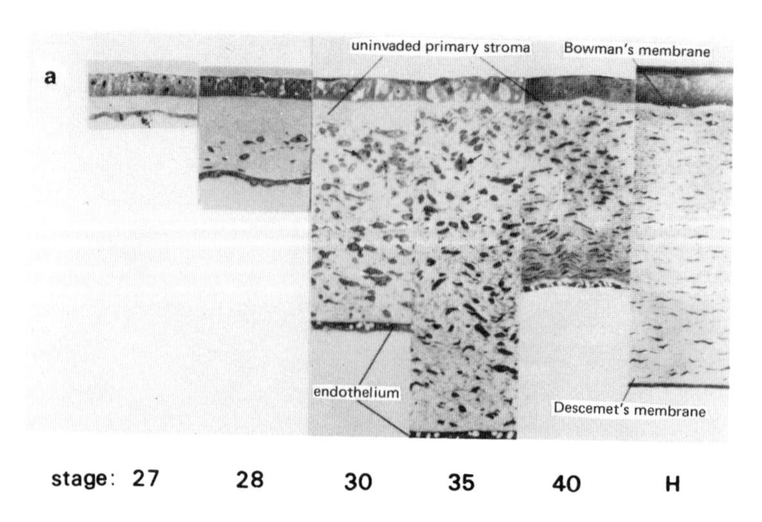
a
uninvaded primary stroma
Bowman's membrane
endothelium
Descemet's membrane
stage: 27
28
30
35
40
H

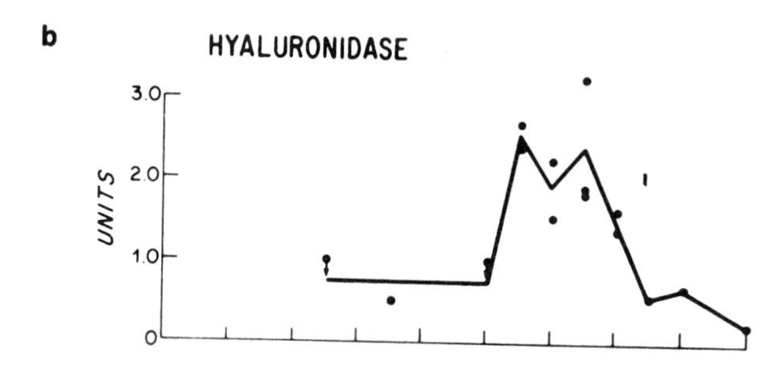
b
HYALURONIDASE
UNITS
3.0
2.0
1.0
0

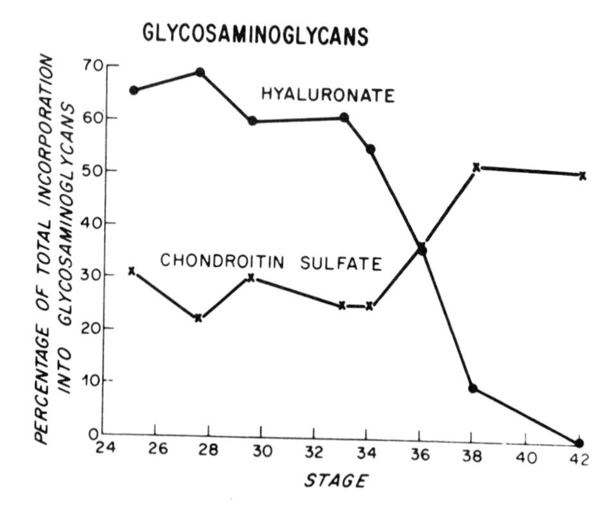
GLYCOSAMINOGLYCANS
PERCENTAGE OF TOTAL INCORPORATION INTO GLYCOSAMINOGLYCANS
HYALURONATE
CHONDROITIN SULFATE
70
60
50
40
30
20
10
0
24
26
28
30
32
34
36
38
40
42
STAGE

By studying hyaluronate metabolism at discrete stages in early development, at times and in tissues where striking morphogenetic cell movements or differentiations are taking place, close correlations between hyaluronate synthesis and cell migration and between hyaluronidase activity and cell differentiation have now been revealed. We have studied three morphogenetic sequences in which a period of extensive cell migration is followed by differentiation of these cells and production of extensive and characteristic extracellular matrix. These are the early stages of development of the chick embryo cornea (Toole and Trelstad, 1971) and vertebral column (Toole, 1972) and of regeneration of the amputated newt limb blastema (Toole and Gross, 1971). Figures 9, 11, and 12 show these tissues before and during migration of the mesenchymal cells to the site of differentiation.

Hay and Revel (1969) have described the course of events during early corneal morphogenesis in the chick embryo in exquisite detail. The corneal fibroblasts derive from the mesenchymal shelf at either edge of the acellular, epithelium-derived, corneal stroma and migrate into this stroma from the end of stage 27 of embryonic development (about 5½ days of incubation; Hamburger and Hamilton, 1951) until approximately stage 35, about 3 days later. The beginning of migration coincides with a marked swelling of the stroma and its cessation with deswelling (Fig. 10A). As shown in Fig. 10, the major GAG component being synthesized by the cornea during the migratory stage is hyaluronate. It is interesting to note that the leading migratory mesenchymal cells always move adjacent to the corneal endothelium (Hay and Revel, 1969) (Figs. 9 and 10), which is the source of hyaluronate (Trelstad *et al.*, 1974; Meier and Hay, 1973). The appearance of hyaluronidase activity in the corneal tissue coincides with the cessation of migration and with deswelling (Toole and Trelstad, 1971) (Fig. 10). Subsequent to deswelling, the cornea becomes transparent (Coulombre and Coulombre, 1958), and at this stage the mesenchymal cells have become corneal fibroblasts (Hay and Revel, 1969) producing large amounts of collagen (Trelstad and Kang, 1974; Trelstad *et al.*, 1974), chondroitin sulfate, and keratan sulfate but very little hyaluronate (Anseth, 1961; Conrad, 1970)

←

Fig. 10. Comparison of (A) morphology and (B) GAG metabolism in embryonic chick cornea. Hyaluronate synthesis is predominant during stromal swelling and the migration of mesenchymal cells into the stroma (stages 27 through 35). Hyaluronidase activity appears at the end of the migratory stage, concomitant with stromal deswelling (stages 35 through 40). A is adapted from micrographs kindly supplied by Dr. Elizabeth Hay (see Hay and Revel, 1969) and B is reproduced with permission of Academic Press from Toole and Trelstad (1971).

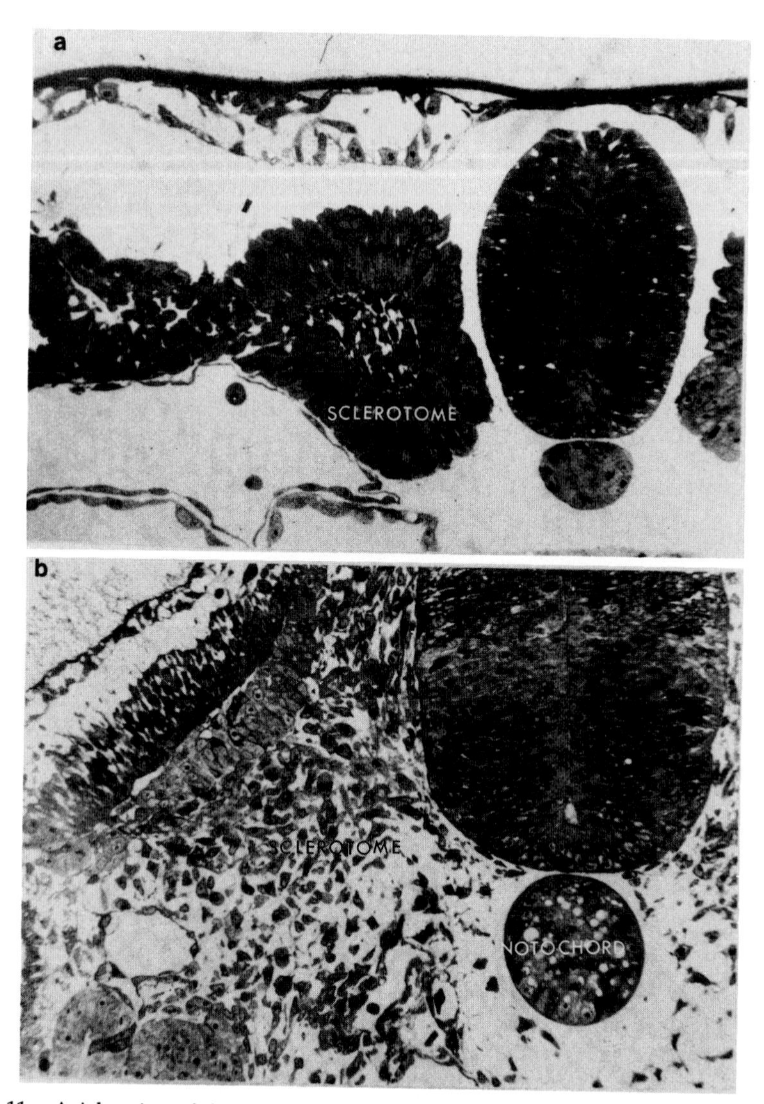

Fig. 11. Axial region of the chick embryo (A) prior to and (B) during migration of the sclerotomal cells from the somite region to the perinotochordal region. ×240. Micrographs kindly supplied by Dr. Robert Trelstad. Reproduced with permission of McGraw-Hill from Hay (1966*b*).

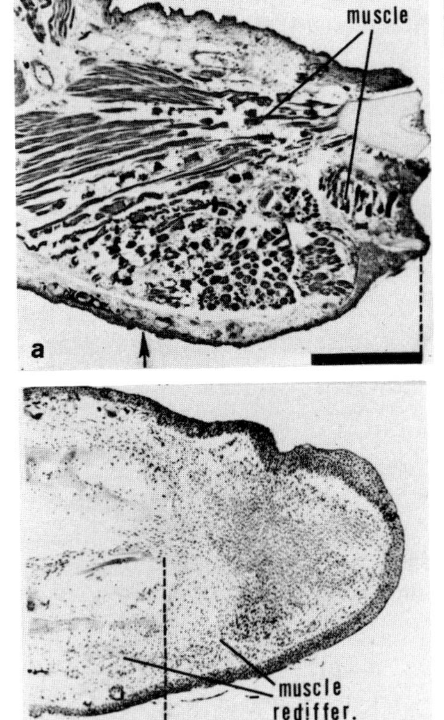

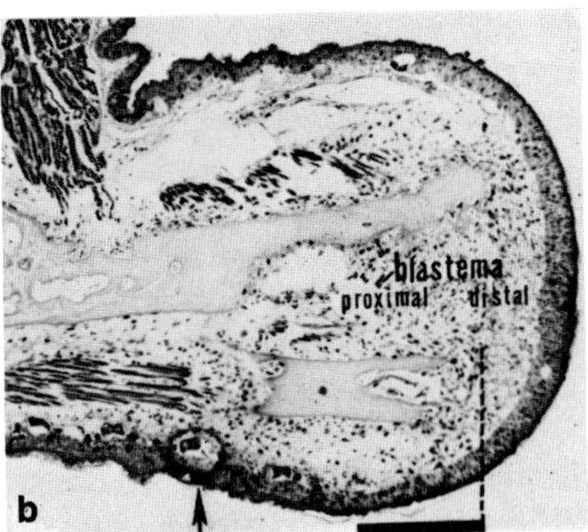

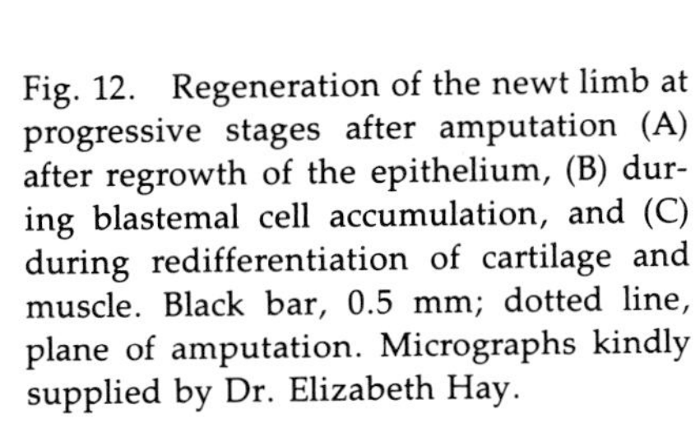

Fig. 12. Regeneration of the newt limb at progressive stages after amputation (A) after regrowth of the epithelium, (B) during blastemal cell accumulation, and (C) during redifferentiation of cartilage and muscle. Black bar, 0.5 mm; dotted line, plane of amputation. Micrographs kindly supplied by Dr. Elizabeth Hay.

(Fig. 10). The change in major GAG from hyaluronate early in development to sulfated GAG subsequently has been confirmed chemically (Praus and Brettschneider, 1970, 1971).

Each somite of the embryonic chick is comprised of three regions: myotome, dermatome, and sclerotome. The sclerotome dissociates to give migratory mesenchymal cells which move toward the notochord, surround it, and form the cartilaginous precursor of the vertebral body (Fig. 11). Kvist and Finnegan (1970*a*,*b*) examined this phenomenon histochemically and chemically with respect to the types of GAG present and to their location. They found that hyaluronate was the major GAG in the matrix surrounding the migrating sclerotomal cells but that at the site of cartilage formation, around the notochord, the amount of chondroitin sulfate, but not hyaluronate, increased dramatically. We examined the production of GAG during these developmental events and obtained similar results (Toole, 1972). In addition, hyaluronidase activity was found to appear precisely at the time that the cells began to produce cartilage matrix after concluding their migration around the notochord.

The formation of the regeneration blastema at the tip of the stump of an amputated newt limb involves (1) degradation and dissolution of the structural elements at the end of the stump (Trampusch and Harrebomee, 1965; Grillo *et al.*, 1968), (2) "redifferentiation" of the released cells to form mesenchymatous blastemal cells (Hay, 1968), and (3) proliferation and migration of these cells to the tip of the stump to form a blastemal outgrowth (Chalkley, 1959; Hay 1966*a*) (Fig. 12). Blastema formation is accompanied by active synthesis of hyaluronate (Fig. 13*b*), at 10–20 times the levels in the normal newt limb or the proximal part of the stump of the amputated limb (Toole and Gross, 1971). The most active area of hyaluronate synthesis corresponds to the most active area of release, proliferation, and migration of blastemal cells—i.e., just proximal to the plane of amputation. Similarly to the two chick embryo systems described above, hyaluronidase activity, decrease in hyaluronate synthesis and increase in chondroitin sulfate synthesis become evident at the end of blastemal formation and at the time of cartilage redifferentiation in the blastema (Toole and Gross, 1971) (Fig.

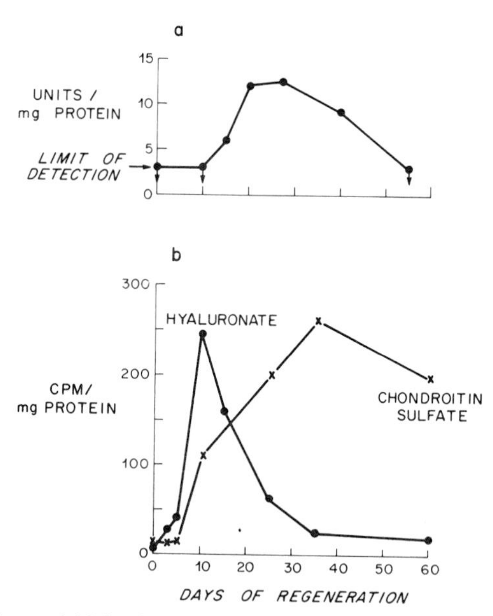

Fig. 13. Comparison of (a) hyaluronidase activity and (b) incorporation of [^{3}H]acetate into hyaluronate and chondroitin sulfate during the course of newt limb regeneration. Reproduced with permission of Academic Press from Toole and Gross (1971).

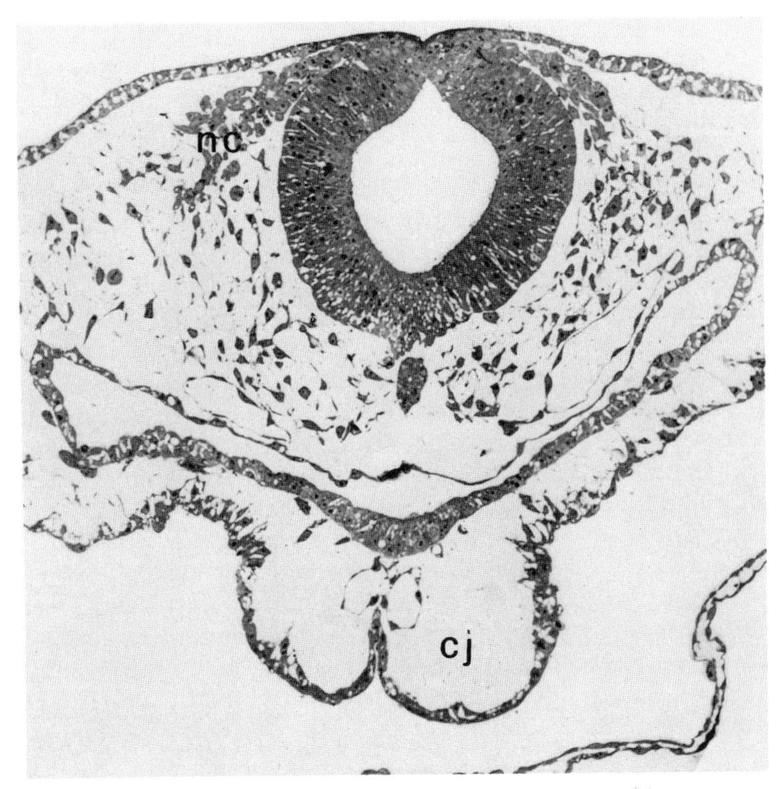

Fig. 14. Cross-section through a stage 10 chick embryo depicting the migrating neural crest cells and the developing heart. Abbreviations: cf, cardiac jelly; nc, neural crest cells. Micrograph kindly supplied by Dr. Elizabeth Hay.

13). Inhibition of blastemal cell accumulation by denervation of the stump shortly after amputation (Singer and Craven, 1948) caused an approximately twofold decrease in hyaluronate synthesis (Smith *et al.*, 1975).

Other investigators have also described the presence of hyaluronate at the time and site of mesenchymal cell migrations during development. Pratt *et al.* (1975) have shown that a major component of the extracellular matrix into which cranial neural crest cells migrate is hyaluronate, and these investigators correlate the production of hyaluronate closely with a large increase in volume of extracellular matrix and the initiation of crest cell migration. Figure 14 shows a cross-section through the anterior region of a stage 10 chick embryo and illustrates the migration of neural crest cells of that region into a large area of previously cell-free matrix, which is apparently filled with hyaluronate. This micrograph also shows the early forming heart. The cardiac jelly contains hyaluronate and, shortly after the stage shown here, will become populated with cushion cells migrating in from the endocardium (Markwald and Adams-Smith, 1972; Manasek *et al.*, 1973).

Invasion by fibroblasts is usually a characteristic of the early stages of repair and remodeling processes in adult organisms. These phenomena are also accompanied by dramatic increases in hyaluronate synthesis—e.g., in tendon regeneration (Dorner, 1968; Reid and Flint, 1974), callus formation in fractured bone (Maurer and Hudack, 1952), granuloma formation (Nemeth-Csoka, 1970), and osteogenesis within muscle implants of insoluble collagen (Iwata and Urist, 1973). Also, in each of these sequences hyaluronate concentration decreases while collagen and sulfated GAG increase subsequent to the migratory stage. Related phenomena may be the high rates of synthesis of hyaluronate *in vitro* during cell division (Morris and Godman, 1960; Morris, 1960; Moscatelli and Rubin, 1975) and by some transformed cells (Ishimoto *et al.*, 1966; Satoh *et al.*, 1973; Makita and Shimojo, 1973). Finally, Fraser *et al.* (1970) have directly observed by microcinematography the movements of cultured synovial fibroblasts within pericellular gel-like investments which are produced by these cells and consist mainly of hyaluronate (Clarris and Fraser, 1968) (Fig. 15), while Terry and Culp (1974) have furnished preliminary evidence, using 3T3 cells, that hyaluronate may

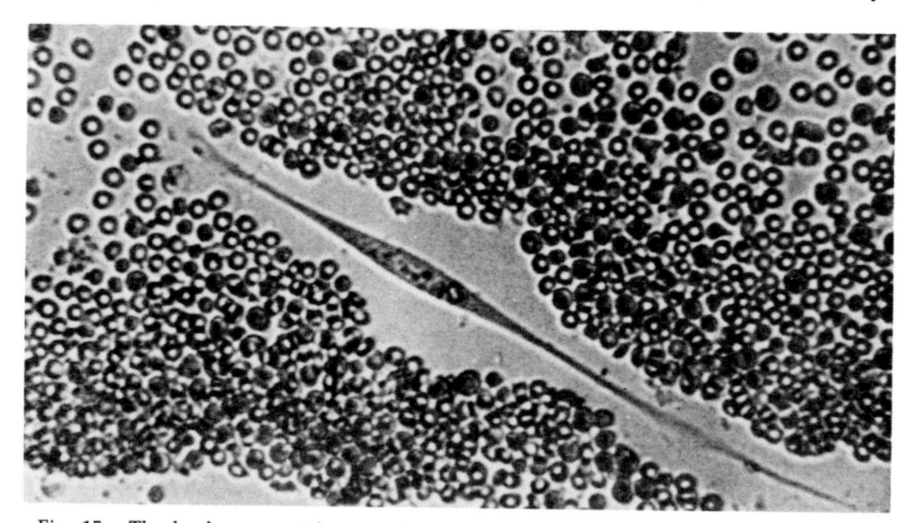

Fig. 15. The hyaluronate-rich, pericellular investment around a synovial fibroblast in culture, demonstrated by exclusion of erythrocytes from this zone. Reproduced with permission of Academic Press from Clarris and Fraser (1968).

be involved in cell–substratum adhesion which is necessary for cell movements.

The data described above demonstrate a striking correlation of hyaluronate synthesis with mesenchymal cell migrations. A second correlation that emerges from some of these studies is that of hyaluronate removal with differentiation of these cells, particularly to form cartilage. It has already been mentioned that hyaluronidase activity becomes measurable subsequent to the phase of cell migration in the newt regeneration blastema and in the chick embryo corneal and axial regions. In the regeneration blastema, hyaluronidase activity was first evident at approximately the same time that redifferentiation of cartilage occurred—i.e., between 15 and 20 days of regeneration—and continued to be present throughout the active stages of cartilage formation, when chondroitin sulfate has become the major GAG being produced (Toole and Gross, 1971) (Fig. 13). This enzyme was found to be approximately 15 times more active with hyaluronate as a substrate than with chondroitin sulfate. However, since there was significant activity toward the latter GAG, the turnover of hyaluronate and chondroitin sulfate was studied at two critical stages: (1) blastema formation, which was maximal at about 10 days of regeneration, when hyaluronate was the major GAG being synthesized and hyaluronidase activity was not evident, and (2) cartilage redifferentiation, which was maximal at 25 days, when chondroitin sulfate was the major GAG and hyaluronidase was maximal. Figure 16 shows the incorporation of [^{3}H]acetate into hyaluronate and chondroitin sulfate between 3 and 48 h after injection of the isotopic precursor into the newts. When animals were injected at 10 days after amputation, the incorporation into both GAG rose to a maximum at about 16 h and then leveled off for the next 32 h. At 25 days a similar pattern occurred for chondroitin sulfate but incorporation into hyaluronate rose for the first 16 h and then dropped of markedly to a low level by 48 h. These results are compatible with preferential action of the hyaluronidase toward hyaluronate during cartilage redifferentiation, as is the known accumulation of chondroitin sulfate as opposed to hyaluronate in mature cartilage. When regeneration is blocked by denervating the amputated limbs at an early stage, the appearance of hyaluronidase activity in the limb is also blocked (Table II). Denervation subsequent to blastema formation does not block regeneration but simply depresses the growth rate of the regenerate (Singer and Craven, 1948). Under the latter conditions, hyaluronidase appears on schedule (Table II), thus enhancing the correlation of hyaluronidase activity with the onset of cartilage differentiation and its absence on inhibition of this event (Smith *et al.*, 1975).

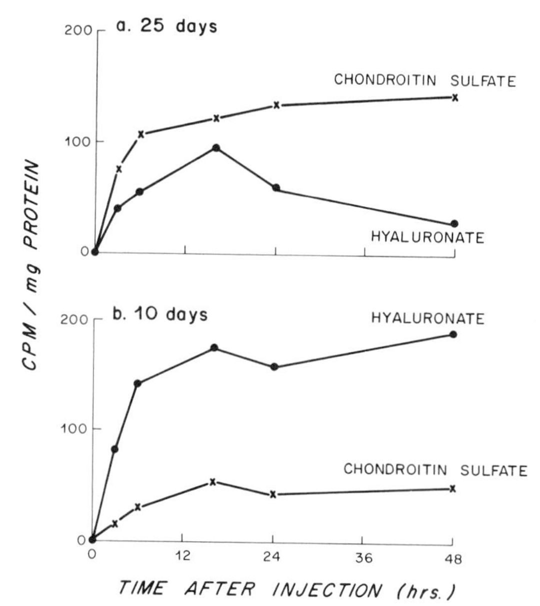

Fig. 16. Time course of incorporation of [^{3}H]acetate into hyaluronate and chondroitin sulfate in newt limb regenerates at (a) 25 days after amputation and (b) 10 days after amputation. Reproduced with permission of Academic Press from Toole and Gross (1971).

In the axial region of the chick embryo, the sclerotomal cell migrations from the somites are not synchronous but proceed in an anterior to posterior sequence, with each successive somite dissociating slightly later than its immediately anterior neighbor (Williams, 1910). However, by examining a restricted length of the trunk (*viz.*, somites 20–32) both histochemically for the appearance of the characteristic metachromasia of cartilage matrix and biochemically for hyaluronidase activity, a very close correlation between the two events was established (Toole, 1972). The chick embryo limb was also examined for the presence of hyaluronidase and synthesis of GAG prior to and at the time of cartilage production (Toole, 1972). In the limb bud, metachromatic cartilage matrix first appears at the stage 25 of development concomitant with a dramatic increase in synthesis and deposition in the matrix of chondroitin sulfate-proteoglycan (Searls, 1965; Medoff, 1967; Huffer, 1970; Levitt and Dorfman, 1974) and cartilage type II collagen, $[\alpha1(\text{II})]_3$ (Linsenmayer *et al.*, 1973*a*,*b*: Linsenmayer, 1974). At precisely this stage of limb development, hyaluronidase activity becomes detectable. It is at this stage also that the precartilage cells of the limb bud become stabilized, in that changes in environment, which at earlier stages repress chondrogenesis, do not now prevent differentiation and production of cartilage matrix (Searls and Janners, 1969; Finch and Zwilling, 1971). Zwilling (1968) has proposed, on the basis of a large body of experimental data of this nature, that early limb development in the chick is comprised of two discrete phases, a morphogenetic phase wherein the mesodermal cells accumulate by proliferation and the future form of the limb is determined and a cytodifferentiative phase wherein these cells differentiate to give cartilage, muscle, connective tissue, etc. Figure 17 depicts this scheme and the relationship of hyaluronidase activity to the two stages. It is readily appreciated that enzyme activity appears at the end of the morphogenetic stage, concomitant with the onset of cytodifferentiation.

As a result of the above correlations, a working hypothesis was formulated which proposed that in addition to the possibility of hyaluronate providing a suitable milieu for cell migration its removal might be necessary for the onset of differentiation. The timing of onset of hyaluronidase activity would then be crucial in the timing of initial differentiation, an important factor since precocious or delayed differentiation leads to aberrant histogenesis.

Table II. Hyaluronidase Activity[a]

Denervation day	Assay day	Activity (units/mg tissue protein)	
		R	D
(a) 10	12	<1	<1
5	15	<1	<1
5	20	8.8	<1
10	20	7.9	<1
5	27	4.6	<1
(b) 15	27	7.3	7.1

[a] Units of activity defined as μg *N*-acetylglucosamine released from 100 μg of hyaluronate in 18 h at 37°C in 0.10M formate/0.15 M NaCl, pH 3.7. Where activity was not detected, the lower limit of detection in the assay is given (i.e., <1). Abbreviations: R, regenerating; D, denervated. (a) Hyaluronidase activity appeared in the normal, regenerating limb (R) subsequent to 15 days postamputation and continued through 27 days. Denervation performed during blastemal cell accumulation (5 or 10 days) resulted in inhibition of the appearance of hyaluronidase activity at 20 through 27 days. (b) Denervation performed subsequent to blastemal cell accumulation but prior to the appearance of hyaluronidase (15 days) did not inhibit the appearance of hyaluronidase activity at 27 days. Results taken from Smith *et al.* (1975).

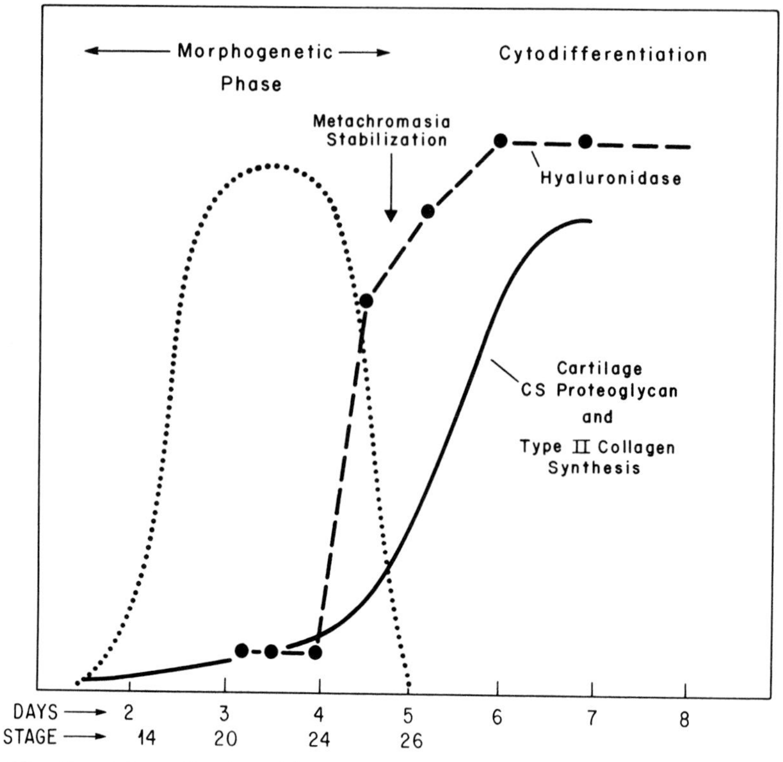

Fig. 17. Correlation of hyaluronidase activity in the developing chick embryo limb with the phase of cytodifferentiation in the scheme of Zwilling (1968). Reproduced with permission of the American Society of Zoologists from Toole (1973*a*).

3.2. *Influence of Hyaluronate on Cells in Vitro*

To test the hypothesis that hyaluronate might influence the course of chondrogenesis *in vivo*, chondrogenic cells were isolated by trypsinization from the somite region of stage 26 chick embryos and treated in culture with hyaluronate (Toole *et al.*, 1972). The somite cells were placed in stationary culture at very high density in order to promote the rapid formation of three-dimensional cartilagelike nodules (Schacter, 1970; Finch and Zwilling, 1971). Addition of hyaluronate to the medium at the beginning of culture prevented this aggregation (Fig. 18). The effect was obtained at concentrations ranging from 1 ng to 500 μg of hyaluronate per milliliter of medium but was not obtained with other polyanions—*viz.*, chondroitin-4- and -6-sulfates, cartilage proteoglycan, heparin, DNA, or RNA. The polymeric nature of hyaluronate was not necessary since the tetrasaccharide (*N*-acetylglucosamine-glucuronic acid)$_2$ exhibited an equivalent inhibitory effect. However, the monosaccharides, alone or mixed together, did not inhibit formation of the nodules (Toole, 1973*a*).

The mechanism by which hyaluronate exerts the above inhibitory action is not known. Examination of cell numbers and [^{3}H]thymidine incorporation into DNA at different times during culture in the presence and absence of hyaluronate showed no significant effect of the polysaccharide on cell proliferation, survival, or plating efficiency (Toole *et al.*, 1972). Attempts to demonstrate a direct effect of hyaluronate on aggregation (Moscona, 1961) or adhesiveness (Steinberg, 1970) of somite cells in swirling cell suspensions have not yielded a positive answer (Toole, unpublished). In fact, experiments with lymphoid cell lines suggest that hyaluronate may induce rather than inhibit aggrega-

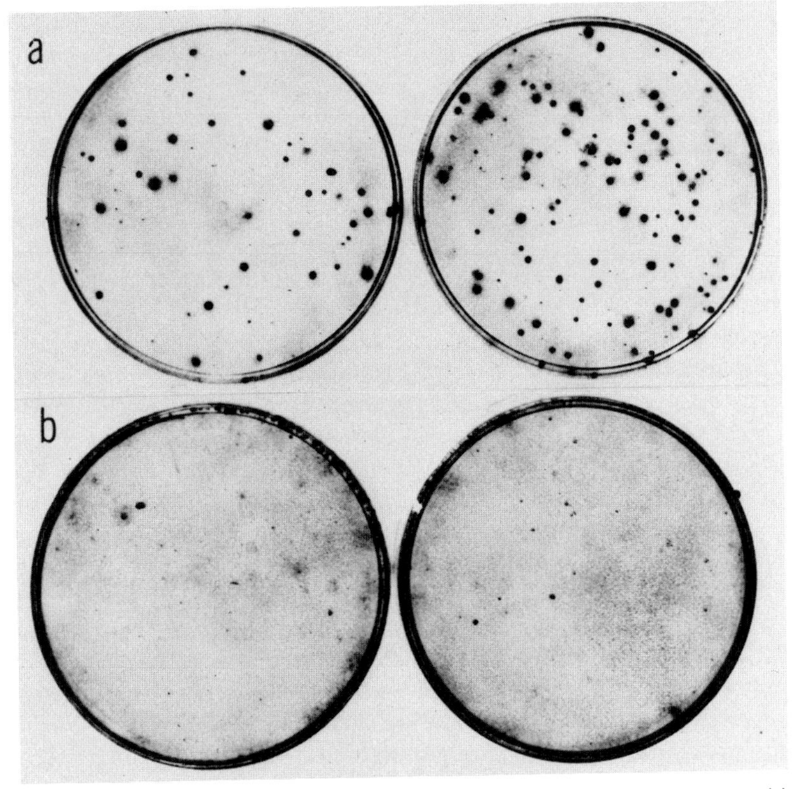

Fig. 18. Appearance of stage 26 chick embryo somite cell cultures after 7 days in (a) the absence and (b) the presence of 0.1 μg/ml hyaluronate. Reproduced with permission of the American Society of Zoologists from Toole (1973*a*).

tion of these particular cells (Pessac and Defendi, 1972; Wasteson *et al.*, 1973).

An obvious potential action of hyaluronate on the chondrogenic cells could be to inhibit matrix deposition, and so to test this the incorporation of labeled precursors into chondroitin sulfate and collagen was measured with and without addition of hyaluronate. No effect on the incorporation of [^{3}H]proline into type I, $[\alpha 1(I)]_2\alpha 2$, or type II, $[\alpha 1(II)]_3$, collagens was obtained, but a twofold reduction of chondroitin sulfate synthesis was evident (Toole, 1973*a*). Other investigators have now demonstrated a similar effect of hyaluronate on proteoglycan synthesis using mature chondrocytes in suspension (Wiebkin and Muir, 1973) or in monolayer cultures (Solursh *et al.*, 1974). In each case, the reduction of chondroitin sulfate synthesis was approximately twofold. Both oligosaccharides and polymeric hyaluronate were efficient inhibitors, although the effective concentrations varied from cell system to system. Figure 19 shows the effect of hyaluronate on the morphology of mature chondrocyte colonies, which is to reduce the amount of visible matrix rather than to inhibit the formation of nodules. An important distinction to make is that the nodules in the high-density somite cell cultures formed very rapidly, presumably by a combination of aggregation and proliferation of the cells, whereas the mature chondrocytes were cultured at clonal densities and the colonies arose solely by cell proliferation (Solursh *et al.*, 1974). The binding characteristics of hyaluronate to proteoglycan subunit and induction of proteoglycan aggregation (Figs. 4 and 5; see Section 2) have several aspects in common with the interaction of hyaluronate with chondrocytes in culture and thus may be involved in the mechanism of hyaluronate inhibition of chondroitin sulfate-proteoglycan synthesis (Wiebkin *et al.*, 1975).

A likely site of action of hyaluronate would be at the cell surface, and several indirect observations do indeed suggest that hyaluronate may become associated with the surface of embryonic mesenchymal cells and fibroblasts. In the case of the somite cells discussed above, the inhibitory effect is expressed if the cells are mixed with hyaluronate briefly, then washed and cultured in hyaluronate-free medium, suggesting the possibility of adherence of the hyaluronate to the cell surface (Toole, 1973*a*). Clarris and Fraser (1968) demonstrated a hyaluronate-rich pericellular zone around synovial fibroblasts *in vitro* by exclusion of particles such as molybdenum disulfide or erythrocytes (Fig. 15). This zone was destroyed by digestion with streptococcal or testicular hyaluronidase but not nucleases, trypsin, neuraminidase, or EDTA. Similar hyaluronate-containing zones or capsules occur around

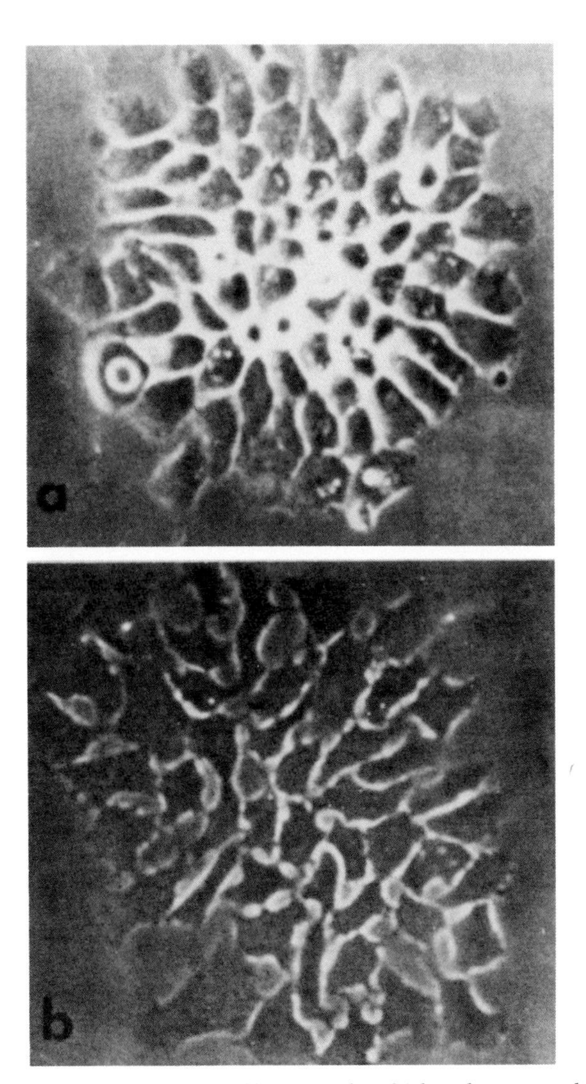

Fig. 19. Phase contrast micrographs of living 13-day chick embryo sternal chondrocytes in (a) the absence and (b) the presence of 200 μg/ml hyaluronate after 5 days of culture. ×200. Reproduced with permission of Academic Press from Solursh *et al*. (1974).

some bacteria (Wilkinson, 1958). Several other investigators (Morris and Godman, 1960; Morris, 1960; Dingle and Webb, 1965; Satoh *et al.*, 1973; Terry and Culp, 1974) have shown that hyaluronate is retained by normal and transformed fibroblasts after washing with physiological buffers or media but is released by trypsin or EDTA treatment subsequent to washing. Also, hyaluronate associated with Rous sarcoma transformed fibroblasts blocks lectin agglutination of these cells (Burger and Martin, 1972) and hyaluronate added to lymphocytes inhibits their stimulation by mitogens (Darzynkiewicz and Balazs, 1971).

The diverse observations presented above indicate that hyaluronate may interact with the surface of various types of mesenchymal cells and in the case of precartilage cells depress the production of a characteristic cartilage component, chondroitin sulfate-proteoglycan. However, the nature of the hyaluronate–cell association, the mechanism of inhibition of proteoglycan synthesis, and the relevance of these to the correlations between hyaluronate removal and cartilage differentiation *in vivo*, presented earlier in this chapter, remain indefinite. Despite this reservation, it seems clear that these studies do provide an interesting base for further investigation of the potential roles of hyaluronate and other GAG and proteoglycans in the control of morphogenesis.

3.3. *Thyroxine Antagonism of the Action of Hyaluronate*

Our interest in thyroxine stems from three types of observations: (1) correlations between thyroxine action and hyaluronate catabolism, (2) the relationship of thyroxine to chondrogenesis, and (3) the dependence of brain development on thyroxine.

Coulombre and Coulombre (1964) demonstrated that dehydration of the cornea and the consequent increase in its transparency which occur at about the fourteenth day of chick embryo development were accelerated by the administration of thyroxine but delayed by a thyroid inhibitor. This period of corneal morphogenesis is associated with hyaluronidase activity and loss of hyaluronate (Toole and Trelstad, 1971), which has in turn been related to degree of hydration (see Section 2). The skin of hypothyroid rats contains elevated levels of hyaluronate, probably due to a decreased rate of degradation (Schiller *et al.*, 1962). This condition was corrected by administration of thyroxine and could not be ascribed to the production of excess thyrotropin. Finally, hyaluronidase is present in, and hyaluronate is removed from, the remodeling backskin of bullfrog tadpoles metamorphosing under the influence of endogenous thyroxine (Lipson *et al.*, 1971). We have extended this last observation by treating tadpoles with exogenously added thyroxine and measuring levels of hyaluronate and hyaluronidase activity in the backskin (Polansky and Toole, 1975). The levels of enzyme activity increased by approximately 100% during treatment with 10^{-7}M L-thyroxine for 10 days. Hyaluronate concentration dropped to about 25% of the level in untreated tadpoles. Similar results were obtained with tadpole brain (see Table IV).

Normal skeletal development is dependent on the action of thyroxine (Sissons, 1971), and chondrogenesis *in vitro* is stimulated by this hormone (Pawelek, 1969). Since hyaluronate has been shown to have the opposite effect *in vitro*—i.e., to depress chondrogenesis—we tested the effect of thyroxine on this inhibitory action of hyaluronate in somite cell cultures and found it to be antagonistic (Toole, 1973*b*) (Table III). An appealing explanation of this effect is that thyroxine induces hyaluronidase activity, thus degrading hyaluronate. However, we have found no evidence for reversal of inhibition by addition of hyaluronidases to hyaluronate-containing media. Also, as indicated earlier, the oligosaccharide breakdown products of vertebrate hyaluronidase treatment of hyaluronate are as efficient in inhibiting chondrogenesis as is polymeric hyaluronate (Toole, 1973*a*; Wiebkin and Muir, 1973; Solursh *et al.*, 1974). Finally, no evidence for hyaluronidase activity in thyroxine-treated cultures has been obtained, even though this hormone is capable of elevating hyaluronidase activities under certain conditions *in vivo* (see above and Table IV). Similar antagonism, as described here for thyroxine, toward the inhibitory effect of hyaluronate on chondrogenesis was obtained with two other hormones known to promote skeletal development or growth: growth hormone and calcitonin (Toole, 1973*b*) (Table III). These two hormones and thyroxine have been shown to stimulate adenosine-3′,5′-cyclic monophosphate (cyclic AMP) levels in other systems (Levey and Epstein, 1969; Moskowitz and Fain, 1970;

Table III. Prevention of Hyaluronate Inhibition of Chondrogenesis by Hormones and Cyclic AMP[a]

Hyaluronate (μg/ml)	Agent	Aggregates per dish (mean)
None	None	18
1	None	1
0.1	None	2
1	Thyroxine (0.01 μM)	16
1	Growth hormone (0.01 μg/ml)	15
1	Calcitonin (0.1 μg/ml)	14
0.1	Cyclic AMP (1 μM)	19
0.1	Theophylline (10 μM)	16

[a] Results taken from Toole (1973*a*,*b*).

Table IV. Thyroxine Stimulation of Tadpole Brain Hyaluronidase in Vivo

	Hyaluronidase activities[a]	
	Controls	Thyroxine treated[b]
	0.73	1.23
	0.75	1.33
	0.68	1.63
	0.79	1.08
	0.49	1.45
		1.24
		1.40
Mean ± SD	0.69 ± 0.12	1.34 ± 0.20

[a] Expressed as μg *N*-acetylhexosamine/mg protein released from 200 μg hyaluronate in 8 h at 37°C in 0.10 M formate 0.15 M NaCl, pH 3.7. [b] L-Thyroxine treatment was for 10 days with 10^{-7} M hormone in the aquarium water.

Heersche *et al.*, 1974; Thanassi and Newcombe, 1974; Marcus, 1975), but cyclic AMP mediation of their action is not established—e.g., the action of thyroxine may be directly on the nucleus (Samuels and Tsai, 1973). Cyclic AMP and theophylline do, however, mimic the action of these hormones in hyaluronate-treated cultures of somite cells (Toole, 1973*b*) (Table III), and so it seems possible that the hormones may overcome the inhibitory action of hyaluronate by stimulating cyclic AMP levels in these cells. In this regard, it is also interesting to note that all three of these hormones and dibutyryl cyclic AMP have been shown to stimulate chondroitin sulfate synthesis in various systems (Pawelek, 1969; Meier and Solursh, 1973; Martin *et al.*, 1969; Goggins *et al.*, 1972).

The role of hyaluronidase *in vivo* remains unclear, but two possibilities are (1) that it may act in concert with β-glucuronidase and β-*N*-acetylglucosaminidase to reduce the hyaluronate to monosaccharides, which do not have the inhibitory effect, or (2) that it simply removes the unneeded polysaccharide subsequent to cell accumulation in order to prevent interference with continued histogenesis. In either case, the level of its activity may also be under the control of thyroxine, as indicated by studies referred to above. The possible interrelationship of thyroxine action, hyaluronidase activity, and GAG levels in developing brain will be discussed in the final section of this chapter.

4. MORPHOGENETIC ROLE OF SULFATED PROTEOGLYCANS

Sulfated GAG and proteoglycans have been implicated in numerous aspects of cell behavior including cytokinesis, cell adhesion, cell migration, epithelial–mesenchymal interactions, and modulation of phenotypic expression. As with hyaluronate, there are diverse indications that under certain conditions sulfated GAG are associated with cell surfaces. Many cultured cell lines synthesize heparan sulfate, a significant proportion of which is associated with the outer cell membrane and requires trypsin treatment for its removal (Kraemer, 1971*a*, *b*). In 3T3 cells, sulfated GAG associated with the cell surface are markedly decreased in amount on transformation by viruses (Roblin *et al.*, 1975; Goggins *et al.*, 1972). Alterations in electrophoretic mobility of cells after treatment with chondroitinases indicate that chondroitin sulfates may also contribute to the surface charge of some cell lines (Suzuki *et al.*, 1970; Kojima and Yamagata, 1971). Addition of exogenous sulfated GAG and proteoglycan to cell lines in culture results in adherence of some of the GAG to the cells, changes in electrophoretic mobility of the cells, and significant alterations in mitotic rate (Lippman, 1968).

Sulfated GAG are synthesized very early in embryonic development and their accumulation has been correlated in particular with early cytokinesis (Kinoshita, 1969; Kosher and Searls, 1973) and cell migrations during gastrulation (Sugiyama, 1972; Kosher and Searls, 1973; Karp and Solursh, 1974). Cyclic changes in the physical state of heparinlike material occur as the first cell divisions take place after fertilization of sea urchin eggs (Kinoshita, 1969), a phenomenon which may relate to the desquamation of heparan sulfate from the surface of cell lines during mitosis in culture (Kraemer and Tobey, 1972). A peak of sulfated GAG synthesis occurs at the time of migration of primary mesechyme cells from the vegetal wall of sea urchin blastula (Karp and Solursh, 1974). This synthesis occurs mainly in the mesenchyme cells (Immers, 1961; Sugiyama, 1972). Deprivation of exogenous sulfate causes aberrant development of sea urchin embryos and particularly affects the migration of primary mesenchyme cells (Immers and Runnstrom, 1965; Sugiyama, 1972; Karp and Solursh, 1974) (Fig. 20). Correlated with this effect is a dramatic inhibition of sulfated GAG synthesis (Karp and Solursh, 1974). However, the GAG which may be involved in these events have not been fully characterized and appear to be different from the known GAG of vertebrate connective tissues. Similar analysis of vertebrate development in the frog embryo has also demon-

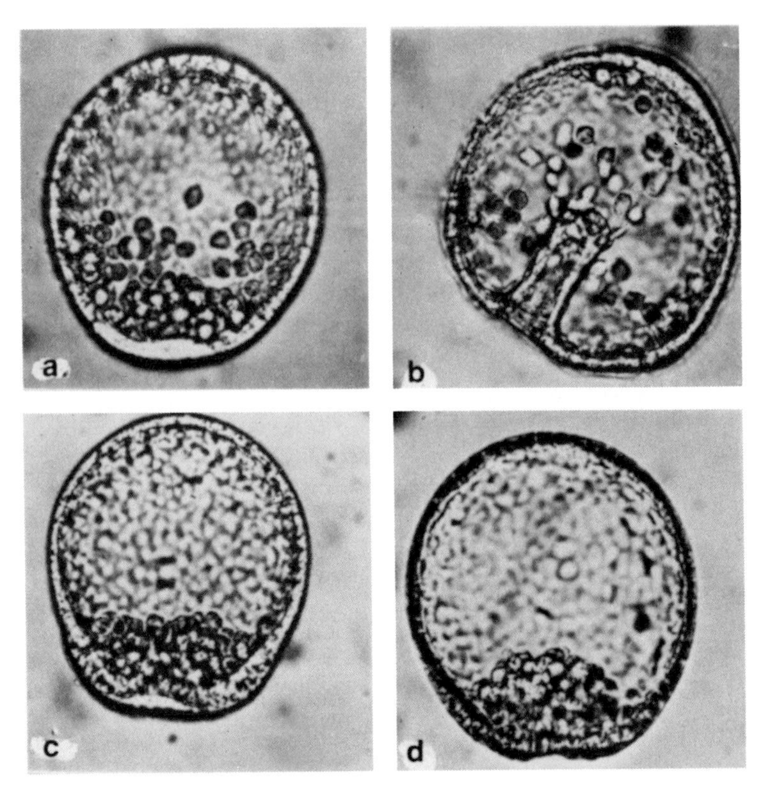

Fig. 20. Sea urchin embryos (*Lytechinus pictus*) raised in complete seawater for (A) 29 h and (B) 38 h after fertilization and in sulfate-free seawater for (C) 29 h and (D) 38 h after fertilization. Reproduced with permission of Academic Press from Karp and Solursh (1974).

strated correlations of sulfated GAG synthesis with early cell cleavage and gastrulation, but the extract nature of these GAG was difficult to determine since the great majority of sulfated GAG in frog embryos is associated with the periphery of yolk platelets (Kosher and Searls, 1973).

Areas of investigation of potential roles for sulfated GAG and their proteoglycans in later stages of development that have been particularly fruitful are (1) the positive feedback effect of sulfated GAG on their own synthesis and (2) their presence and role in basement membrane during epithelial–mesenchymal cell interactions.

Nevo and Dorfman (1972) demonstrated that addition of sulfated polyanions to suspensions of cultured chondrocytes caused an approximately fivefold stimulation of chondroitin sulfate production by these cells. For several years, Lash and coworkers (Lash *et al.*, 1957; Lash, 1968; Ellison and Lash, 1971; Gordon and Lash, 1974) have been investigating the nature of the stimulating or stabilizing influence imparted by notochord and spinal cord on somite chondrogenesis. Recently they have found that this influence is closely matched by replacement of the notochord or spinal cord with cartilage chondroitin sulfate-proteoglycan. Explanted somites produced significantly greater amounts of chondroitin sulfate in the presence of exogenous proteoglycan than in its absence (Fig. 21) and cartilage matrix was visibly increased (Kosher *et al.*, 1973). In addition, they have demonstrated that treatment of notochord with chondroitinase or testicular hyaluronidase removes its enhancing effect on chondrogenesis (Lash and Kosher, 1975). Meier and Hay (1974) have performed similar experiments with the chick embryo corneal epithelium, which produces the early corneal stroma. They find that synthesis of both GAG products of this epithelium—i.e., chondroitin sulfate and heparan sulfate—is stimulated by either of these two GAG but not by other GAG—e.g., dermatan sulfate, keratan sulfate, or hyaluronate.

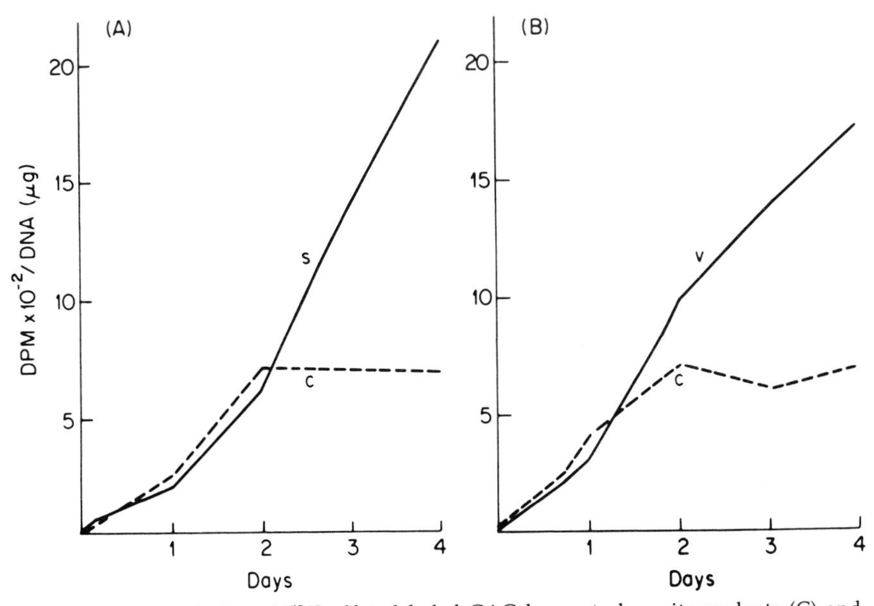

Fig. 21. Accumulation of [^{35}S]sulfate-labeled GAG by control somite explants (C) and by explants supplied with 200 μg/ml exogenous proteoglycan. A: Exogenous proteoglycan derived from sternal cartilage (S). B: Exogenous proteoglycan derived from vertebral cartilage (V). Reproduced with permission of Academic Press from Kosher *et al.* (1973).

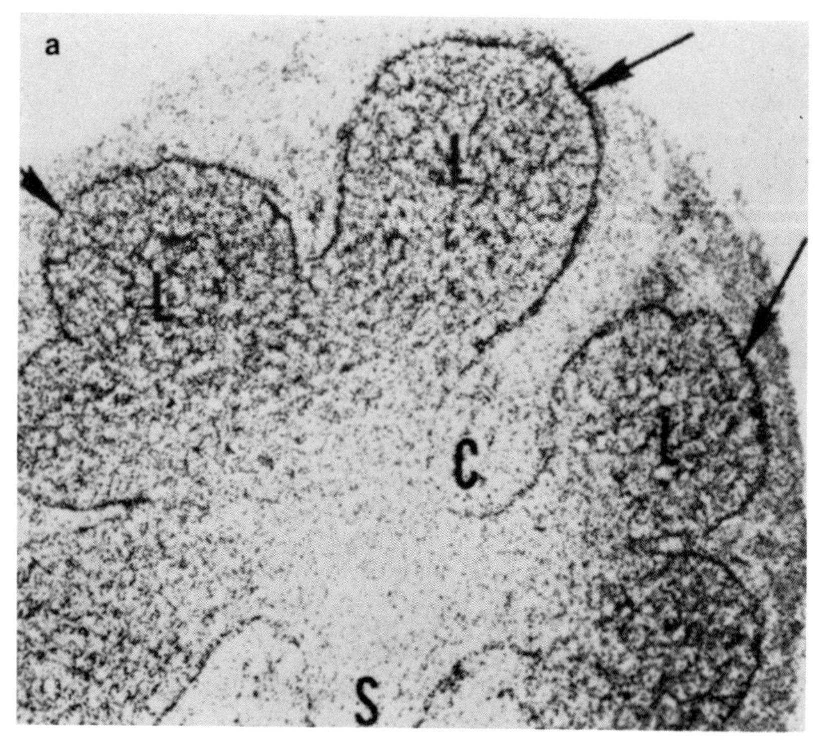

Fig. 22. Localization of GAG at the epithelial surface during submandibular salivary gland morphogenesis. a: Autoradiogram of a 13¼-day mouse embryo salivary gland incubated for 2 h with [^{3}H]glucosamine. The label is maximally localized in GAG at the distal surfaces of the growing lobules (arrows) as opposed to the interlobular clefts (C) and

The interface between epithelium and mesenchyme has been established as crucial in the control of several morphogenetic sequences in the embryo (Grobstein, 1967). The basement membranes underlying epithelia are thought to be important in this control. Until recently attention has been focused on the role of the collagenous element of this structure (Kallman and Grobstein, 1965; Grobstein and Cohen, 1965; Wessells and Cohen, 1968; Bernfield, 1970), but sulfated GAG are now recognized as a prominent component during morphogenesis (Bernfield and Banerjee, 1972; Trelstad *et al.*, 1974; Hay and Meier, 1974) (Figs. 7 and 22). The GAG in the corneal basement membrane is chiefly chondroitin sulfate, organized in a regular array in two planes on either side of the electron-dense basal lamina (Trelstad *et al.*, 1974) (Fig. 7). Bernfield and coworkers (Bernfield and Banerjee, 1972; Bernfield *et al.*, 1972, 1973) have correlated GAG deposition in the basement membrane with branching of embryonic salivary gland

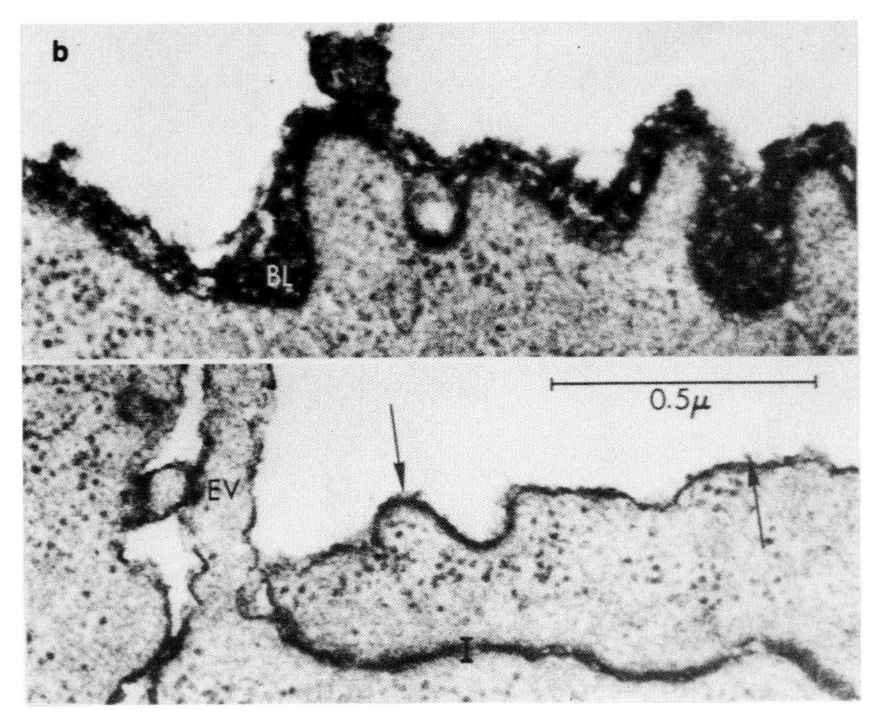

the stalk (S). ×145. Reproduced with permission of American Society of Zoologists from Bernfield *et al.* (1973). b: Electron micrographs of ruthenium red stained epithelial basement membrane (BL) before (top micrograph) and after (bottom micrograph) treatment with testicular hyaluronidase. ×74,000. Reproduced with permission of Rockefeller Press from Bernfield *et al.* (1972).

explants and have demonstrated the dependence of this morphogenetic process on the presence of these GAG. They propose that the interaction of chondroitin sulfate-proteoglycan and collagen molecules might control fibrogenesis in presumptive clefts, endowing these regions with stability (Fig. 22).

Although the above evidence indicates that sulfated GAG undoubtedly are involved in morphogenesis, little is known about the native structure of the proteoglycans of embryonic tissues and the relationship of this structure to influences on cell behavior.

5. HYALURONATE AND BRAIN MORPHOGENESIS

Immature developing brain contains greater amounts of GAG (Fig. 23) (Singh and Bacchawat, 1965, 1968; Young and Custod, 1972; Polansky *et al.*, 1974; Krusius *et al.*, 1974; Margolis *et al.*, 1975), larger

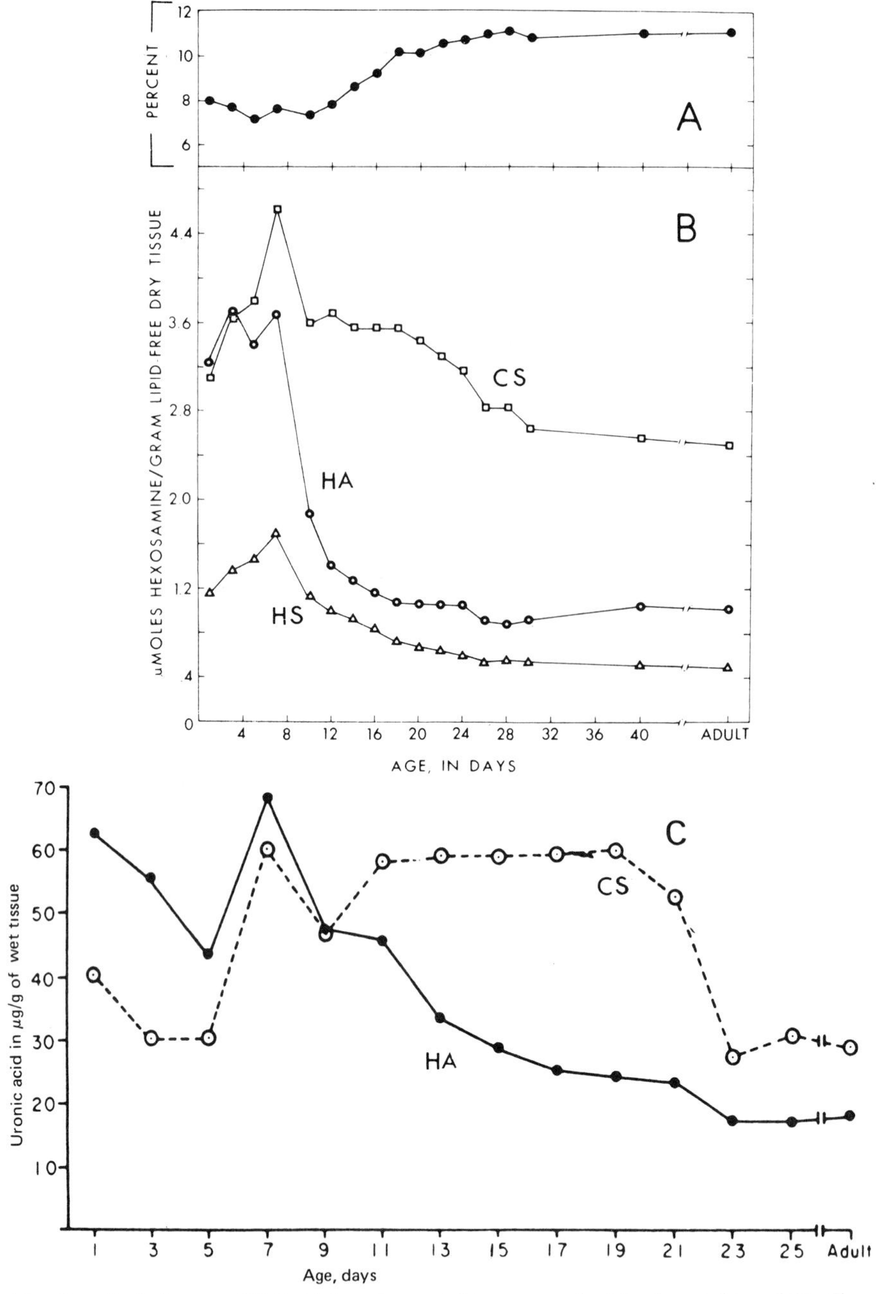

Fig. 23. Variations in (A) lipid-free dry weight as percent of fresh weight and (B, C) GAG content during postnatal development of rat brain. Abbreviations: HA, hyaluronate; CS, chondroitin sulfate; HS, heparan sulfate. A and B reproduced with permission of American Chemical Society from Margolis *et al*. (1975). C reproduced with permission of Pergamon Press from Singh and Bacchawat (1965).

areas of intercellular space (Fig. 1) (Bondareff and Pysh, 1968), and higher water content (DeSouza and Dobbing, 1971; Young and Custod, 1972; Polansky *et al.*, 1974; Margolis *et al.*, 1975) (Fig. 23) than mature brain.

The GAG components of brain are hyaluronate, chondroitin sulfate, and heparan sulfate (Szabo and Roboz-Einstein, 1962; Singh and Bacchawat, 1965; Margolis, 1967; Singh *et al.*, 1969; Margolis and Atherton, 1972), and all three are produced in both neurons and glial cells (Margolis and Margolis, 1974). The decrease in hyaluronate content during brain development is considerably more marked than in the other GAG (Fig. 23). Hyaluronate content has been related to water content in many other systems (Fessler, 1960; Rienits, 1960; Szirmai, 1966; Ogston, 1966; Toole and Trelstad, 1971) and thus may dictate the high degree of hydration of immature brain and in turn the size and properties of the intercellular spaces. Correlations have been noted above (Section 3) between active hyaluronate synthesis, high levels of hydration, increased matrix volume, and cell migration in several morphogenetic systems—e.g., the embryonic cornea (Toole and Trelstad, 1971) and neural crest (Pratt *et al.*, 1975). We feel it is likely then that the relatively high concentration of hyaluronate in developing brain is related to the extensive neuronal migrations that characterize development of the multilayered segments of the brain (Jacobson, 1970; Sidman and Rakic, 1973).

Chondroitin sulfate and heparan sulfate also undergo changes in concentration during brain development, although they are less dramatic than for hyaluronate (Fig. 23). The clearest demonstrations of sulfated GAG involvement in morphogenesis that have been derived from other biological systems (see Section 4) related to (1) the enhancement of matrix deposition by positive feedback effects and (2) the mediation of the basement membrane in epithelial–mesenchymal interactions. However, these two aspects probably do not relate directly to morphogenesis of the central nervous system, which (1) is relatively poor in structured matrix and (2) does not contain basement membranes except underlying blood capillary endothelia. Sulfated GAG could of course be involved in these ways in the formation of other structures associated with the various components of the nervous systems—e.g., in the meninges, endoneurium, Schwann cell basement membranes, and other cellular sheaths. The association of chondroitin sulfate and heparan sulfate with the surface of various cells (see Section 4 for references) suggests potential roles in binding of divalent cations, in cell–cell adhesion, or even in cell migrations. However, these possibilities are still very conjectural. A transient increase in all three brain GAG has been observed during the stage of active myelination in brain—e.g., at approximately 7 days postnatal development in the rat (Singh and Bacchawat, 1965; Young and Custod, 1972; Margolis *et al.*, 1975) (Fig. 23). However, no GAG could be found in myelin on direct analysis (Margolis, 1967).

We have proposed that the synthesis and enzymatic degradation of hyaluronate may be involved in the temporal control of cell migration and cell differentiation in several developing or remodeling tissues (see Section 3). The developmental sequences, whose study has led to this postulate, can be separated into (1) a morphogenetic phase during which cells accumulate by proliferation and migration at a suitable location and in the required number and (2) a phase of overt differentiation. The former phase is characterized by hyaluronate production, the latter by hyaluronidase activity (e.g., see Figs. 10, 13, and 17). During chick brain development, a different situation is encountered where both the concentration of hyaluronate and hyaluronidase activity measured in whole brain are high during early development of the organ and decrease concurrently at hatching (Polansky *et al.*, 1974) (Fig. 24).

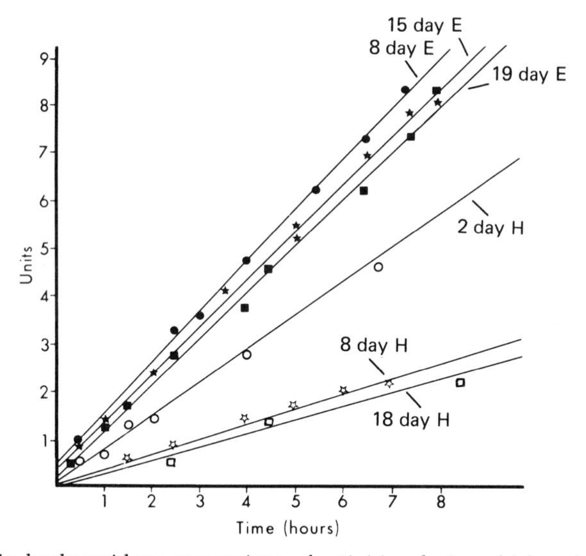

Fig. 24. Brain hyaluronidase: comparison of activities during chick embryo development and after hatching. Extracts of embryo brain (E) contain higher enzymatic activities than the hatched chick (H). The slope of each line represents the rate of enzymatic activity expressed as micrograms of *N*-acetylhexosamine released from 200 μg of hyaluronate per milligram of brain tissue protein. Reproduced with permission of the American Association for the Advancement of Science from Polansky *et al.* (1974).

However, brain development is not divided into two distinct phases of cell accumulation and differentiation but is comprised of continuous, overlapping sequences of neuronal proliferations, migrations, and differentiations (Jacobson, 1970; LaVail and Cowan, 1971*a,b;* Sidman and Rakic, 1973). Thus in the chick both hyaluronate content and hyaluronidase activity are greater during these sequences, which continue throughout most of embryonic life in the chick, than subsequent to hatching, when most of the neuronal migrations have ceased (Hanaway, 1967; LaVail and Cowan, 1971*a,b*). Thus it seems an attractive possibility that, by analogy to the other systems described above (Section 3), hyaluronate may prevent precocious neuronal differentiation as well as facilitate migration. Investigation of the temporal and spatial distribution in brain of cells producing hyaluronate and cells exhibiting hyaluronidase activity will be critical in evaluating this hypothesis.

In Section 3 I have outlined the possible relationship between the morphogenetic action of thyroxine and hyaluronate metabolism. Normal brain development is dependent on the action of thyroxine. Studies of cerebellar development in hypothyroid rats have indicated that lack of thyroxine results in delayed and aberrant differentiation of Purkinje and external granule cells as well as delayed migration of the latter (LeGrand, 1971; Hamburgh *et al.*, 1971). Since the nonmigrating external granule cells continued to divide in the hypothyroid rat, Hamburgh *et al.*, (1971) proposed that thyroxine might act as a "timer" for turning off proliferation and initiating differentiation necessary for subsequent migration of the differentiated cells. A somewhat analogous experimental system for examining thyroxine effects is found in the bullfrog tadpole, where the cerebellum is very immature. Administration of thyroxine stimulates neuronal differentiation (Kollros and McMurray, 1956; Pesetsky, 1966) and induces the formation and subsequent migration of the external granule layer (Gona, 1973). In preliminary studies we have found that thyroxine also causes a twofold enhancement of hyaluronidase activity in tadpole brain (Table IV) (Polansky and Toole, 1975), thus highlighting the potential interrelationship between thyroxine and hyaluronate metabolism in morphogenesis.

Acknowledgments

I am most grateful to Drs. Romaine Bruns, Jerome Gross, Thomas Linsenmayer, Jon Polansky, Gerald Smith, and Robert Trelstad for their critical reading of the manuscript, for their helpful discussions, and for their fruitful collaborations. I also wish to thank the several people who have allowed me to reproduce data from their publications. The author is an Established Investigator of the American Heart Association (No. 73–138). The original research described in this chapter was supported by grants from the American Heart Association (No. 73. 757), the National Foundation—March of Dimes (No. CRBS-312), and the United States Public Health Service (AM 3564). This is publication No. 661 of the Lovett Memorial Group for the Study of Disease Causing Deformities.

6. REFERENCES

Anseth, A., 1961, Glycosaminoglycans in the developing corneal stroma, *Expl. Eye Res.* **1:**116.

Arnott, S., Guss, J. M., Hukins, D. W., and Mathews, M. B., 1973, Mucopolysaccharides: Comparison of chondroitin sulfate conformations with those of related polyanions, *Science* **180:**743.

Aronson, N. N., and Davidson, E. A., 1967*a*, Lysosomal hyaluronidase from rat liver. I. Preparations, *J. Biol. Chem.* **242:**437.

Aronson, N. N., and Davidson, E. A., 1967*b*, Lysosomal hyaluronidase from rat liver. II. Properties, *J. Biol. Chem.* **242:**441.

Ashhurst, D. E., and Costin, N. M., 1971, Insect mucosubstances. II. Mucosubstances of the central nervous system, *Histochem. J.* **3:**297.

Atkins, E. D., Phelps, C. F., and Sheehan, J. K., 1972, The conformation of mucopolysaccharides: Hyaluronates, *Biochem. J.* **128:**1255.

Atkins, E. D., Hardingham, T. E., Isaac, D. H., and Muir, H, 1974, X-ray fibre diffraction of cartilage proteoglycan aggregates containing hyaluronic acid, *Biochem. J.* **141:**919.

Bernfield, M. R., 1970, Collagen synthesis during epithelio-mesenchymal interactions, *Dev. Biol.* **22:**213.

Bernfield, M. R., and Banerjee, S. D., 1972, Acid mucopolysaccharide (glycosaminoglycan) at the epithelial–mesenchymal interface of mouse embryo salivary glands, *J. Cell Biol.* **52:**664.

Bernfield, M. R., Banerjee, S. D., and Cohn, R. H., 1972, Dependence of salivary epithelial morphology and branching morphogenesis upon acid mucopolysaccharide–protein (proteoglycan) at the epithelial surface, *J. Cell Biol.* **52:**674.

Bernfield, M. R., Cohn, R. H., and Banerjee, S. D., 1973, Glycosaminoglycans and epithelial organ formation, *Am. Zool.* **13:**1067.

Bondareff, W., and Narotsky, R., 1972, Age changes in the neuronal microenvironment, *Science* **176:**1135.

Bondareff, W., and Pysh, J. J., 1968, Distribution of the extracellular space during postnatal maturation of rat cerebral cortex, *Anat. Rec.* **160:**773.

Bradbury, M. W., Villamil, M., and Kleiman, C. R., 1968, Extracellular fluid, ionic distribution and exchange in isolated frog brain, *Am. J. Physiol.* **214:**643.

Breen, M., Weinstein, H. G., Johnson, R. L., Veis, A., and Marshall, R. T., 1970, Acidic glycosaminoglycans in human skin during fetal development and adult life, *Biochim. Biophys. Acta* **201:**54.

Breen, M., Johnson, R. L., Sittig, R. A., Weinstein, H. G., and Veis, A., 1972, The acidic glycosaminoglycans in human fetal development and adult life: Cornea, sclera and skin, *Connect. Tissue Res.* **1**:291.

Burger, M. M., and Martin, G. S., 1972, Agglutination of cells transformed by Rous sarcoma virus by wheat germ agglutinin and concanavalin A, *Nature (London) New Biol.* **237**:9.

Chalkley, D. T., 1959, The cellular basis of limb regeneration, in: *Regeneration in Vertebrates* (C. S. Thornton, ed.), pp. 34–58, University of Chicago Press, Chicago.

Clarris, B. J., and Fraser, J. R., 1968, On the pericellular zone of some mammalian cells *in vitro*, *Exp. Cell Res.* **49**:181.

Cleland, R. L., 1970, Molecular weight distribution in hyaluronic acid, in: *Chemistry and Molecular Biology of the Intercellular Matrix,* Vol. 2 (E. A. Balazs, ed.), pp. 733–742, Academic Press, New York.

Comper, W. D., and Preston, B. N., 1974, Model connective tissue systems: A study of polyion-mobile ion and of excluded-volume interactions of proteoglycans, *Biochem. J.* **143**:1.

Conrad, G. W., 1970, Collagen and mucopolysaccharide biosynthesis in the developing chick cornea, *Dev. Biol.* **21**:292.

Coulombre, A. J., and Coulombre, J. L., 1958, Corneal development. I. Corneal transparency, *J. Cell. Comp. Physiol.* **51**:1.

Coulombre, A. J., and Coulombre, J. L., 1964, Corneal development. III. The role of the thyroid in dehydration and the development of transparency, *Exp. Eye Res.* **3**:105.

Custod, J. T., and Young, I. J., 1968, Cat brain mucopolysaccharides and their *in vivo* hyaluronidase digestion, *J. Neurochem.* **15**:809.

Darzynkiewicz, A., and Balazs, E. A., 1971, Effect of connective tissue intercellular matrix on lymphocyte stimulation. I. Suppresion of lymphocyte stimulation by hyaluronic acid, *Exp. Cell Res.* **66**:113.

Davidson, E. A., 1970, Glycoprotein and mucopolysaccharide hydrolysis, in: *Metabolic Conjugation and Metabolic Hydrolysis,* Vol. 1, (W. H. Fishman, ed.), pp. 327–353, Academic Press, New York.

Dea, I. C., Moorhouse, R., Rees, D. A., Arnott, S., Guss, J. M., and Balazs, E. A., 1973, Hyaluronic acid: A novel, double helical molecule, *Science* **179**:560.

DeLuca, S., Richmond, M. E., and Silbert, J. E., 1973, Biosynthesis of chondroitin sulfate: Sulfation of the polysaccharide chain, *Biochemistry* **12**:3911.

DeSouza, S. W., and Dobbing, J., 1971, Cerebral edema in developing brain. 1. Normal water and cation content in developing rat brain and postmortem changes,*Exp. Neurol.* **32**:431.

Dingle, J. T., and Webb, M., 1965, Mucopolysaccharide metabolism in tissue culture, in: *Cells and Tissues in Culture,* Vol. 1 (E. N. Willmer, ed.), pp. 353–396, Academic Press, New York.

Dingle, J. T., Barrett, A. J., and Weston, P. D., 1971, Cathepsin D—Characteristics of immunoinhibition and the confirmation of a role in cartilage breakdown, *Biochem. J.* **123**:1.

Disalvo, J., and Schubert, M., 1966, Interaction during fibril formation of soluble collagen with cartilage protein polysaccharide, *Biopolymers* **4**:247.

Dorner, R. W., 1968, Changes in glycosaminoglycan composition associated with maturation of regenerating rabbit tendon, *Arch. Biochem. Biophys.* **128**:34.

Dunstone, J. R., 1962, Ion-exchange reactions between acid mucopolysaccharides and various cations, *Biochem. J.* **85**:336.

Eisenstein, R., Larsson, S., Sorgente, N., and Kuettner, K. E., 1973, Collagen–proteoglycan relationships in epiphyseal cartilage, *Am. J. Pathol.* **73**:443.

Ellison, M. L., and Lash, J. W., 1971, Environmental enhancement of *in vitro* chondrogenesis, *Dev. Biol.* **26:**486.

Fessler, J., 1960, A structural function of mucopolysaccharide in connective tissue, *Biochem. J.* **76:**124.

Finch, R. A., and Zwilling, E., 1971, Culture stability of morphogenetic properties of chick limb-bud mesoderm, *J. Exp. Zool.* **176:**397.

Fransson, L. A., 1970, Structure and metabolism of the proteoglycans of dermatan sulfate, in: *Chemistry and Molecular Biology of the Intercellular Matrix,* Vol. 2 (E. A. Balazs, ed.), pp. 823–842, Academic Press, New York.

Fraser, J. R., Clarris, F. J., and Kont, L. A., 1970, The morphology and motility of human synovial cells and their pericellular gels: A time-lapse microcinematographic study, *Aust. J. Biol. Sci.* **23:**1297.

Gerber, B. R., and Schubert, M., 1964, The exclusion of large solutes by cartilage protein polysaccaride, *Biopolymers* **2:**259.

Goetinck, P. F., Pennypacker, J. P., and Royal, P. D., 1974, Proteochondroitin sulfate synthesis and chondrogenic expression, *Exp. Cell Res.* **87:**241.

Goggins, J. F., Johnson, G. S., and Pastan, I., 1972, The effect of dibutyryl cyclic adenosine monophosphate on synthesis of sulfated acid mucopolysaccharides by transformed cells, *J. Biol. Chem.* **247:**5759.

Gona, A. G., 1973, Effects of thyroxine, thyrotropin, prolactin, and growth hormone on the maturation of the frog cerebellum, *Exp. Neurol.* **38:**494.

Gordon, J. S., and Lash, J. W., 1974, *In vitro* chondrogenesis and cell viability, *Dev. Biol.* **36:**88.

Gregory, J. D., 1973, Multiple aggregation factors in cartilage proteoglycan, Biochem. J. **133:**383.

Grillo, H. C., Lapiere, C. M., Dresden, M. H., and Gross, J., 1968, Collagenolytic activity in regenerating forelimbs of the adult newt *(Triturus viridescens), Dev. Biol.* **17:**571.

Grobstein, C., 1967, Mechanisms of organogenetic tissue interaction, *Natl. Cancer Inst. Monogr.* **26:**279.

Grobstein, C., and Cohen, J., 1965, Collagenase: Effect on the morphogenesis of embryonic salivary epithelium *in vitro, Science* **150:**626.

Hamburger, F., and Hamilton, H. L., 1951, A series of normal stages in the development of the chick embryo, *J. Morphol.* **88:**49.

Hamburgh, M., Mendoza, L. A., Burkart, J. F., and Weil, F., 1971, in: *Hormones in Development* (M. Hamburgh and E. J. Barrington, eds.), pp. 403–415, Appleton-Century-Crofts, New York.

Hanaway, J., 1967, Formation and differentiation of the external granular layer of the chick cerebellum, *J. Comp. Neurol.* **131:**1.

Hardingham, T. E., and Muir, H., 1972, The specific interaction of hyaluronic acid with cartilage proteoglycans, *Biochim. Biophys. Acta* **279:**401.

Hardingham, T. E., and Muir, H., 1974, Hyaluronic acid in cartilage and proteoglycan aggregation, *Biochem. J.* **139:**565

Hascall, V. C., and Heinegard, D., 1974*a*, Aggregation of cartilage proteoglycans. I. The role of hyaluronic acid, *J. Biol. Chem.* **249:**4232.

Hascall, V. C., and Heinegard, D., 1974*b*, Aggregation of cartilage proteoglycans. II. Oligosaccharide competitors of the proteoglycan–hyaluronic acid interaction, *J. Biol. Chem.* **249:**4242.

Hascall, V. C., and Heinegard, D., 1975, The structure of cartilage proteoglycans, in: *Extracellular Matrix Influences on Gene Expression* (H. C. Slavkin and R. Greulich, eds.), pp. 423–434, Academic Press, New York.

Hascall, V. C., and Sajdera, S. W., 1970, Physical properties and polydispersity of proteoglycan from bovine nasal cartilage, *J. Biol. Chem.* **245:**4920.
Hay, E. D., 1966*a*, *Regeneration,* Holt, Rinehart and Winston, New York.
Hay, E. D., 1966*b*, Embryologic origin of tissues, in: *Histology* (R. O. Greep, ed.), pp. 56–73, McGraw-Hill, New York.
Hay, E. D., 1968, Dedifferentiation and metaplasia in vertebrate and invertebrate regeneration, in: *Stability of the Differentiated State* (H. Ursprung, ed.), pp. 85–108, Springer, New York.
Hay, E. D., 1970, Regeneration of muscle in the amputated amphibian limb, in: *Regeneration of Striated Muscle, and Myogenesis* (A. Mauro, S. A., Shafiq, and A. T. Milhorat. eds), pp. 3–24, Excerpta Medica, Amsterdam.
Hay, E. D., 1973, Origin and role of collagen in the embryo, *Am. Zool.* **13:**1085.
Hay, E. D., and Meier, S., 1974, Glycosaminoglycan synthesis by embryonic inductors: Neural tube, notochord, and lens, *J. Cell Biol.* **62:**889.
Hay, E. D., and Revel, J. P., 1969, *Fine Structure of the Developing Avian Cornea* (A. Wolsky and P. S. Chem, eds.), Vol. I of *Monographs in Developmental Biology,* Karger, Basel.
Heersche, J. N., Marcus, R., and Aurbach, G. D., 1974, Calcitonin and the formation of 3′, 5′-AMP in bone and kidney, *Endocrinology* **94:**241.
Heinegard, D., and Hascall, V. C., 1974, Aggregation of cartilage proteoglycans. III. Characteristics of the proteins isolated from trypsin digests of aggregates, *J. Biol. Chem.* **249:**4250.
Hirano, S., and Meyer, K., 1971, Enzymatic degradation of corneal and cartilagenous keratosulfates, *Biochem. Biophys. Res. Commun.* **44:**1371.
Hopwood, J. J., Fitch, F. W., and Dorfman, A., 1974, Hyaluronic acid synthesis in a cell-free system from rat fibrosarcoma, *Biochem. Biophys. Res. Commun.* **61:**583.
Hovingh, P., and Linker, A., 1970, The enzymatic degradation of heparin and heparitin sulfate. III. Purification of a heparitinase and a heparinase from flavobacteria, J. Biol. Chem. **245:**6170.
Huffer, E. S., 1970, Glycosaminoglycans in the cartilage of developing chick embryo limbs, *Calc. Tissue Res.* **6:**55.
Immers, J., 1961, Comparative study of the localization of incorporated ^{14}C-labeled amino acids and $^{35}SO_4$ in the sea urchin ovary, egg and embryo, *Exp. Cell Res.* **24:**356.
Immers, J., and Runnstrom, J., 1965, Further studies of the effects of deprivation of sulfate on the early development of the sea urchin *Paracentrosus lividus, J. Embryol. Exp. Morphol.* **14:**289.
Ishimoto, N., Temin, H. M., and Strominger, J. L., 1966, Studies of carcinogenesis by avian sarcoma viruses. II. Virus-induced increase in hyaluronic acid synthetase in chicken fibroblasts, *J. Biol. Chem.* **241:**2052.
Iwata, H., and Urist, M. R., 1973, Hyaluronic acid production and removal during bone morphogenesis in implants of bone matrix in rats, *Clin. Orthop. Rel. Res.* **90:**236.
Jacobson, M., 1970, *Developmental Neurobiology,* Holt, Rinehart and Winston, New York.
Kallman, F., and Grobstein, C., 1965, Source of collagen at epitheliomesenchymal interfaces during inductive interaction, *Dev. Biol.* **11:**169.
Karp, G. C., and Solursh, M., 1974, Acid mucopolysaccharide metabolism, the cell surface, and primary mesenchyme cell activity in the sea urchin embryo, *Dev. Biol.* **41:**110.
Katchalsky, A., 1964, Polyelectrolytes and their biological interactions, *Biophys. J. Suppl.* **4:**9.
Kempson, G. E., Muir, H., Swanson, S. A., and Freeman, M. A., 1970, Correlations

between stiffness and the chemical constituents of cartilage on the human femoral head, *Biochim. Biophys. Acta* **215**:70.
Kinoshita, S., 1969, Periodical release of heparin-like polysaccharide within cytoplasm during cleavage of sea urchin egg, *Exp. Cell Res.* **56**:39.
Kojima, K., and Yamagata, T., 1971, Glycosaminoglycans and electrokinetic behavior of rat ascites hepatoma cells, *Exp. Cell Res.* **67**:142.
Kollros, J. J., and McMurray, V. M., 1956, The mesencephalic V nucleus in anurans. II. The influence of thyroid hormone on cell size and cell number, *J. Exp. Zool.* **131**:1.
Konigsberg, I. R., 1970, The relationship of collagen to the clonal development of embryonic skeletal muscle, in: *Chemistry and Molecular Biology of the Intercellular Matrix*, Vol. 3 (E. A. Balazs, ed.), pp. 1779–1810, Academic Press, New York.
Kosher, R. A., and Searls, R. L., 1973, Sulfated mucopolysaccharide synthesis during the development of *Rana pipiens*, *Dev. Biol.* **32**:50.
Kosher, R. A., Lash, J. W., and Minor, R. R. , 1973, Environment enhancement of *in vitro* chondrogenesis. IV. Stimulation of somite chondrogenesis by exogenous chondromucoprotein, *Dev. Biol.* **35**:210.
Kraemer, P. M., 1971*a*, Heparan sulfates of cultured cells. I. Membrane associated and cell-sap species in Chinese hamster cells, *Biochemistry* **10**:1437.
Kraemer, P. M., 1971*b*, Heparan sulfates of cultured cells. II. Acid-soluble and precipitable species of different cell lines, *Biochemistry* **10**:1445.
Kraemer, P. M., and Tobey, R. A., 1972, Cell-cycle dependent desquamation of heparan sulfate from the cell surface, *J. Cell Biol.* **55**:713.
Krusius, T., Finne, J., Karkkainen, J., and Jarnefelt, J., 1974, Neutral and acidic glycopeptides in adult and developing rat brain, *Biochim. Biophys. Acta* **365**:80.
Kvist, T. N., and Finnegan, C. V., 1970*a*, The distribution of glycosaminoglycans in the axial region of the developing chick embryo. I. Histochemical analysis, *J. Exp. Zool.* **175**:221.
Kvist, T. N., and Finnegan, C. V., 1970*b*, The distribution of glycosaminoglycans in the axial region of the developing chick embryo. II. Biochemical analysis, *J. Exp. Zool.* **175**:241.
Lash, J. W., 1968, Chondrogenesis: Genotypic and phenotypic expression. *J. Cell. Physiol.* **72**:Suppl. 1, 35–46.
Lash. J. W., and Kosher, R. A., 1975, Perinotochordal proteoglycans and somite chondrogenesis, in: *Extracellular Matrix Influences on Gene Expression* (H. C. Slavkin and R. Greulich, eds.), pp. 671–676, Academic Press, New York.
Lash, J., Holtzer, S., and Holtzer, H., 1957, An experimental analysis of the development of the spinal column, *Exp. Cell Res.* **13**:292.
Laurent, T. C., 1964, The interaction between polysaccharides and other macromolecules. 9. The exclusion of molecules from hyaluronic acid gels and solutions, *Biochem. J.* **93**:106.
Laurent, T. C., 1966, *In vitro* studies on the transport of macromolecules through the connective tissue, *Fed. Proc.* **25**:1128.
Laurent, T. C., 1970, Structure of hyaluronic acid, in: *Chemistry and Molecular Biology of the Intercellular Matrix*, Vol. 2 (E. A. Balazs, ed.), pp. 703–732. Academic Press, New York.
Laurent, T. C., and Ogston, A. G., 1963, The interaction between polysaccharides and other macromolecules. 4. The osmotic pressure of mixtures of serum albumin and hyaluronic acid, *Biochem. J.* **89**:249.
Laurent, T. C., Ryan, M., and Pietruszkiewicz, A., 1960, Fractionation of hyaluronic acid: The polydispersity of hyaluronic acid from the bovine vitreous body, *Biochim. Biophys. Acta* **42**:476.

Laurent, T. C., Bjork, I., Pietruskiewicz, A., and Persson, H., 1963, On the interaction between polysaccarides and other macromolecules. II. The transport of globular particles through hyaluronic acid solutions, *Biochim. Biophys. Acta* **78:**351.

LaVail, J. H., and Cowan, W. M., 1971*a*, The development of the chick optic tectum. I. Normal morphology and cytoarchitectonic development, *Brain Res.* **28:**391.

LaVail, J. H., and Cowan, W. M., 1971*b*, The development of the chick optic tectum. II. Autoradiographic studies, *Brain Res.* **28:**421.

LeGrand, J., 1971, Comparative effects of thyroid deficiency and undernutrition on maturation of the nervous system and particularly on myelination in the young rat, in: *Hormones in Development* (M. Hamburgh and E. J., Barrington, eds.), pp. 381–390 Appleton-Century-Crofts, New York.

Levey, G. S., and Epstein, S. E., 1969, Myocardial adenylcyclase: Activation by thyroid hormone and evidence for two adenyl cyclase systems, *J. Clin. Invest.* **48:**1663.

Levitt, D., and Dorfman, A., 1974, Concepts and mechanisms of cartilage differentiation, in: *Current Topics in Development Biology,* Vol. 8 (A. Moscona, ed.), pp. 103–149, Academic Press, New York.

Lindahl, U., 1970, Structure of heparin, heparan sulfate and their proteoglycans, in: *Chemistry and Molecular Biology of the Intercellular Matrix,* Vol. 2 (E. A. Balazs, ed.), pp. 943–960, Academic Press, New York.

Linsenmayer, T. F., 1974, Temporal and spatial transitions in collagen types during embryonic chick limb development. II. Comparison of the embryonic cartilage collagen molecule with that from adult cartilage, *Dev. Biol.* **40:**372.

Linsenmayer, T. F., Trelstad, R. L., Toole, B. P., and Gross, J., 1973*a*, The collagen of osteogenic cartilage in the embryonic chick, *Biochem. Biophys. Res. Commun.* **52:**870.

Linsenmayer, T. F., Toole, B. P., and Trelstad, R. L., 1973*b*, Temporal and spatial transitions in collagen types during embryonic chick limb development, *Dev. Biol.* **35:**232.

Lippman, M., 1968, Glycosaminoglycans and cell division, in: *Epithelial–Mesenchymal Interactions* (R. Fleischmajer and R. E. Billingham, eds.), pp. 208–229, Williams and Wilkins, Baltimore.

Lipson, M. J., Cerskus, R. A., and Silbert, J. E., 1971, Glycosaminoglycans and glycosaminoglycan degrading enzyme of *Rana catesbeiana* back skin during late stages of metamorphosis, *Dev. Biol.* **25:**198.

Loewi, G., and Meyer, K., 1958, The acid mucopolysaccharides of embryonic skin, *Biochim. Biophys. Acta* **27:**453.

Lowther, D. A., Preston, B. N., and Meyer, F. A., 1970, Isolation and properties of chondroitin sulphates from bovine heart valves, *Biochem. J.* **118:**595.

Makita, A., and Shimojo, H., 1973, Polysaccarides of SV40-transformed green monkey kidney cells, *Biochim. Biophys. Acta* **304:**571.

Manasek, F. J., Reid, M., Vinson, W., Seyer, J., and Johnson, R., 1973, Glycosaminoglycan synthesis by the early embryonic chick heart, *Dev. Biol.* **35:**332.

Marcus, R., 1975, Cyclic nucleotide phosphodiesterase from bone: Characterization of the enyzyme and studies of inhibition by thyroid hormones, *Endocrinology* **96:**400.

Margolis, R. U., 1967, Acid mucopolysaccharides and proteins of bovine whole brain, white matter and myelin, *Biochim. Biophys. Acta* **141:**91.

Margolis, R. U., and Atherton, D. M., 1972, The heparan sulfate of rat brain, *Biochim. Biophys. Acta* **273:**368.

Margolis, R. U., and Margolis, R. K., 1974, Distribution and metabolism of mucopolysaccharides and glycoproteins in neuronal perikarya, astrocytes and oligodendroglia, *Biochemistry* **13:**2849.

Margolis, R. U., Margolis, R. K., Chang, L. B., and Preti, C., 1975, Glycosaminoglycans of brain during development, *Biochemistry* **41**:85.

Markwald, R. R., and Adams-Smith, W. N., 1972, Distribution of mucosubstances in the developing rat heart, *J. Histochem. Cytochem.* **20**:896.

Martin, T. J., Harris, G. S., Melick, R. A., and Fraser, J. R., 1969, Effect of calcitonin on glycosaminoglycan synthesis of embryo calf bone cells *in vitro*, *Experientia* **25**:375.

Mathews, M. B., 1965, The interaction of collagen and acid mucopolysaccharides: A model for connective tissue, *Biochem. J.* **96**:710.

Mathews, M. B., and Decker, L., 1968, The effect of acid mucopolysaccarides and acid mucopolysaccharide-proteins on fibril formation from collagen solutions, *Biochem. J.* **109**:517.

Maurer, P. H., and Hudack, S. S., 1952, Isolation of hyaluronic acid from callus tissue during early healing, *Arch. Biochem.* **38**:49.

Medoff, J., 1967, Enzymatic events during cartilage differentiation in the chick embryonic limb bud, *Dev. Biol.* **16**:118.

Meier, S., and Hay, E. D., 1973, Synthesis of sulfated glycosaminoglycans by embryonic corneal epithelium, *Dev. Biol.* **35**:318.

Meier, S., and Hay, E. D., 1974, Stimulation of extracellular matrix synthesis in the developing cornea by glycosaminoglycans, *Proc. Natl. Acad. Sci. USA* **71**:2310.

Meier, S., and Solursh, M., 1973, Mediation of growth hormone-enhanced expression of the cartilage phenotype *in vitro* by the availability of the essential amino acid valine, *Dev. Biol.* **30**:290.

Meyer, K., Hoffman, P., and Linker, A., 1960, Hyaluronidases, in: *The Enzymes*, Vol. 4 (P. D. Boyer, H. Lardy, and K. Myrback, eds.), pp. 447–460, Academic Press, New York.

Meyer, F. A., Preston, B. N., and Lowther, D. A., 1969, Isolation and properties of hyaluronic acid from bovine heart valves, *Biochem. J.* **113**:559.

Meyer, F. A., Comper, W . D., and Preston, B. N., 1971, Model connective tissues systems: A physical study of gelatin gels containing proteoglycans, *Biopolymers* **10**:1351.

Minns, R. J., Soden, P. D., and Jackson, D. S., 1973, The role of the fibrous components and ground substance in the mechanical properties of biological tissues: A preliminary investigation, *J. Biomech.* **6**:153.

Morris, C. C., 1960, Quantitative studies on the production of acid mucopolysaccharides by replicate cell cultures of rat fibroblasts, *Ann. N.Y. Acad. Sci.* **86**:878.

Morris, C. C., and Godman, G. C., 1960, Production of acid mucopolysaccharides by fibroblasts in cell cultures, *Nature (London)* **188**:407.

Moscatelli, D., and Rubin, H., 1975, Increased hyaluronic acid production on stimulation of DNA synthesis in chick embryo fibroblasts, *Nature (London)* **254**:65.

Moscona, A. A., 1960, Patterns and mechanisms of tissue reconstruction from dissociated cells, in: *Developing Cell Systems and Their Control* (D. Rudnick, ed.), pp. 45–70, Ronald Press, New York.

Moscona, A. A., 1961, Rotation-mediated histogenetic aggregation of dissociated cells: A quantifiable approach to cell interactions *in vitro*, *Exp. Cell Res.* **22**:455.

Moskowitz, J., and Fain, J. N., 1970, Stimulation by growth hormone and dexamethasone of labelled cyclic adenosine 3′,5′-monophosphate accumulation by white fat cells, *J. Biol. Chem.* **245**:1101.

Muir, H., 1969, The structure and metabolism of mucopolysaccharides (glycosaminoglycans) and the problem of the mucopolysaccharidoses, *Am. J. Med.* **47**:673.

Mustafa, M., and Kamat, D. N., 1973, Mucopolysaccharide histochemistry of *Musca domestica*. VII. The brain, *Acta Histochem.* **45:**254.

Nakao, K., and Bashey, R. I., 1972, Fine structure of collagen fibrils as revealed by ruthenium red, *Exp. Mol. Pathol.* **17:**6.

Nanto, V., 1969, Electrophoretic analysis of acidic glycosaminoglycans and its application to the developing chick embryo, *Ann. Acad. Sci. Fenn. Ser. A5 (Medica)* **144:**1.

Nemeth-Csoka, M., 1970, Importance of sulphated acid mucopolysaccharides for fibrillogenesis in carrageenin granulomata of rats at different ages, *Exp. Gerontol.* **5:**67.

Neufeld, E. F., 1974, The biochemical basis for mucopolysaccharidoses and mucolipidoses, in: *Progress in Medical Genetics,* Vol. 10 (A. G. Steinberg and A. G. Bearn, eds.), pp. 81–101, Grune and Stratton, New York.

Nevis, A. H., and Collins, G. H., 1967, Electrical impedance and volume changes in brain during development, *Brain Res.* **5:**57.

Nevo, Z., and Dorfman, A., 1972, Stimulation of chondromucoprotein synthesis in chondrocytes by extracellular chondromucoprotein, *Proc. Natl. Acad. Sci. USA* **69:**2069.

Obrink, B., 1972, Isolation and partial characterization of a dermatan sulfate proteoglycan from pig skin, *Biochim. Biophys. Acta* **264:**354.

Obrink, B., 1973*a*, A study of the interactions between monomeric tropocollagen and glycosaminoglycans, *Eur. J. Biochem.* **33:**387.

Obrink, B., 1973*b*, The influence of glycosaminoglycans on the formation of fibers from monomeric tropocollagen *in vitro, Eur. J. Bioch.* **34:**129.

Ogston, A. G., 1966, On water binding, *Fed. Proc.* **25:**986.

Ogston, A. G., 1970, The biological functions of the glycosaminoglycans, in: *Chemistry and Molecular Biology of the Intercellular Matrix,* Vol. 3 (E. A. Balazs, ed.), pp. 1231–1240, Academic Press, New York.

Ogston, A. G., and Phelps, C. F., 1961, The partition of solutes between buffer solutions and solutions containing hyaluronic acid, *Biochem. J.* **78:**827.

Ogston, A. G., and Sherman, T. F., 1961, Effects of hyaluronic acid upon diffusion of solutes and flow of solvent, *J. Physiol. (London)* **156:**67.

Ogston, A. G., and Stanier, J. E., 1953, The physiological function of hyaluronic acid in synovial fluid; viscous, elastic and lubricant properties, *J. Physiol. (London)* **119:**244.

Ohya, T., and Kaneko, Y., 1970, Novel hyaluronidase from *Streptomyces, Biochim. Biophys. Acta* **198:**607.

Palmoski, M. J., and Goetinck, P. F., 1972, Synthesis of proteochondroitin sulfate by normal, nanomelic and 5-bromodeoxyuridine-treated chondrocytes in cell culture, *Proc. Natl. Acad. Sci. USA* **69:**3385.

Palmoski, M., Khosla, R., and Brandt, K., 1974, Small proteoglycans of cartilage: Confirmation of their presence by non-disruptive extraction, *Biochim. Biophys. Acta* **372:**171.

Pawelek, J. M., 1969, Effects of thyroxine and low oxygen tension on chondrogenic expression in cell culture, *Dev. Biol.* **19:**52.

Pesetsky, I., 1966, The role of the thyroid in the development of Mauthner's neuron: A karyometric study in thyroidectomized anuran larvae, *Z. Zellforsch.* **75:**138.

Pessac, B., and Defendi, V., 1972, Cell aggregation: Role of acid mucopolysaccharides, *Science* **175:**898.

Podrazky, V., Steven, F. S., Jackson, D. S., Weiss, J. B., and Leibovich, S. J., 1971, Interaction of tropocollagen with protein polysaccharide complexes: An analysis of the ionic groups responsible for interaction, *Biochim. Biophys. Acta* **229:**690.

Polansky, J., and Toole, B. P., 1975, unpublished results.
Polansky, J., Toole, B. P., and Gross, J., 1974, Brain hyaluronidase: Changes in activity during chick development, *Science* **183:**862.
Pratt, R. M., Larsen, M. A., and Johnston, M. C., 1975, Migration of cranial neural crest cells in a cell-free hyaluronate-rich matrix, *Dev. Biol.* **44:**298–305.
Praus, R., and Brettschneider, I., 1970, Presence of a non-sulphated glucosaminoglycan in embryonic cornea, *FEBS Lett.* **6:**221.
Praus, R., and Brettschneider, I., 1971, Glycosaminoglycans in the developing chicken cornea, *Ophthalmol. Res.* **2:**367.
Preston, B. N., and Snowden, J. M., 1972, Model connective tissue systems: The effect of proteoglycans on the diffusional behavior of small non-electrolytes and microions, *Biopolymers* **11:**1627.
Preston, B. N., Davies, M., and Ogston, A. G., 1965, The composition and physicochemical properties of hyaluronic acids prepared from ox synovial fluid and from a case of mesothelioma, *Biochem. J.* **96:**449.
Quintarelli, G., Vocaturo, A., Bellocci, M., Roden, L., Iffolito, E., and Baker, J. R., 1974, Preliminary ultrastructural demonstration of hyaluronic acid–proteoglycan interaction in cartilage matrix, *Am. J. Anat.* **140:**433.
Reid, T., and Flint, M. H., 1974, Changes in glycosaminoglycan content of healing rabbit tendon, *J. Embryol. Exp. Morphol.* **31:**489.
Rienits, K. G., 1960, The acid mucopolysaccharides of the sexual skin of apes and monkeys, *Biochem. J.* **74:**27.
Robinson, H. C., Telser, A., and Dorfman, A., 1966, Studies on biosynthesis of the linkage region of chondroitin sulfate–protein complex, *Proc. Natl. Acad. Sci.* **56:**1859.
Roblin, R., Albert, S. O., Gelb, N. A., and Black, P. H., 1975, Cell surface changes correlated with density-dependent growth inhibition: Glycosaminoglycan metabolism in 3T3, SV3T3, and Con A selected revertant cells, *Biochemistry* **14:**347.
Roden, L., 1970*a*, The structure and metabolism of the proteoglycans of chondroitin sulfates and keratan sulfate, in: *Chemistry and Molecular Biology of the Intercellular Matrix* Vol. 2 (E. A. Balazs, ed.), pp. 797–821, Academic Press, New York.
Roden, L., 1970*b*, Biosynthesis of acidic glycosaminoglycans (mucopolysaccharides), in: *Metabolic Conjugation and Metabolic Hydrolysis*, Vol. 2 (W. H. Fishman, ed.), pp. 345–442, Academic Press, New York.
Rosenberg, L., Hellmann, W., and Kleinschmidt, A. K., 1970, Macromolecular models of protein polysaccarides from bovine nasal cartilage based on electron microscopic studies, *J. Biol. Chem.* **245:**4123.
Saito, H. Yamagata, T., and Suzuki, S., 1968, Enzymatic methods for the determination of small quantities of isomeric chondroitin sulfates, *J. Biol. Chem.* **243:**1536.
Sajdera, S. W., and Hascall, V. C., 1969, Proteinpolysaccharide complex from bovine nasal cartilage: A comparison of low and high shear extraction procedures, *J. Biol. Chem.* **244:**77.
Sajdera, S. W., Hascall, V. C., Gregory, J. D., and Dziewiatkowski, D. D., 1970, The proteoglycans of bovine nasal cartilage: Structure of the aggregate, in: *Chemistry and Molecular Biology of the Intercellular Matrix,* Vol. 2 (E. A. Balazs, ed.), pp. 851–858, Academic Press, New York.
Samuels, H. H., and Tsai, J. S., 1973, Thyroid hormone action in cell culture: Demonstration of nuclear receptors in intact cells and isolated nuclei, *Proc. Natl. Acad. Sci.* **70:**3488.
Sapolsky, A. I., Howell, D. S., and Woessner, J. F., 1974, Neutral proteases and cathepsin D in human articular cartilage, *J. Clin. Invest.* **53:**1044.

Satoh, C., Duff, R., Rapp, F., and Davidson, E. A., 1973, Production of mucopolysaccharides by normal and transformed cells, *Proc. Natl. Acad. Sci.* **70**:54.

Schacter, L. P., 1970, Effect of conditioned media on differentiation in mass cultures of chick limb bud cells. I. Mophological effects, *Exp. Cell Res.* **63**:19.

Schiller, S., Slover, G. A., and Dorfman, A., 1962, Effect of the thyroid gland on metabolism of acid mucopolysaccharides in skin, *Biochim. Biophys. Acta* **58**:27.

Schubert, M., and Hamerman, D., 1964, The functioning of the diffuse macromolecules of joints, *Bull. Rheum. Dis.* **14**:345.

Schubert, M., and Hammerman, D., 1968, *A Primer on Connective Tissue Biochemistry*, Lea and Febiger, Philadelphia.

Schwartz, N. B., Roden, L., and Dorfman, A., 1974, Biosynthesis of chondroitin sulfate: Interaction between xylosyltransferase and galactosyltransferase, *Biochem. Biophys. Res. Commun.* **56**:717.

Schwartz, N. B., Dorfman, A., and Roden, L., 1975, Role of enzyme–enzyme interactions in the organization of multi–enzyme systems, in: *Extracellular Matrix Influences on Gene Expression* (H. C. Slavkin and R. Greulich, eds.), pp. 197–208, Academic Press, New York.

Scott, J. E., 1968, Ion binding in solutions containing acid mucopolysaccharides, in: *The Chemical Physiology of Mucopolysaccharides* (G. Quintarelli, ed.), pp. 171–187, Little, Brown, Boston,.

Searls, R. L., 1965, An autoradiographic study of the uptake of S^{35}-sulfate during the differentiation of limb bud cartilage, *Dev. Biol.* **11**:155.

Searls, R. L., and Janners, M. Y., 1969, The stabilization of cartilage properties in the cartilage-forming mesenchyme of the embryonic chick limb, *J. Exp. Zool.* **170**:365.

Serafini-Fracassini, A., Wells, P. J., and Smith, J. W., 1970, in: *Chemistry and Molecular Biology of the Intercellular Matrix,* Vol. 2 (E. A. Balazs, ed.), pp. 1201–1215, Academic Press, New York.

Sidman, R. L., and Rakic, P., 1973, Neuronal migration with special reference to developing human brain: A review, *Brain Res.* **62**:1.

Silbert, J. E., and DeLuca, S., 1970, Degradation of glycosaminoglycans by tadpole tissue: Differences in activity toward chondroitin 4-sulfate and chondroitin 6-sulfate, *J. Biol. Chem.* **245**:1506.

Silpananta, P., Dunstone, J. R., and Ogston, A. G., 1968, Fractionation of a hyaluronic acid preparation in a density gradient: Some properties of the hyaluronic acid, *Biochem. J.* **109**:43.

Singer, M., and Craven, L., 1948, The growth and morphogenesis of the regenerating forelimb of adult *Triturus* following denervation at various stages of development, *J. Exp. Zool.* **108**:279.

Singh, M., and Bacchawat, B. K., 1965, The distribution and variation with age of different uronic acid-containing mucopolysaccharides in brain, *J. Neurochem.* **12**:519.

Singh, M., and Bacchawat, B. K., 1968, Isolation and characterization of glycosaminoglycans in human brain of different age groups, *J. Neurochem.* **15**:249.

Singh, M., Chandrasekaran, E. V., Cherian, R., and Bacchawat, B. K., 1969, Isolation and characterization of glycosaminoglycans in brain of different species, *J. Neurochem.* **16**:1157.

Sissons, H. A., 1971, The growth of bone, in: *The Biochemistry and Physiology of Bone,* Vol. 3 (G. H. Bourne, ed.), pp. 145–180, Academic Press, New York.

Smith, G. N., Toole, B. P., and Gross, J., 1975, Hyaluronidase activity and glycosaminoglycan synthesis in the amputated newt limb: Comparison of denervating, nonregenerating limbs with regenerates, *Dev. Biol.* **43**:221–232.

Smith, J. W., and Frame, J., 1969, Observations on the collagen and protein–polysaccharide complex of rabbit corneal stroma, *J. Cell Sci.* **4**:421.

Solursh, M., Vaerewyck, S. A., and Reiter, R. S., 1974, Depression by hyaluronic acid of glycosaminoglycan synthesis by cultured chick embryo chondrocytes, *Dev. Biol.* **41**:233.

Steinberg, M. S., 1970, Does differential adhesion govern the self-assembly of tissue structure? Equilibrium configurations and the emergence of a hierarchy among populations of embryonic cells, *J. Exp. Zool.* **173**:395.

Stoolmiller, A. C., and Dorfman, A., 1969, The biosynthesis of hyaluronic acid by *Streptococcus, J. Biol. Chem.* **244**:236.

Sugiyama, K., 1972, Occurrence of mucopolysaccharides in the early development of the sea urchin embryo and its role in gastrulation, *Dev. Growth Differ.* **14**:63.

Suzuki, S., Kojima, K., and Utsumi, K. R., 1970, Production of sulfated mucopolysaccharides by established cell lines of fibroblastic and nonfibroblastic origin, *Biochim. Biophys. Acta* **222**:240.

Swann, D. A., 1969, Hyaluronic acid: Structure of the macromolecule in the connective tissue matrix, *Biochem. Biophys. Res. Commun.* **35**:571.

Szabo, M. M., and Roboz-Einstein, E., 1962, Acidic polysaccharides in the central nervous system, *Arch. Biochem. Biophys.* **98**:406.

Szirmai, J. A., 1966, Effect of steroid hormones on the glycosaminoglycans of target connective tissue, in: *The Amino Sugars,* Vol. 2B (R. W. Jeanloz, and E. A. Balazs, eds.), pp. 129–154, Academic Press, New York.

Telser, A., Robinson, H. C., and Dorfman, A., 1966, The biosynthesis of chondroitin sulfate, *Arch. Biochem. Biophys.* **116**:458.

Terry, A. H., and Culp. L. A., 1974, Substrate-attached glycoproteins from normal and virus-transformed cells, *Biochemistry* **13**:414.

Thanassi, N. M., and Newcombe, D. S., 1974, Cyclic AMP and Thyroid hormone: Inhibition of epiphyseal cartilage cyclic 3′,5′-nucleotide phosphodiesterase activity by L-triiodo-thyronine, *Proc. Soc. Exp. Biol. Med.* **147**:710.

Toole, B. P., 1972, Hyaluronate turnover during chondrogenesis in the developing chick limb and axial skeleton, *Dev. Biol.* **29**:321.

Toole, B. P., 1973*a*, Hyaluronate and hyaluronidase in morphogenesis and differentiation, *Am. Zool.* **13**:1061.

Toole, B. P., 1973*b*, Hyaluronate inhibition of chondrogenesis: Antagonism of thyroxine, growth hormone and calcitonin, *Science* **180**:302.

Toole, B. P., and Gross, J., 1971, The extracellular matrix of the regenerating newt limb: Synthesis and removal of hyaluronate prior to differentiation, *Dev. Biol.* **25**:57.

Toole, B. P., and Lowther, D. A., 1967, Precipitation of collagen fibrils *in vitro* by protein polysaccarides, *Biochem. Biophys. Commun.* **29**:515.

Toole, B. P., and Lowther, D. A., 1968*a*, Dermatan sulfate-protein: Isolation from and interaction with collagen, *Arch. Biochem. Biophys.* **128**:567.

Toole, B. P., and Lowther, D. A., 1968*b*, The effect of chondroitin sulfate-protein on the formation of collagen fibrils *in vitro, Biochem. J.* **109**:857.

Toole, B. P., and Trelstad, R. L., 1971, Hyaluronate production and removal during corneal development in the chick, *Dev. Biol.* **26**:28.

Toole, B. P., Jackson, G., and Gross, J., 1972, Hyaluronate in morphogenesis: Inhibition of chondrogenesis *in vitro, Proc. Natl. Acad. Sci.* **69**:1384.

Trampusch, H. A., and Harrebomee, A. E., 1965, Dedifferentiation, a prerequisite of

regeneration, in: *Regeneration in Animals and Related Problems* (V. Kiortsis and H. A. Trampusch, eds.), pp. 341–374, North-Holland, Amsterdam.
Trelstad, R. L., 1975, Collagen fibrillogenesis *in vitro* and *in vivo:* The existence of unique aggregates and the special state of the fibril end, in: *Extracellular Matrix Influences on Gene Expression* (H. C. Slavkin and R. Greulich, eds.), pp. 331–340, Academic Press, New York.
Trelstad, R. L., and Kang, A. H., 1974, Collagen heterogeneity in the avian eye: Lens, vitreous body, cornea and sclera, *Exp. Eye Res.* **18:**395.
Trelstad, R. L., Hayashi, K., and Toole, B. P., 1974, Epithelial collagens and glycosaminoglycans in the embryonic cornea: Macromolecular order and morphogenesis in the basement membrane, *J. Cell Biol.* **62:**815.
Tsiganos, C. P., Hardingham, T. E., and Muir, H., 1971, Proteoglycans of cartilage: An assessment of their structure, *Biochim. Biophys. Acta* **229:**529.
Vaes, G., 1967, Hyaluronidase activity in lysosomes of bone tissue, *Biochem. J.* **103:**802.
Van Harreveld, A., Crowell, J., and Malhotra, S. K., 1965, A study of extracellular space in central nervous tissue by freeze-substitution, *J. Cell Biol.* **25:**117.
Van Harreveld, A., Dafny, N., and Khattab, F. I., 1971, Effects of calcium on the electrical resistance and the extracellular space of cerebral cortex, *Exp. Neurol.* **31:**358.
Varma, R., Varma, R. S., Allen, W. S., and Wardi, A. H., 1974, On the carbohydrate–protein linkage group in vitreous humor hyaluronate, *Biochim. Biophys. Acta* **362:**584.
Wang, H. H., and Adey, W. R., 1969, Effects of cations and hyaluronidase on cerebral electrical impedance, *Exp. Neurol.* **25:**70.
Wasteson, A., Lindahl, V., and Hallen, A., 1972, Mode of degradation of the chondroitin sulphate proteoglycan in rat costal cartilage, *Biochem. J.* **130:**729.
Wasteson, A., Westermark, B., Lindahl, V., and Ponten, J., 1973, Aggregation of feline lymphoma cells by hyaluronic acid, *Int. J. Cancer,* **12:**169.
Watson, E. M., and Pierce, R. H., 1949, The mucopolysaccharide content of skin in localized (pretibial) myxedema, *Am. J. Clin. Pathol.* **19:**442.
Weinstein, H., Sachs, C. R., and Schubert, M., 1963, Proteinpolysaccharide in connective tissue: Inhibition of phase separation, *Science* **142:**1073.
Weiss, P., 1958, Cell contact, *Int. Rev. Cytol.* **7:**391.
Wessells, N. K., and Cohen, J. H., 1968, Effects of collagenase on developing epithelia *in vitro*: Lung, ureteric bud, and pancreas, *Dev. Biol.* **18:**294.
Wiebkin, O. W., and Muir, H., 1973, The inhibition of sulphate incorporation in isolated adult chondrocytes by hyaluronic acid, *FEBS Lett.* **37:**42.
Wiebkin, O. W., Hardingham, T. E., and Muir, H., 1975, Hyaluronic acid–proteoglycan interaction and the influence of hyaluronic acid on proteoglycan synthesis by chondrocytes from adult cartilage, in: *Extracellular Matrix Influences on Gene Expression* (H. C. Slavkin and R. Greulich, eds.), pp. 209–224, Academic Press, New York.
Wilkinson, J. F., 1958, The extracellular polysaccharides of bacteria, *Bacteriol. Rev.* **22:**46.
Williams, L. W., 1910, The somites of the chick, *Am. J. Anat.* **11:**55.
Young, I. J., and Custod, J. T., 1972, Isolation of glycosaminoglycans and variation with age in the feline brain, *J. Neurochem.* **19:**923.
Zugibe, F. T., 1962, The demonstration of the individual acid mucopolysaccharides in human aortas, coronary arteries and cerebral arteries. II. Identification and significance with aging, *J. Histochem. Cytochem.* **10:**448.
Zwilling, E., 1968, Morphogenetic phases in development, *Dev. Biol. Suppl.* **2:**184.

11

Cell Surface Carbohydrate-Binding Proteins: Role in Cell Recognition

SAMUEL H. BARONDES and STEVEN D. ROSEN

1. INTRODUCTION

The complexity of cellular associations in the nervous system suggests a high level of development of cellular recognition processes. To understand these processes, it will be necessary to isolate the specific molecules that mediate cellular recognition. This task might be particularly difficult in the nervous system, where cellular interactions appear to be so complicated. Therefore, studies with simpler biological systems seem preferable to lay the groundwork. Although additional mechanisms may be required to provide the graded specifications required for precise interneuronal interactions, it is likely that the general mechanisms employed in simpler systems will be conserved in some form.

Our understanding of the principles of cellular recognition is extremely limited. However, a number of ideas have been developed over the years based on a general understanding of molecular associations, the composition of cell surfaces, and studies of cell adhesion in various systems. These ideas can be summarized as follows: (1) There are discrete molecular entitites on the surface of cells which recognize specific molecules on the surface of certain other cells. (2) The recogni-

SAMUEL H. BARONDES and STEVEN D. ROSEN · Department of Psychiatry, University of California, San Diego, School of Medicine, La Jolla, California.

tion molecules on the surfaces of two interacting cells are probably complementary and associate in a noncovalent manner—i.e., the association is like that of an antigen to an antibody; this is often called the Tyler–Weiss hypothesis (Tyler, 1946; Weiss, 1947). (3) One or both members of the recognition pair are probably proteins because of the many specific recognition sites that can be created from proteins. (4) Carbohydrates available on cell surfaces in large amounts and in great diversity (Winzler, 1970; Rambourg, 1971; Nicolson, 1974) may be involved in cellular recognition processes; one possibility supported by studies in several systems (to be discussed later) is that cellular recognition may occur by specific association between an oligosaccharide receptor (in glycoprotein or glycolipid) on the surface of one cell with a protein specific for this oligosaccharide on the surface of another cell. (5) Some of the recognition molecules, although associated with membranes might be, in Singer's (1974) terminology, "peripheral proteins" that can be dissociated from the cell surface and isolated in aqueous solution without addition of detergents. The sponge aggregation factor (Humphreys, 1963; Moscona, 1963) would be in this category. Other components might be "integral proteins" that are embedded in the lipid bilayer and require detergents for solubilization.

The major purpose of this chapter is to describe recent studies of cellular association in cellular slime molds, which provide experimental support for each of these ideas. The major conclusion from this work is that cellular association in these organisms is mediated by the interaction between carbohydrate-binding proteins and specific oligosaccharide receptors that are both present on the cell surface. Preliminary evidence with several species of slime molds suggests that both the carbohydrate-binding proteins and the oligosaccharide receptors are species specific, providing a potential mechanism to account for the species-specific sorting out observed in slime molds (Raper and Thom, 1941; Shaffer, 1957; Bonner and Adams, 1958). We will then briefly review evidence from other experimental systems that suggests involvement of a mechanism of this type in several kinds of cellular interactions. Then we will consider the implications of this work for an analysis of the molecular basis of neuronal recognition.

2. ADVANTAGES OF CELLULAR SLIME MOLDS FOR STUDIES OF CELL RECOGNITION

The life cycle of cellular slime molds such as *Dictyostelium discoideum* has been described in detail (Bonner, 1967; Gerisch, 1968; Sussman, 1975; Loomis, 1975). The organisms may be found as unicellular amoebae. The cells consume bacteria and divide every 3 h. In this condition the cells are called *vegetative cells* and this part of their life cycle is called the *nonsocial phase*, since there is no tendency for the amoebae to associate with one another. As long as ample food is available, the amoebae remain in this state. However, when the food supply is exhausted the amoebae differentiate into an aggregation-competent state over the course of about 9–12 h after food deprivation. During this phase of development, a chemotactic system draws the cells together. They are then capable of forming stable intercellular contacts. A multicellular structure (pseudoplasmodium) containing up to 10^5 cells is assembled. In the next 12 h this aggregate undergoes further differentiation, culminating in the formation of a fruiting body with a spore cap that contains spores. The spores are in a dormant state and are resistant to inclement environments. However, if spores are exposed to a favorable environment they transform into amoebae, and a new life cycle is begun.

There are several advantages in using cellular slime molds for studies of specific cellular associations: (1) There are a number of species of cellular slime molds that have been shown to display species-specific cellular association—i.e., mixtures of cells from different species "sort out" into separate colonies each composed of slime mold cells of a single species (Raper and Thom, 1941; Shaffer, 1957; Bonner and Adams, 1958). Sorting out is presumably due, at least in part, to species-selective intercellular affinities. Thus this system provides the opportunity for studying the molecular basis of selective cellular cohesion. (2) A large number of identical cells can be raised in culture. Therefore, the problem of cellular heterogeneity which would complicate studies of cellular recognition in a complex tissue is eliminated. (3) The culture conditions are simple and resemble the natural environment of these organisms. Therefore, slime mold cells retain cellular recognition properties under defined conditions where cells from higher organisms might lose them. (4) The cells can be isolated in a noncohesive form (as vegetative cells) and at various stages during development of cohesiveness.

Beug, Gerisch, and colleagues (Beug *et al.*, 1970, 1971, 1973*a*,*b*) have demonstrated the usefulness of cellular slime molds for studies of cellular association. They have carried out elegant immunological studies comparing vegetative slime mold amoebae with cohesive cells. They raised high-affinity antibodies to cohesive *D. discoideum* cells in rabbits and degraded the resultant immunoglobulins to univalent antibodies (Fab fragments). After adsorption with vegetative cells, the unadsorbed Fab fragments were shown to block cohesion of *D.*

discoideum. This result indicates that new antigens have appeared on the surface of cohesive slime mold cells and that these antigens, termed *contact sites*, presumably mediate cell cohesion. The contact sites were shown to be species specific with respect to *Polysphondylium pallidum*. Because the antibodies specific for the cohesive cells are presumably heterogeneous, it would be difficult to use them as a tool for isolation of the specific substances involved in cohesion. Despite this limitation, the work demonstrated that discrete substances (identified as antigens) appear on the cell surface as they become cohesive and that some of these substances play a role in cohesiveness.

3. DEVELOPMENTALLY REGULATED CARBOHYDRATE-BINDING PROTEINS IN CELLULAR SLIME MOLDS

3.1. *Parallel Appearance of Carbohydrate-Binding Proteins and Cohesiveness*

A possible approach to the mechanism of cell cohesion in cellular slime molds was suggested by the finding (Rosen, 1972) that extracts of cohesive *D. discoideum* cells contained a substance that agglutinated formalinized erythrocytes that had been treated with tannic acid. Appearance of this agglutinin in soluble extracts of two slime mold species, *D. discoideum* (Rosen *et al.*, 1973) and *P. pallidum* (Rosen *et al.*, 1974), paralleled the differentiation of these cells from a vegetative to a cohesive form (Fig. 1). It was subsequently shown that it is not necessary to treat the erythrocytes with either tannic acid or formaldehyde to demonstrate this agglutination activity, but generally we use formalinized erythrocytes of the appropriate type for these agglutination assays since they are a more stable and thus more convenient reagent than fresh erythrocytes. Appearance of the agglutinin and development of cell cohesiveness were blocked by cycloheximide, a protein synthesis inhibitor (Rosen *et al.*, 1973).

Having demonstrated these developmentally regulated agglutinins, we then sought to determine their mode of action. Were these simply nonspecific "sticky" substances or were they substances whose agglutination activity could be explained? An explanation was suggested by the finding that agglutination activity could be blocked by specific sugars. For example, the agglutination of formalinized sheep erythrocytes by *D. discoideum* agglutinin was inhibited by *N*-acetyl-D-galactosamine (Table I). Several other monosaccharides, including D-galactose, also inhibited agglutination but were less potent. Still others

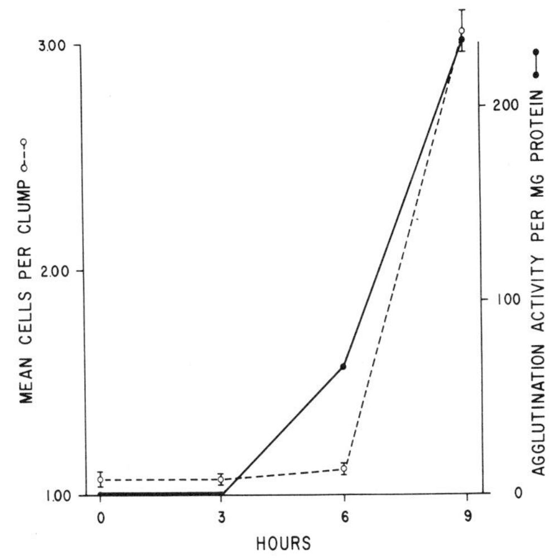

Fig. 1. Development of cohesiveness and agglutination activity with differentiation of *D. discoideum*. Vegetative *D. discoideum* cells were separated from bacteria at time 0 and differentiated on Millipore pads. At the indicated times, cells were assayed for cohesiveness (mean cells per clump) with a quantitative cohesiveness assay and aliquots were extracted and assayed for agglutination activity with formalinized sheep erythrocytes. For details, see Rosen *et al.* (1973).

(e.g., *N*-acetyl-D-glucosamine) were totally inactive when tested at concentrations up to 150 mM. Agglutination activity of extracts of *P. pallidum* was also inhibited by specific saccharides (Table I), but the relative potency of the saccharides as inhibitors was different from that observed with the agglutinin from *D. discoideum* (Rosen *et al.*, 1974). These findings suggested that the agglutinins were like the carbohydrate-binding proteins called lectins (Lis and Sharon, 1973) isolated from a wide variety of plant and animal sources, which agglutinate erythrocytes and other cell types. Like the plant lectins, the slime mold agglutinins are apparently multivalent carbohydrate-binding proteins. They cause agglutination by binding to oligosaccharides on the surfaces of adjacent cells. Synthesis of these slime mold agglutinins as the cells became cohesive suggested that they might function to agglutinate slime mold cells by a similar mechanism. Evidence for this will be considered later.

3.2. *Purification of Discoidin and Pallidin*

Because of its affinity for D-galactose, the agglutinin from *D. discoideum* was adsorbed by a Sepharose column (made up of a polymer of

Table I. Sugar Effects on Agglutination of Erythrocytes by Agglutinins from P. pallidum and D. discoideum[a]

Sugar	Sugar concentration (mM) for 50% inhibition of agglutination	
	P. pallidum agglutinin	*D. discoideum* agglutinin
Lactose	1.6	12.5
α-Methyl-D-galactose	6.2	6.2
D-Galactose	6.2	25
N-Acetyl-D-galactosamine	25	1.6
D-Fucose	25	3.1
L-Fucose	50	12.5
3-*O*-Methyl-D-glucose	100	1.6
α-Methyl-D-glucose	≥100	12.5
β-Methyl-D-glucose	≥100	100
D-Glucose	≥100	≥100
D-Glucosamine	≥100	≥100
N-Acetyl-D-glucosamine	≥100	≥100
D-Mannose	≥100	≥100
α-Methyl-D-mannose	≥100	≥100

[a] Agglutination of formalinized human erythrocytes by serial dilutions of slime mold agglutinins and the sugar concentration that produced 50% inhibition of agglutination were determined as described in Rosen *et al.* (1974).

alternating D-galactose and 3,6-anhydro-L-galactose) whereas other proteins in the extract were not adsorbed. The adsorbed agglutinin could then be eluted from the column with D-galactose (Fig. 2). The purified protein, named *discoidin* (Simpson *et al.*, 1974), was subsequently resolved into two proteins (discoidin I and discoidin II) by chromatography on DEAE-cellulose (Frazier *et al.*, 1975). Discoidin I, the major protein, is a tetramer with subunit molecular weight of about 26,000 and discoidin II is a tetramer with subunit molecular weight of about 24,000. Discoidin I and discoidin II differ in amino acid composition, peptide maps, and their relative agglutination activity, when tested with sheep or rabbit erythrocytes (Frazier *et al.*, 1975). They also differ in the relative efficacy of different carbohydrates in blocking their agglutination activity. They may play different roles in the differentiation of *D. discoideum* since they show a different pattern of developmental regulation. Both are absent in vegetative cells, but discoidin II approaches its maximum level at 9 h after food deprivation, whereas the level of discoidin I increases considerably between 9 and 12 h (Frazier *et al.*, 1975), at which time it is about 9 times as abundant as discoidin II.

Purification of the agglutinin from *P. pallidum* was achieved by adsorption of the agglutinin from a crude extract with formalinized human erythroyctes and elution of the agglutinin with D-galactose (Rosen *et al.*, 1974; Simpson *et al.*, 1975). The purified protein, named *pallidin*, contained only one subunit on polyacrylamide gel electrophoresis in sodium dodecylsulfate (Fig. 3). Its subunit molecular weight is about 25,000 daltons and the weight average molecular weight of pallidin is about 250,000 daltons, suggesting that it is a decamer. Pallidin has a different isoelectric point, amino acid composition, and peptide map than discoidin I or II. The relative potency of monosaccharides and disaccharides in inhibiting pallidin's erythrocyte agglutination activity is also different from that found with discoidins I and II.

Both discoidin and pallidin are prominent cell proteins. They each comprise about 1–2% of the total soluble protein of cohesive cells. We estimate that there are about 2×10^6 molecules of discoidin (I and II) per cohesive *D. discoideum* cell and a similar number of pallidin molecules per cohesive *P. pallidum* cell. Neither purified discoidin nor purified pallidin contains detectable covalently bound carbohydrate (Simpson *et al.*, 1974, 1975).

4. CELL SURFACE LOCATION OF DISCOIDIN AND PALLIDIN

If discoidin and pallidin are involved in cell cohesion, they must be present on the cell surface. Two types of experiments indicate that these agglutinins are indeed present on the surface of cohesive cells: (1)

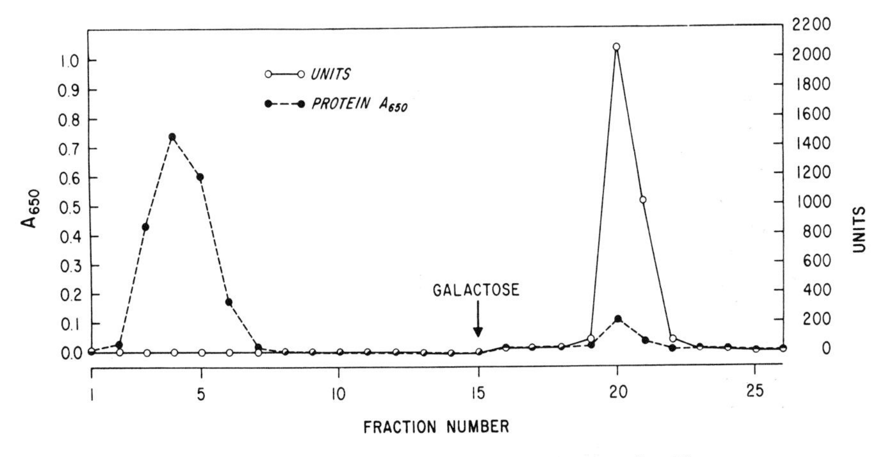

Fig. 2. Discoidin purification by affinity chromatography. *D. discoideum* extracts were applied to a Sepharose 4B column. All the agglutination activity bound to the column and was eluted by washing with 0.3 M galactose. For details, see Simpson *et al.* (1974).

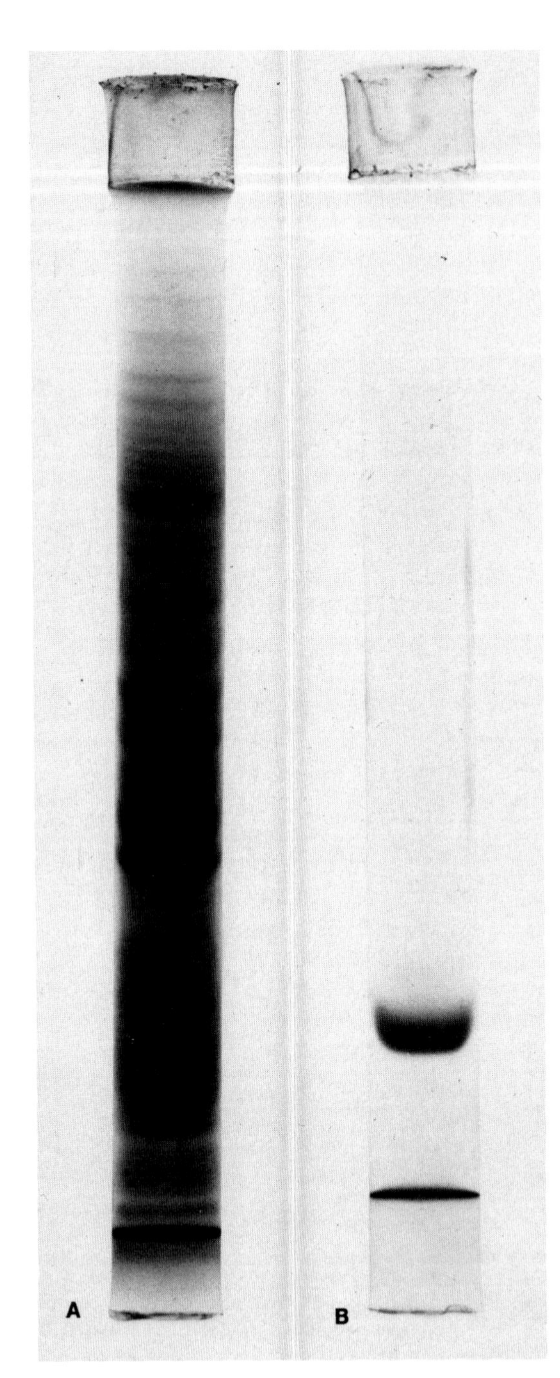

Fig. 3. Polyacrylamide gel electrophoresis in a sodium dodecylsulfate system of (A) a soluble extract of cohesive *P. pallidum* cells and (B) pallidin (the broad single band) that had been purified by adsorption of the soluble extract to formalinized erythrocytes followed by elution of the adsorbed protein with galactose. The fine band near the bottom of the gel is cytochrome C which was added as an internal standard. For details, see Simpson *et al.* (1975).

cohesive cells form rosettes with erythrocytes which can be blocked by appropriate sugars (Rosen *et al.*, 1973, 1974) and (2) antibodies against purified discoidin or pallidin bind to the cell surface (Chang *et al.*, 1975, and unpublished).

For the rosette experiments, formalinized sheep erythrocytes were mixed with slime mold cells. The erythrocytes showed little binding to vegetative *D. discoideum* cells but substantially more binding to cohesive *D. discoideum* cells (Rosen *et al.*, 1973). The binding of erythrocytes to cohesive *D. discoideum* cells is specific since it could be blocked by *N*-acetyl-D-galactosamine, a specific sugar that interacts with purified discoidin but not by a nonspecific sugar, such as *N*-acetyl-D-glucosamine (Rosen *et al.*, 1973). Similarly, with cohesive *P. pallidum* cells rosette formation could be blocked with lactose or D-galactose but not with D-glucose. Scanning electron micrographs of the association of formalinized sheep erythrocytes with cohesive *P. pallidum* cells and the effects of galactose and glucose are shown in Fig. 4.

Using antibodies to purified discoidin or pallidin, we demonstrated the location of these proteins on the surface of cohesive slime mold cells using both immunofluorescent and immunoferritin labeling techniques (Chang *et al.*, 1975, and unpublished). For example, vegetative *D. discoideum* cells incubated with rabbit antibody to discoidin followed by reaction with fluorescein-labeled goat anti-rabbit immunoglobulin showed no fluorescence, whereas cohesive cells showed substantial fluorescence. This result was found whether these cells were examined in a living state after differentiation in suspension culture or after fixation *in situ* following differentiation on a glass slide. Gentle fixation conditions were used to preserve the reactivity of discoidin with antibodies. When cells differentiated for 12 h on glass and then fixed were reacted with the antibodies, fluorescent staining was diffusely distributed over the surface of the cells with no preferential localization at sites of intercellular contact (Fig. 5A). When the cells were differentiated in suspension culture, they had a round shape. When reacted with antibodies as above, these cells initially showed diffuse fluorescent antibody staining over their surface. If the cells were kept at room temperature after reaction with antibody, patches of surface antigen were found. Eventually, one fluorescent cap formed per cell (Fig. 5B), indicating that the agglutinin can move laterally in the plane of the membrane, as has been observed with some surface components in other cell types (Nicolson, 1974). When cohesive cells fixed *in situ* were reacted with rabbit anti-discoidin immunoglobulin followed by ferritin-labeled goat anti-rabbit immunoglobulin, cell surface location of discoidin was again shown (Fig. 5C). Similar studies (Chang *et al.*, unpublished) with pallidin show cell surface location of

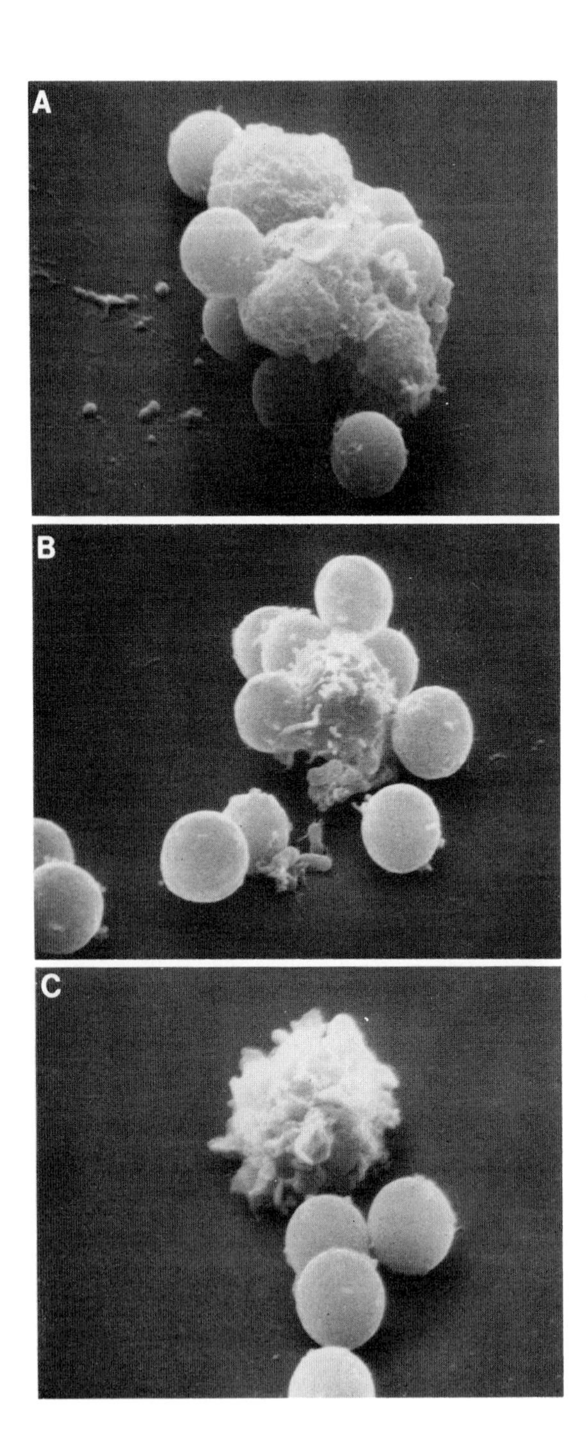
A
B
C

this protein. In addition, when the glutaraldehyde-fixed cells were treated with acetone to render the surface membranes permeable to immunoglobulins, considerable pallidin was observed intracellularly. Based on this and other evidence, it appears that only a small percentage of pallidin and discoidin is present on the cell surface.

5. DEVELOPMENTAL REGULATION OF CELL SURFACE RECEPTORS FOR DISCOIDIN AND PALLIDIN

For slime mold agglutinins to act as cellular cohesion substances, it is necessary not only that the agglutinins be present on the cell surface but also that the cell surface contain receptors for these carbohydrate-binding proteins. To evaluate this, we prepared glutaraldehyde-fixed *D. discoideum* cells at a number of stages in differentiation and determined the agglutinability of these cells with discoidin, pallidin, and several lectins purified from plants. The glutaraldehyde fixation conditions used here were sufficient to inactivate the endogenous carbohydrate-binding protein but have no effect on oligosaccharides. We found that high concentrations of discoidin I or discoidin II were required to agglutinate fixed vegetative *D. discoideum* cells but that the concentration required for agglutination diminished very markedly as the cells became cohesive (Reitherman *et al.*, 1975). Cells fixed 3 h after food deprivation were slightly agglutinable, those fixed 6 h after food deprivation were more agglutinable, and those fixed 9 h after food deprivation required a much smaller amount of discoidin I or II to produce agglutination. The amount of discoidin I or II required to produce agglutination of fixed slime mold cells at various stages of differentiation is shown in Fig. 6. The increased agglutinability of these cells with differentiation paralleled the endogenous synthesis of discoidin by these cells and the development of cohesiveness (Fig. 7). The agglutinability of fixed *D. discoideum* cells by pallidin also increased with differentiation, but the level of agglutination was less than with discoidin I or II (Reitherman *et al.*, 1975). Studies with wheat germ agglutinin (WGA), a plant lectin that reacts with *N*-acetyl-D-glucosamine residues (Lis and Sharon, 1973), indicate a specificity in the increased agglutinability

Fig. 4. Scanning electron micrographs of living cohesive *P. pallidum* cells that were mixed with formalinized sheep erythrocytes in the presence and absence of sugars. (A) In the absence of sugar, the sheep erythrocytes bound to the surface of the *P. pallidum* cells. (B) Addition of 0.15 M D-glucose had little or no effect on this binding. (C) Addition of D-galactose blocked the binding of erythrocytes to the surface of the slime mold cells. This experiment was conducted in collaboration with C.-M. Chang.

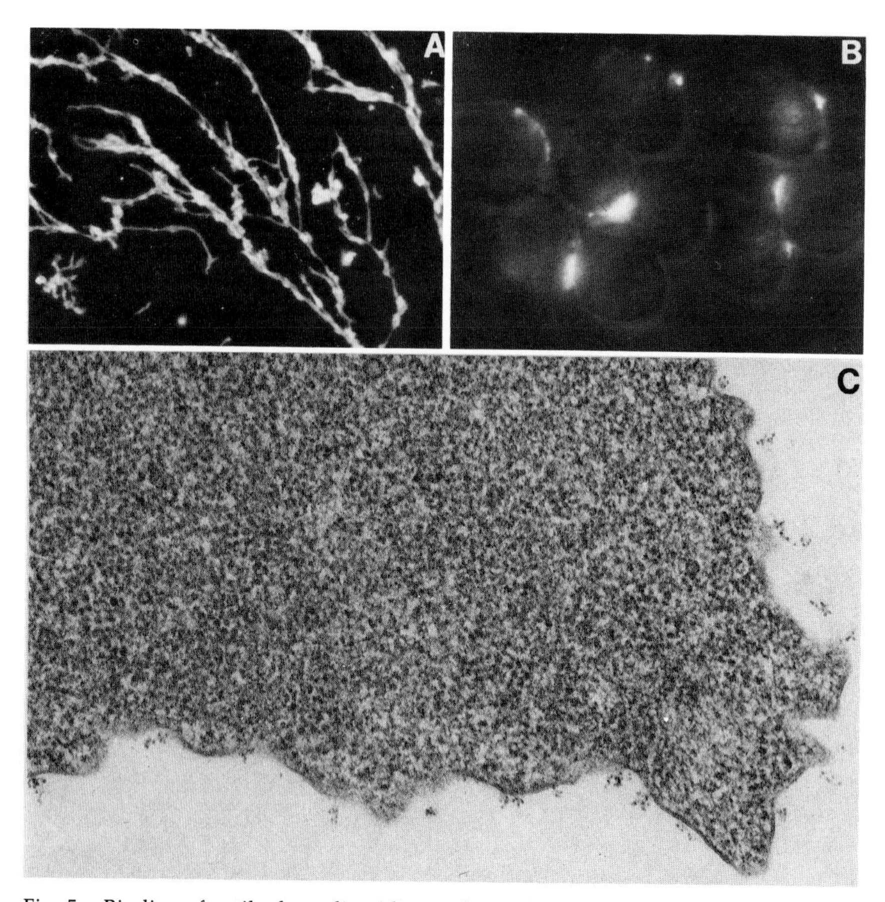

Fig. 5. Binding of antibody to discoidin on the surface of cohesive *D. discoideum* cells. (A) Streams of *D. discoideum* cells differentiated on a glass surface were fixed *in situ* by brief treatment with glutaraldehyde and reacted first with an antibody to discoidin raised in rabbits followed by fluorescein-conjugated goat anti-rabbit immunoglobulin. Fluorescent labeling is observed over the entire surface of the streaming cells. No staining was observed when control rabbit serum was used in place of antibody. (B) *D. discoideum* cells were differentiated in suspension in the absence of bacteria, reacted sequentially with the two antibodies as in (A), and then maintained at room temperature for 3 h. Note the fluorescent caps indicating mobility of the antigen (discoidin) in the plane of the cell surface membrane. (C) Electron micrograph of a cohesive *D. discoideum* cell fixed briefly with glutaraldehyde followed by reaction with antibody to discoidin and ferritin-conjugated goat anti-rabbit immunoglobulin. Note the ferritin labeling at the surface of the cell. For details of all these experiments, see Chang *et al.* (1975).

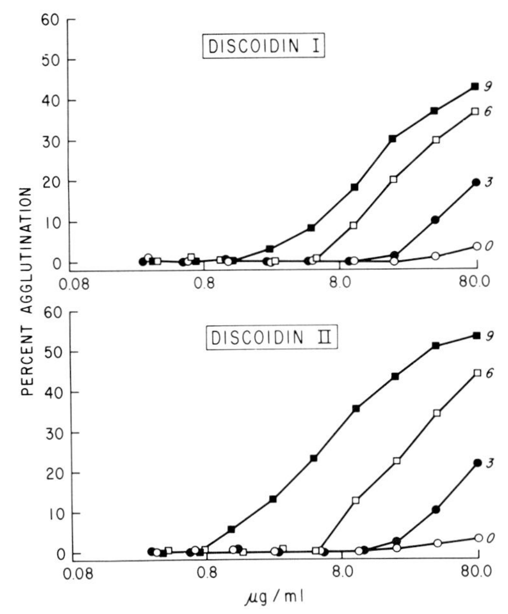

Fig. 6. Increased agglutinability of *D. discoideum* cells by discoidins I and II with differentiation. *D. discoideum* cells were separated from bacteria at time 0 and then differentiated for up to 9 h in suspension. At the indicated times, cells were fixed with glutaraldehyde and their agglutinability by discoidins I and II was determined using a quantitative agglutination assay. Effects of a series of concentrations of discoidins I and II on agglutination of the cells at different stages of differentiation are shown. For details, see Reitherman *et al.* (1975).

with differentiation. In contrast to the slime mold lectins, the concentration of WGA required to agglutinate glutaraldehyde-fixed vegetative *D. discoideum* cells was considerably *smaller* than that required to agglutinate fixed cohesive *D. discoideum* cells (Reitherman *et al.*, 1975).

Unlike *D. discoideum* cells, *P. pallidum* cells that were fixed with glutaraldehyde were not agglutinated by pallidin. However, *P. pallidum* cells that were heat-treated (60°C for 10 min) were agglutinated by added pallidin (Rosen *et al.*, 1974) (Fig. 8). Pallidin markedly enhanced the agglutination of the cells above the endogenous cohesiveness that survived heat treatment. Both the residual cohesiveness and the agglutination produced by added pallidin were inhibited by D-galactose but not by the nonspecific sugar D-glucose. Thus *P. pallidum* cells, like *D. discoideum* cells, contain cell surface receptors for their respective carbohydrate-binding proteins.

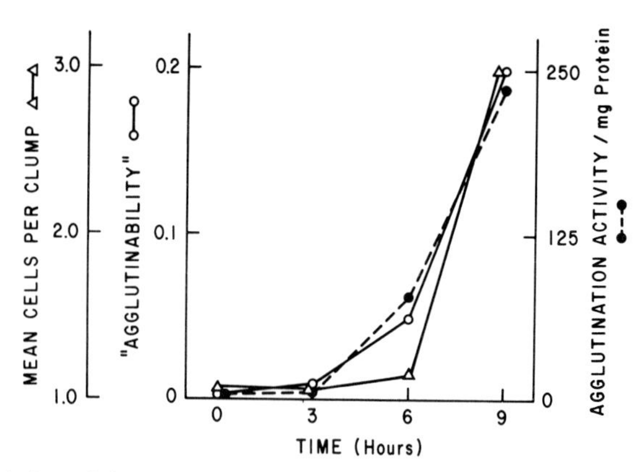

Fig. 7. Correlation of changes in agglutinability, cohesiveness, and carbohydrate-binding protein with differentiation of *D. discoideum* cells. *D. discoideum* cells were differentiated for the indicated times, and at each time cohesiveness of the living cells, levels of agglutination activity (discoidin), and agglutinability of cells by discoidin after glutaraldehyde fixation are shown. The data on cohesiveness (cells per clump) and agglutination activity are the same as those shown in Fig. 1. The data on agglutinability were derived from the experiment shown in Fig. 6 in which we plotted the relative concentrations of discoidin I required to produce 20% agglutination of the fixed *D. discoideum* cells under the assay conditions.

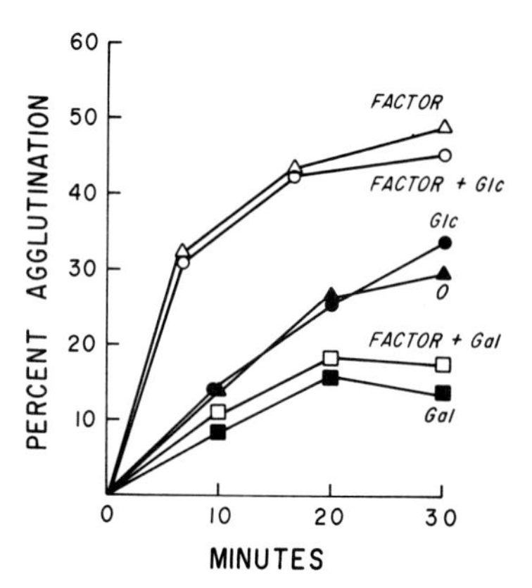

Fig. 8. Agglutination of heat-treated (60°C for 10 min) *P. pallidum* cells by pallidin. Heat-treated cohesive *P. pallidum* cells were gyrated for indicated times in the presence or absence of pallidin (designated *FACTOR* in the figure), 0.2 M D-glucose (Glc), or 0.2 M D-galactose (Gal). The experiments in which no factor was added are shown by closed symbols and the experiments with added factor are designated by open symbols. Note that with this degree of heat treatment the cells remain somewhat cohesive since they agglutinate in the absence of added pallidin. This endogenous cohesiveness is inhibited by galactose but not glucose. Agglutination is markedly augmented by addition of pallidin and this augmented agglutination is also inhibited by galactose but not by glucose. For details, see Rosen *et al.* (1974).

6. BLOCKING OF SLIME MOLD COHESION BY SPECIFIC SUGARS

Given the presence of both carbohydrate-binding proteins and receptors on the surface of cohesive cells, it still remains to be shown that the interaction of these substances mediates cell cohesion. If the interaction of the carbohydrate-binding proteins and the receptors mediates cell cohesion, then cohesion should be blocked by appropriate saccharides. This result would occur only if saccharides were added in concentrations sufficient to block the high affinity of carbohydrate-binding proteins for cell surface receptors, as will be discussed below. This condition has been successfully met in studies with cohesive *P. pallidum* cells (Rosen *et al.*, 1974). When these cells were gyrated in a suspension, they aggregated. The formation of aggregates was determined quantitatively using a Coulter particle counter by determining the disappearance of single cells from a gyrated suspension. Addition of lactose or D-galactose, sugars which have a relatively high affinity for pallidin when compared with other simple sugars, blocked the cohesion of heat-treated or living *P. pallidum* cells in this assay at substantially lower concentrations than D-mannose or D-glucose, which have relatively low affinity for pallidin (Fig. 9). These studies indicate that interaction between pallidin and oligosaccharide receptors on the surface of cohesive *P. pallidum* cells mediates the agglutination of these cells as determined in this assay. Presumably, a similar reaction plays a role in aggregation under natural conditions. In similar studies with *D. discoideum* cells (Rosen *et al.*, 1973), very high concentrations of *N*-acetyl-D-galactosamine or D-galactose were required to block aggregation of cohesive cells, and at these concentrations the osmotic effect of the sugars (which cause cell shrinkage) might have been responsible, since D-glucose was also effective. However, recent unpublished experiments with an axenically grown mutant of *D. discoideum* suggest that cohesion of living cells of the mutant can be blocked with much lower concentrations of specific sugars, whereas similar concentrations of other sugars are ineffective.

7. SPECIFICITY OF CARBOHYDRATE-BINDING PROTEINS FROM OTHER SPECIES OF CELLULAR SLIME MOLDS

As mentioned above, when slime mold cells of two different species are mixed and allowed to differentiate, the species segregate (Raper and Thom, 1941; Shaffer, 1957; Bonner and Adams, 1958). Whereas all the details of segregation have not yet been analyzed, the general

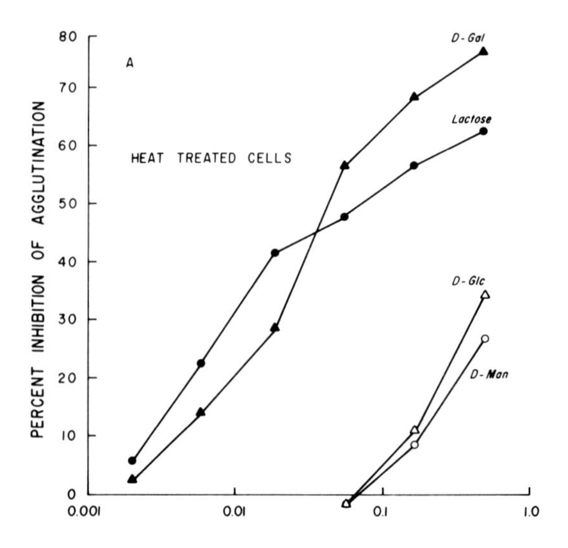

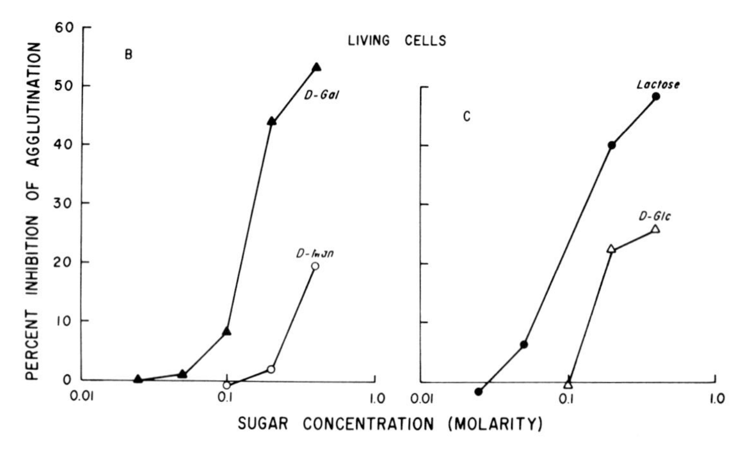

Fig. 9. Effect of sugars on agglutination of (A) heat-treated (51°C for 10 min) or (B, C) normal *P. pallidum* cells. Both the heat-treated and living cells were cohesive and could be agglutinated by gyration. Agglutination was monitored with a quantitative assay. Results show the concentration of sugars required to inhibit agglutination of both heat-treated and living cells. Note that D-galactose and lactose, which interact well with pallidin, are much more potent inhibitors than D-glucose or D-mannose. Note also that the concentrations of sugars required to inhibit agglutination of the living cells are higher than those required to produce comparable inhibition with the heat-treated cells. Purified pallidin retains about 50% of its erythrocyte agglutination activity when heated at 51°C for 10 min. For details, see Rosen *et al.* (1974).

result is that fruiting bodies contain cells of only one species. Do the interactions of specific carbohydrate-binding proteins and specific oligosaccharide receptors of each species play a role in species-specific sorting out? The studies reported already indicate that pallidin and discoidin are different not only in several physical-chemical properties but also in their interaction with a series of simple sugars. The differences in the carbohydrate-binding of pallidin and discoidin could mediate species-specific sorting out of these two species. To extend our studies, we determined that four other species of cellular slime molds also contain agglutinins (Rosen *et al.*, 1975). Using both semiquantitative and quantitative agglutination assays, we determined the relative efficacy of a large number of simple sugars and derivatives as inhibitors of hemagglutination by these six agglutinins. The complete results of these studies have been presented in detail (Rosen *et al.*, 1975). Some critical information is summarized in Table II. Using this test, we found that the agglutinins from the six species of cellular slime mold were readily discriminable, with the exception of those of *D. discoideum* and *P. violaceum*, which show only slight differences with the sugars tested. In general, then, the carbohydrate-binding specificity of the carbohydrate-binding proteins from these species is sufficiently different so that it might be a critical factor underlying species-specific sorting out. Other factors may also play a major role in sorting out. For example, it is known that the chemotactic signal for aggregation in *D. discoideum* and several other Dictyosteliaceae is cyclic AMP (Bonner *et al.*, 1972). In contrast, the chemotactic

Table II. Relative Concentrations of Sugars Required to Inhibit Erythrocyte Agglutination Activity from Six Slime Mold Species[a]

Sugar	Relative sugar concentration for 50% inhibition (lactose = 1)					
	D. purpureum	*D. mucoroides*	*D. discoideum*	*D. rosarium*	*P. violaceum*	*P. pallidum*
Me-α-D-Glc	200	4.6	15	≥21	13	≥25
Me-α-D-Gal	18	2.6	2.0	2.3	3.3	3.5
Me-β-D-Gal	5.0	3.1	0.74	1.3	1	4.1
D-GalNAc	20	0.71	0.15	0.75	0.33	≥25
Lac	1	1	1	1	1	1

[a] Agglutination activity in the presence and absence of a series of sugar concentrations was determined with fresh erythrocytes using a quantitative hemagglutination assay. For details, see Rosen *et al.* (1975).

signal in *P. pallidum* is apparently not cyclic AMP (Bonner *et al.*, 1972). Differences in chemotactic signaling could obviously contribute to sorting out. The different rates of differentiation of slime mold species upon food deprivation may also play a role in this process.

8. SPECIES SPECIFICITY OF CELL SURFACE RECEPTORS ON *P. PALLIDUM* AND *D. DISCOIDEUM*

In order for cell surface carbohydrate-binding proteins and oligosaccharide receptors to mediate species-specific sorting out, it is necessary not only that the carbohydrate-binding proteins be different but also that there be discriminable receptors on the cell surface of each species with a higher affinity for the carbohydrate-binding protein of the same species. To test this prediction, we determined the affinity of pallidin, discoidin I, and discoidin II for receptors on the surface of glutaraldehyde-fixed differentiated *D. discoideum* and *P. pallidum* cells. We also determined the affinity of these lectins for the surface of glutaraldehyde-fixed vegetative *D. discoideum* cells and glutaraldehyde-fixed erythrocytes. In addition, we determined the affinity of a plant lectin, concanavalin A, for glutaraldehyde-fixed cohesive *D. discoideum* cells. All these studies were done by adsorbing solutions containing a fixed concentration of the carbohydrate-binding protein with varying concentrations of glutaraldehyde-fixed cells. By determining the residual carbohydrate-binding protein after adsorption in the supernatant using a quantitative hemagglutination assay, we determined the amount of agglutinin bound by various concentrations of the cell type (Reitherman *et al.*, 1975). Given this information, the association constant of each carbohydrate-binding protein for each of the cell types and also the number of receptor sites could be calculated (Reitherman *et al.*, 1975).

Fixed cohesive (9 h) *D. discoideum* cells bound discoidin I or discoidin II with high affinity. In both cases, the association constant was in the range of 10^9 M^{-1} (Table III). In contrast, the association constant of these lectins for fixed cohesive *P. pallidum* cells was an order of magnitude lower. With pallidin, the association constant for fixed differentiated *P. pallidum* cells was in the range of 4×10^9 M^{-1} and an order of magnitude lower for fixed differentiated *D. discoideum* cells, indicating clear species specificity. Specificity was also shown by the finding that the association constant of discodin I for fixed rabbit erythrocytes was 3 orders of magnitude lower than that for fixed differentiated *D. discoideum* cells and that the association constant of concanavalin A for fixed

Table III. Association Constants and Number of Receptors per Cell[a]

Cell type	Discoidin I	Discoidin II	Pallidin	Concanavalin A
		$K_\alpha(M^{-1})$		
D.d. (0)	5×10^7	6.5×10^7	1.3×10^8	
D.d. (dif.)	1.3×10^9	1.5×10^9	5.0×10^8	1.8×10^6
P.p. (dif.)	1.6×10^8	1.9×10^8	3.8×10^9	
Rabbit RBC	1.3×10^6			
		Binding sites per cell		
D.d. (0)	3×10^5	3×10^5	1.4×10^5	
D.d. (dif.)	5×10^5	4×10^5	2.7×10^5	4×10^5
P.p. (dif.)	3×10^5	3×10^5	3.2×10^5	
Rabbit RBC	9×10^5			

[a] All cells were fixed with glutaraldehyde. D.d. (0), vegetative *D. discoideum* cells; D.d. (dif.) and P.p. (dif.), differentiated, cohesive *D. discoideum* and *P. pallidum* cells. For details, see Reitherman *et al.* (1975).

differentiated *D. discoideum* cells was 3 orders of magnitude lower than the association constant of discoidin I or II for these cells. The association constant of discoidin I or II for fixed vegetative *D. discoideum* cells was about twentyfold lower than for fixed cohesive *D. discoideum* cells, confirming our conclusion (from the developmental study of *D. discoideum* agglutinability by discoidin) that the receptors are developmentally regulated. These studies indicate that as slime mold cells become cohesive, high-affinity receptors appear on their surface, and these receptors have a higher affinity for the carbohydrate-binding protein from the same species than for that from another species of slime mold. The generality of these findings is under investigation.

9. EVIDENCE FOR CELL SURFACE PROTEIN–CARBOHYDRATE INTERACTIONS IN OTHER SYSTEMS

All our findings are consistent with the hypothesis that in differentiation from a vegetative to a cohesive state the cell surface of cellular slime molds is modified by addition of cell surface species-specific carbohydrate-binding proteins as well as molecules that contain species-specific oligosaccharide receptors. The evidence presently available provides strong support for the hypothesis that cellular recognition in cellular slime molds is mediated by interaction of these substances. This general class of interactions also appears to mediate other types of cell–cell associations.

Specifically, interactions of this type involving viruses, bacteria, and unicellular amoebae have been observed. Thus it has been known for many years that interaction of influenza virus with erythrocytes or other eukaryotic cells is mediated by the association of a viral protein with sialic acid residues on the cell surface (Schulze, 1973; Laver, 1973). Many other examples of carbohydrate-binding components on the surface of viruses are known. These components apparently mediate viral–host cell attachment. Evidence has been presented that the symbiotic association of specific nitrogen-fixing bacteria with specific plants is due to the interaction of lectins on the roots with oligosaccharides of the bacterial cell surface (Hamblin and Kent, 1973; Bohlool and Schmidt, 1974). In other cases, carbohydrate-binding proteins present on bacteria associate with oligosaccharides on the surface of cells such as erythrocytes (Gesner and Thomas, 1966) and intestinal mucosal cells (Gibbons *et al.*, 1975), leading to infection of the cells. Finally, unicellular *Acanthamoeba castelanii* apparently contains a carbohydrate binding protein on its cell surface that mediates association with specific other cells such as yeast and fungi and facilitates phagocytosis (Brown *et al.*, 1975).

These results indicate that interaction between cell surface carbohydrate-binding proteins and oligosaccharide receptors is a common biological phenomenon that extends from viruses to bacteria to simple eukaryotes. The major physiological functions identified so far include microbe–host cell association, symbiosis, and phagocytosis. Our results with slime molds suggest that carbohydrate-binding proteins on cell surfaces also have been adapted for an additional significant biological process—cell recognition in the construction of a homotypic multicellular organism. We suggest that the proposed mechanism of cell adhesion in the slime molds involving specific carbohydrate binding–oligosaccharide interactions on cell surfaces may also have been adapted for cellular recognition phenomena in higher systems. This view is an extension of previous speculations (reviewed in Barondes, 1970).

10. IMPLICATIONS FOR NEURONAL RECOGNITION

If indeed it can be proved that species-specific cellular recognition and cohesion in cellular slime molds are mediated by the interaction between cell surface carbohydrate-binding proteins and oligosaccharide receptors, what relevance does this have for the problem of cellular recognition in the nervous system? It seems very unlikely that the wiring diagram of brain will be established exclusively by a mechanism like that which we have identified in the cellular slime molds. Other processes, considered by other contributors to this volume, will undoubtedly prove to be important (see also Sidman, 1974). Chemotactic signals like those which lead to aggregation of slime mold cells are likely to be found. Guidance of neurons to their destination by reversible interaction with specific glial cells has been shown (Rakic, 1974). Differentiative processes which determine the size and branching pattern of neurons undoubtedly play a role; and the differentiative processes that determine the neuron's neurotransmitter and its pattern of cell surface receptors for neurotransmitters might also prove relevant. Factors that control the timing of cell maturation are also likely to be important. A nerve terminal that arrives at a potential target cell before another one has a competetive advantage which may be decisive. Therefore, identification of specific cell adhesion molecules in the nervous system, should this occur, will not be sufficient to explain the differentiation of this complex organ. The thing that is particularly appealing about identification of molecules like those we have found in slime molds is that, once identified, they provide both insights and reagents for further investigation. The history of molecular biology has amply proved that analysis of processes is tremendously facilitated by identification of relevant substances.

How then might substances like those apparently used in specific cell cohesion in slime molds mediate specific connections in the nervous system? In his analysis of the development of retinal–tectal connections, Sperry (1963, 1965) concluded by postulating the existence of specific chemicals on the surface of retinal cells that are recognized by appropriate tectal cells. How could the types of processes believed to be operative in cellular recognition in slime molds lead to the ordered development of retinal–tectal connections?

One way in which this might be achieved is by the specification of each cell in the retina and tectum to synthesize a unique cell surface protein which could interact only with the unique complementary protein. This seems unlikely since there are millions of retinal and tectal cells, which exceeds the number of genes for specific proteins in vertebrate DNA (unless there were some type of genetic splicing like that sometimes proposed to explain antibody diversity). A more likely possibility is that a *gradation of relative affinities* produces specific retinal–tectal associations. One way to think of this is in terms of an inductive gradient that spreads from a source (e.g., the midline) in an embryo and sets up a graded signal across both the retina and tectum (see Jacobson, this volume). The signal could either be the concentration of the inductive substance or a secondary change that it produces.

Systematic variation in the intensity of a signal could lead to a permanent graded expression of genes that control properties of the cell surface, as argued previously (Barondes, 1970). For example, there could be graded expression in both the retina and tectum of two genes—one that controls the synthesis of a specific carbohydrate-binding protein and the other that controls the synthesis of a specific glycoprotein receptor. The amount of glycoprotein receptor synthesized per cell could be regulated most easily by control of the synthesis of the polypeptide backbone, but control of the synthesis of a specific glycosyltransferase that confers receptor specificity might also be effective. In this model, the inductive gradient would have opposite effects on the genes controlling carbohydrate-binding protein and receptor so that cells rich in receptor would be poor in carbohydrate-binding protein and conversely. Both the retinal and tectal cells close to the source of the inductive gradient might therefore have a large amount of the carbohydrate-binding protein on their surface and a small amount of the receptor, and those far from the source of the gradient might be rich in the receptor and poor in the carbohydrate-binding protein. The cells might then associate in a manner whereby nerve endings rich in the carbohydrate-binding protein but poor in the glycoprotein receptor preferred target cells that were rich in the receptor glycoprotein but poor in the carbohydrate-binding protein. In this way, cells in the retina close to the source of the inductive gradient would make synapses with cells in the optic tectum far from the source during the time of the inductive gradient. Similar inductive gradients could also determine the synthesis of another pair of complementary cell surface proteins in the perpendicular (e.g., dorsal–ventral) dimension. The overall results would be consistent with the observations reviewed by Jacobson in this volume. The role of carbohydrate–protein interactions is compatible not only with our observations on slime molds but also with studies on vertebrate cell cohesion reviewed by other contributors to this volume.

A number of experimental approaches might be applied in a search for a mechanism of this type. One possible tactic is to attempt to stain the surface of nerve endings from retinal ganglion cells or dendrites from retinal tectal cells with specific reagents that would discriminate a gradient of a specific oligosaccharide receptor. With a great deal of luck, one of the many known carbohydrate-binding proteins that have been purified from plant and animal sources (Lis and Sharon, 1973) might have a specific high affinity for the particular configuration of the appropriate oligosaccharide receptor. If this were so, one could show that the labeled carbohydrate-binding protein (e.g., labeled with fluorescein or ferritin) bound more to cell surfaces in one region compared to another. We have developed a general method for purification of carbohydrate-binding proteins which could provide a very large battery of reagents of this type (Reitherman *et al.*, 1974). If graded amounts of a specific oligosaccharide receptor were present on nerve endings of retinal ganglion cells or on tectal dendrites as a function of their anatomical position, and if the right carbohydrate-binding protein were used, the gradient could be demonstrable. It remains to be seen whether systems will be found in which these conditions are met.

Another possible approach is to look for carbohydrate-binding proteins in extracts of developing brain using erythrocyte agglutination as a screening assay. This approach was successful with the cellular slime molds because oligosaccharides on the surface of appropriate erythrocyte types were sufficiently similar to the oligosaccharide receptors on the surface of cohesive slime mold cells to react with them. Since the slime mold proteins are polyvalent, they could be detected by erythrocyte agglutination. Carbohydrate-binding proteins on the surfaces of cells of the developing brain might also agglutinate erythrocytes (but they could be monovalent and not be identified as agglutinins). In an attempt to apply this approach, we have identified an agglutinin of formalinized chicken erythrocytes in aqueous extracts of developing chick brain (Barondes *et al.*, 1974). The activity of this agglutinin per milligram of soluble brain protein is high between 6 and 10 days of embryonic chick development and declines thereafter. It has been partially purified by gel filtration followed by equilibrium sedimentation in cesium chloride. We have been unable to find (with the saccharides we have used in an attempt to inhibit agglutination) evidence that the substance in question is a carbohydrate-binding protein. The substance binds to formalinized chicken erythrocytes, but the nature of this binding is not known.

The agglutination activity of this developmentally regulated brain agglutinin is contained in a proteoglycan complex consisting of about 75% carbohydrate and 25% protein. Major carbohydrate constituents are glycosaminoglycans, tentatively identified as hyaluronic acid and chondroitin sulfate, although other types of carbohydrates are also present. However, purified glycosaminoglycans and trypsin-treated proteoglycan are inactive as agglutinins in our assay.

It seems most likely then that the substance we have identified may be, at least in part, the native form of the glycosaminoglycans discussed by Toole in this volume that appear to play a role in regulation of migration of cells or neurites in brain development. Why we should have been led to identify developmentally regulated proteoglycans as agglutinins of formalinized erythrocytes is presently unclear. However, this approach to the problem of cell cohesion in vertebrate systems still merits further exploration. In this regard, the observation of a carbohydrate-binding protein in electric organ tissue that agglutinates trypsin-treated rabbit erythrocytes and of a similar agglutinin in embryonic chick muscle (Teichberg *et al.*, 1975) seems very encouraging.

ACKNOWLEDGMENTS

The research program in this laboratory is supported by Grant MH18282 from the United States Public Health Service and by a grant from the Alfred P. Sloan Foundation.

11. REFERENCES

Barondes, S. H., 1970, Brain glycomacromolecules and interneuronal recognition, in: *The Neurosciences, Second Study Program* (F. O. Schmitt, ed.), pp. 747–760, Rockefeller University Press, New York.

Barondes, S. H., Rosen, S. D., Simpson, D. L., and Kafka, J. A., 1974, Agglutinins of formalinized erythrocytes: Changes in activity with development of *Dictyostelium discoideum* and embryonic chick brain, in: *Dynamics of Degeneration and Growth in Neurons* (K. Fuxe, L. Olson, and Y. Zotterman, eds.), pp. 449–454, Pergamon Press, Oxford.

Beug, H., Gerisch, G., Kampff, S., Riedel, V., and Cremer, G., 1970, Specific inhibition of cell contact formation in *Dictyostelium* by univalent antibodies, *Exp. Cell Res.* **63**:147.

Beug, H., Gerisch, G., and Müller, E., 1971, Cell dissociation: Univalent antibodies as a possible alternative to proteolytic enzymes, *Science* **173**:742.

Beug, H., Katz, F. E., Stein, A., and Gerisch, 1973*a*, Quantitation of membrane sites in aggregating *Dictyostelium* cells by use of tritiated univalent antibody, *Proc. Natl. Acad. Sci. USA* **70**:3150.

Beug, H., Katz, F. E., and Gerisch, G., 1973*b*, Dynamics of antigenic membrane sites relating to cell aggregation in *Dictyostelium discoideum*, *J. Cell Biol.* **56**:647.

Bohlool, B. B., and Schmidt, E. L., 1974, Lectins: A possible basis for specificity in the rhizobium–legume root nodule symbiosis, *Science* **185**:269.

Bonner, J. T., 1967, *The Cellular Slime Molds*, 2nd ed., Princeton University Press, Princeton, N.J.

Bonner, J. T., and Adams, M. S., 1958, Cell mixtures of different species and strains of cellular slime moulds, *J. Embryol. Exp. Morphol.* **6**:346.

Bonner, J. T., Hall, E. M., Noller, S., Oleson, F. B., Jr., and Roberts, A. B., 1972, Synthesis of cyclic AMP and phosphodiesterase in various species of cellular slime molds and its bearing on chemotaxis and differentiation, *Dev. Biol.* **29**:402.

Brown, R. C., Bass, H., and Coombs, J. P., 1975, Carbohydrate binding proteins involved in phagocytosis by *Acanthamoeba*, *Nature (London)* **254**:434.

Chang, C.-M., Reitherman, R. W., Rosen, S. D., and Barondes, S. H., 1975, Cell surface localization of discoidin, a developmentally regulated carbohydrate-binding protein from *Dictyostelium discoideum*, *Exp. Cell Res.* **95**:136.

Frazier, W. A., Rosen, S. D., Reitherman, R. W., and Barondes, S. H., 1975, Purification and comparison of two developmentally regulated lectins from *Dictyostelium discoideum*: Discoidin I and II, *J. Biol. Chem.* **250**:7714.

Gerisch, G., 1968, Cell aggregation and differentiation in *Dictyostelium*, in: *Current Topics in Developmental Biology*, Vol. 3 (A. Monroy and A. Moscona, eds.), pp. 157–197, Academic Press, New York.

Gesner, B. M., and Thomas, L., 1966, Sialic acid binding sites: Role in hemagglutination by *Mycoplasma galliseptium*, *Science* **151**:590.

Gibbons, R. A., Jones, G. W., and Sellwood, R., 1975, An attempt to identify the intestinal receptor for the K88 adhesin by means of a haemagglutinin inhibition test using glycoproteins and fractions from sow colustrum, *J. Gen. Microbiol.* **86**:228.

Hamblin, J., and Kent, S. P., 1973, Possible role of phytohaemagglutinin in *Phaseolus vulgaris* L., *Nature (London) New Biol.* **245**:28.

Humphreys, T., 1963, Chemical dissolution and *in vitro* reconstruction of sponge cell adhesions. I. Isolation and functional demonstration of the components involved, *Dev. Biol.* **8**:27.

Laver, W. G., 1973, The polypeptides of influenza, *Advan. Virus Res.* **18**:57.

Lis, H., and Sharon, N., 1973, The biochemistry of plant lectins (phytohemagglutinins), *Ann. Rev. Biochem.* **42**:541.

Loomis, W. F., Jr., 1975, *Dictyostelium discoideum, a Developmental System*, Academic Press, New York.

Moscona, A. A., 1963, Studies on cell aggregation: Demonstrations of materials with selective cell-binding activity, *Proc. Natl. Acad. Sci. USA* **49**:742.

Nicolson, G. L., 1974, The interaction of lectins with animal cell surfaces, *Int. Rev. Cytol.* **39**:89.

Rakic, P., 1974, Mode of cell migration to the superficial layers of fetal monkey neocortex, *J. Comp. Neurol.* **145**:61.

Rambourg, A., 1971, Morphological and histochemical aspects of glycoproteins at the surface of animal cells, *Int. Rev. Cytol.* **31**:57.

Raper, K. B., and Thom, C., 1941, Interspecific mixtures in the Dictyosteliaceae, *Am. J. Bot.* **28**:69.

Reitherman, R. W., Rosen, S. D., and Barondes, S. H., 1974, Lectin purification using formalinized erythrocytes as a general affinity adsorbent, *Nature (London)* **248**:599.

Reitherman, R. W., Rosen, S. D., Frazier, W. A., and Barondes, S. H., 1975, Cell surface species-specific high affinity receptors for discoidin: Developmental regulation in *D. discoideum*, *Proc. Natl. Acad. Sci. USA* **72**:3541.

Rosen, S. D., 1972, A possible assay for intercellular adhesion molecules, Ph.D. thesis, Cornell University.

Rosen, S. D., Kafka, J., Simpson, D. L., and Barondes, S. H., 1973, Developmentally-regulated, carbohydrate-binding protein in *Dictyostelium discoideum*, *Proc. Natl. Acad. Sci. USA* **70**:2554.

Rosen, S. D., Simpson, D. L., Rose, J. E., and Barondes, S. H., 1974, Carbohydrate-binding protein from *Polysphondylium pallidum* implicated in intercellular adhesion, *Nature (London)* **252**:128.

Rosen, S. D., Reitherman, R. W., and Barondes, S. H., 1975, Distinct lectin activities from six species of cellular slime molds, *Exp. Cell Res.* **95**:159.

Schulze, I. T., 1973, Structure of the influenza virion, *Advan. Virus Res.* **18**:1.

Shaffer, B. M., 1957, Properties of slime-mould amoebae of significance for aggregation, *Q. J. Microsc. Sci.* **98**:377.

Sidman, R. L., 1974, Contact interaction among developing mammalian brain cells, in: *The Cell Surface in Development* (A. A. Moscona, ed.), pp. 221–253, Wiley, New York.
Simpson, D. L., Rosen, S. D., and Barondes, S. H., 1974, Discoidin, a developmentally regulated carbohydrate-binding protein from *Dictyostelium discoideum*: Purification and characterization, *Biochemistry* **13:**3487.
Simpson, D. L., Rosen, S. D., and Barondes, S. H., 1975, Pallidin: Purification and characterization of a carbohydrate-binding protein from *Polysphondylium pallidum* implicated in intercellular adhesion, *Biochim. Biophys. Acta* **412:**109.
Singer, S. J., 1974, The molecular organization of membranes, *Annu. Rev. Biochem.* **43:**805.
Sperry, R. W., 1963, Chemoaffinity in the orderly growth of nerve fiber patterns and connections, *Proc. Natl. Acad. Sci. USA* **50:**703.
Sperry, R. W., 1965, Embryogenesis of behavioral nerve nets, in: *Organogenesis* (R. L. DeHaan and H. Ursprung, ed.), pp. 161–186, Holt, Rinehart and Winston, New York.
Sussman, M., 1975, The genesis of multicellular organization and the control of gene expression in *D. discoideum*, in: *Progress in Molecular and Subcellular Biology*, Vol. IV (F. Hahn ed.), Springer, Berlin.
Teichberg, V. I., Silman, I., Beitsch, D., and Resheff, D., 1975, A β-D-galactoside binding protein from electric organ tissue of *Electrophorus electricus, Proc. Natl. Acad. Sci. USA* **72:**1383.
Tyler, A., 1946, An auto-antibody concept of cell structure, growth and differentiation, *Growth* **10:**7.
Weiss, P., 1947, The problem of specificity in growth and development, *Yale J. Biol. Med.* **19:**235.
Winzler, R. J., 1970, Carbohydrates in cell surfaces, *Int. Rev. Cytol.* **29:**77.

Note Added in Proof to Chapter 5

In collaboration with Dr. Marie-France Maylié-Pfenninger, these lectin studies have meanwhile been extended with a battery of well-characterized ferritin conjugates (Maylié-Pfenninger *et al.*, 1975) specific for galactosyl (ferritin-ricin, F-R), *N*-acetyl-glucosaminyl (ferritin-wheat germ agglutinin, F-WGA), fucosyl (ferritin-lotus lectin, F-LTL) and glucosyl/mannosyl residues (ferritin-concanavalin A, F-ConA). An investigation of the carbohydrate composition of cultured nerve cell surfaces (cf. Pfenninger and Maylié-Pfenninger, 1975) was possible only after the careful removal of serum glycoproteins from the cultures. This was achieved by preincubation in hapten-free medium in which the serum had been replaced by 1% bovine serum albumin. Subsequently the cultures were aldehyde-fixed or cooled to 2–4°C before the labeling experiments were carried out.

In brief, the results are the following: In all parts of rat superior cervical ganglion neurons, the amount of F-ConA binding is relatively small and no fucosyl residues accessible to F-LTL are detectable. However, high densities of F-R and F-WGA binding are observed, even on the growth cones. The levels of nonspecific binding in these experiments are negligible. So far, there is no indication of a lectin receptor density gradient in parallel to the distribution of intramembranous particles along the nerve fiber axis. This indicates the presence in the membrane of carbohydrate-bearing components which are independent from intramembranous particles and stresses the complexity of the chemistry of the growth cone membrane surface (cf. discussions on pp. 170 and 171). Indeed, comparative studies on cultured growth cones from rat spinal cord, superior cervical and dorsal root ganglia with F-WGA and F-R have revealed differential concentrations of lectin-accessible *N*-acetyl-glucosaminyl and galactosyl residues in a way which permits the clear-cut distinction of the three types of growth cones: whereas superior cervical ganglion cones exhibit large quantities of both carbohydrate moieties, dorsal root ganglion cones contain a high density of F-WGA receptors but only a small number of F-R binding sites. Quite different still are growth cones from spinal cord cultures which bind virtually no F-R but exhibit a modest number of F-WGA binding sites. Thus a *differential surface chemistry* has been established in these three *different types of growth cones.* These findings may serve as a potential basis for the *hypothesis* that coding for neuronal recognition may lie within membrane surface components, especially carbohydrates.

REFERENCES

Maylié-Pfenninger, M. F., Palade, G. E., and Jamieson, J. D., 1975, Interaction of lectins with the surface of dispersed pancreatic cells, *J. Cell Biol.* **67**:333a.

Pfenninger, K. H., and Maylié-Pfenninger, M. F., 1975, Distribution and fate of lectin binding sites on the surface of growing neuronal processes. *J. Cell Biol.* **67**:332a.

Index